T0318686

URBAN WATER DISTRIBUTION NETWORKS

URBAN WATER DISTRIBUTION NETWORKS

Assessing Systems Vulnerabilities, Failures, and Risks

Symeon E. Christodoulou
University of Cyprus

Michalis Fragiadakis
National Technical University of Athens

Agathoklis Agathokleous
University of Cyprus

Savvas Xanthos
City University of New York

Butterworth-Heinemann is an imprint of Elsevier
The Boulevard, Langford Lane, Kidlington, Oxford OX5 1GB, United Kingdom
50 Hampshire Street, 5th Floor, Cambridge, MA 02139, United States

Library of Congress Cataloging-in-Publication Data
A catalog record for this book is available from the Library of Congress

British Library Cataloguing-in-Publication Data
A catalogue record for this book is available from the British Library

ISBN: 978-0-12-813652-2

Publisher: Matthew Deans
Acquisition Editor: Ken McCombs
Editorial Project Manager: Anna Valutkevich
Production Project Manager: Vijayaraj Purushothaman
Designer: Mark Rogers

Typeset by VTeX

CONTENTS

LIST OF FIGURES

LIST OF TABLES

ABOUT THE AUTHORS

Dr. Symeon E. Christodoulou is an alumnus of Columbia University, New York City (USA), where he first studied on an academic scholarship from the Agency for International Development (AID) towards a degree in civil engineering (BSc 1991), and then on academic and research fellowships towards postgraduate degrees in civil engineering (MSc 1992, MPhil 1996, PhD 1998).

In the period of 1993–1995 he worked as a project engineer on the reconstruction of the Columbus Avenue (New York City) and the Gowanus Expressway and Prospect Expressway (New York City). Subsequently (1995–1998), he was employed by the firm of O'Brien-Kreitzberg Inc., the largest pure construction management firm in the USA, and worked on the Newark International Airport's Automated People Mover System (1995–1996), and the campus development program for Columbia University (1996–1998), with a budget of 500 million and 600 million USD, respectively.

Upon completion of his doctoral studies, and after several years of professional career in construction engineering and management, Dr. Christodoulou joined the academic staff of Polytechnic University (New York City) as an Assistant Professor and academic coordinator of the construction engineering program, at the university's Department of Civil Engineering (1998–2003).

In 2004, Dr. Christodoulou joined the faculty of the University of Cyprus as an Assistant Professor at the Department of Civil and Environmental Engineering (2004–2009). Concurrently, he taught for the graduate program of the Democritus University of Thrace (Greece) as a visiting professor in an interdisciplinary academic program funded by the European Commission. In 2009 Dr. Christodoulou was elected as an Associate Professor at the University of Cyprus, and subsequently served as Head of the Department of Civil and Environmental Engineering (2011–2013) and as Vice Dean of Engineering (2014–2017). He is also a founding member and serves on the Board of Directors for the Nireas International Water Research Center (Nireas-IWRC), a research center established through funding by the European Regional Development Fund and the Republic of Cyprus. In 2017, Dr. Christodoulou was promoted to the rank of Professor.

Dr. Christodoulou is the author of numerous scientific publications, the recipient of research funding in excess of three million euro (including research grants from USA's National Science Foundation, the European Science Foundation, the European Commission, the European Territorial Cooperation Programme, Cyprus's Research Promotion Foundation, and New York City's Metropolitan Transit Authority). Further, he is the recipient of two international research awards (London 1999, Athens 2015), whilst he also served (2008–2012) as the Republic of Cyprus's national representative to the European Commission's COST Domain Committee on transport and urban development.

Finally, Dr. Christodoulou has served as an Associate Editor for the *European Water* journal, and as an Editorial Board member for the journals of *Information Technology in Construction* (ITCon) and *Smart Cities*, and he serves as scientific reviewer for over 40 scientific journals, among which the journals of *Construction Engineering and Management* (ASCE), *Computing in Civil Engineering* (ASCE), *Professional Issues in Engineering* (ASCE), *Construction Management and Economics* (Taylor and Francis), *Automation in Construction* (Elsevier), and *Water Resources Management* (Springer).

Dr. Michalis Fragiadakis is Assistant Professor at the School of Civil Engineering of the National Technical University of Athens (NTUA), Greece. He received his PhD from NTUA and holds MSc degrees in "Earthquake Engineering and Structural Dynamics" from Imperial College, London and also in "Structural Analysis and Design" from NTUA. He has served as adjunct lecturer at the Department of Civil Engineering of the University of Thessaly for 4.5 years and he has also worked as a postdoctoral researcher at the University of Cyprus and the University of Pavia. His teaching includes structural analysis and earthquake engineering courses at both under- and post-graduate levels. His research activity is focused on structural dynamics, earthquake engineering and on the seismic risk assessment of structures and lifelines. His interests also extend to computational methods in structural analysis and on the application of novel soft-computing methodologies for structural analysis and design. Dr. Fragiadakis is a member of the Hellenic Society for Earthquake Engineering (HSTAM) and has been involved in various national and international research projects. He has coorganized three international conferences (COMPDYN 2009, 2011, 2017) and several mini-symposia and special sessions in international conferences. He has authored and coauthored more than 145 research publications: 2 edited books, 36 journal papers, 17 book chapters and 90 papers in international conferences. His work has been extensivelly disseminated and has received a large number of citations.

Dr. Agathoklis Agathokleous is currently a Postdoctoral Associate of Nireas-IWRC's *EUPALINOS – Construction Engineering and Water Distribution Networks Management Laboratory*. Dr. Agathokleous has a Higher National Diploma in Civil Engineering from Cyprus's Higher Technical Institute (2001), an MEng in Civil Engineering from the University of Surrey, UK (2005) and an MSc in Structural Engineering from the same university (2006). In 2007 he joined the Department of Civil and Environmental Engineering of the University of Cyprus as a researcher and PhD candidate, from where he graduated, receiving his PhD in December of 2014. Since January 2012, Dr. Agathokleous is a member of the EUPALINOS Lab research team, and upon completion of his PhD he became the Head of the Lab's WDN management research group.

His research interests focus on the rehabilitation and asset management of urban water distribution networks, and his research work has been on risk analysis, decision support systems and sensor development for the sustainable management of urban water distribution networks. He has worked on a number of research projects funded by the Cyprus Research Promotion Foundation and the European Regional Development Fund (ERDF), and has published extensively on his research work through several journal papers and proceedings of international conferences.

Dr. Savvas Xanthos has a background in Mechanical Eng., with specialization in experimental fluid mechanics (PhD 2004, City University, NYC). His research field has been primarily in Fluid Mechanics, specializing in shock wave and expansion wave interaction with turbulence, hot wire anemometry techniques using 3D velocity, and fully 3D vorticity probes. He has been involved in blast wave mitigation utilizing shock wave propagation, Laser Doppler Velocimetry (LDV) measurements and Particle Image Velocimetry (PIV) techniques used for verification purposes in applied CFD projects of concern. In addition, he has also shown interest in the field of biomedical engineering and especially applying experimental procedures as well as finite element method (FEM) solutions to verifying various theoretical models for atherosclerosis in human arteries. Finally, in the recent years he has served as a research consultant on the development of a three-dimensional CFD model investigating the hydrodynamic behavior of activated sludge in Final Clarifiers of Water Pollution Control Plants.

This model has been supplemented with a decade's worth of bench scale experiments and data and is considered one of the most comprehensive three dimensional models available. This CFD model is now being used as a tool for Final Settling Tank optimization studies for several municipalities in the USA especially in the states of NY and NJ. He has collaborated in many different types of grants under a wide inter disciplinary area of mechanical, biomedical and civil engineering projects and has been an author and reviewer of multiple scientific publications relating to the aforementioned areas of interest, some of which have won different awards from the scientific community. Dr. Xanthos successfully completed a one-year research placement at the University of Cyprus on a Fulbright Scholarship from the USA and was consequently employed at the EUPALINOS Lab from 2011 to 2014 on two externally-funded projects related to water distribution networks (Nireas-IWRC) and building energy efficiency (ISES). In 2014 Dr. Xanthos returned to the USA and lives in NY.

PREFACE

Lifeline systems, such as water distribution networks, are of critical importance to the uninterrupted provision of services and thus to the resiliency of a city. The term "lifelines" refers to essential utility and transportation systems that serve communities across all jurisdictions and locales. These systems share the attributes of being distributed systems, rather than isolated facilities; and of providing products or services that are transferred through networks that often cross legal and jurisdictional boundaries [6]. In effect, lifelines are the infrastructure systems that support urban life and because of their interconnected nature, once a lifeline system is damaged by an earthquake or other natural hazard, other lifeline systems (or infrastructure components) may malfunction as well. The term "resilience" refers to a system's (or city's) capacity to recover quickly and effectively from a catastrophic event.

Forecasting, managing, and reducing the risk of lifelines are actions of paramount importance to Authorities and citizens alike. As a result, lifeline owners have made significant investments in designing, constructing, and retrofitting their systems to reduce the risk of damage in both normal and abnormal operating conditions, as well as in continuously monitoring the performance of the systems. The latter typically refers to seismic risk and to the effects of an earthquake on a lifeline. However, the seismic performance standards for lifelines vary widely and are not tied to generally-applicable public policies for reducing risk or for ensuring community resilience in the event of a major earthquake.

The work included in the pages of this book discusses several of the facets of water distribution network (WDN) vulnerability, providing a methodology for system-wide assessment of WDNs based on component analysis, network topology, and hydraulic analysis, utilizing methods such as statistics, survival analysis, data mining, Monte Carlo simulation, time series and spatio-temporal analysis. Further, the effects of a network's past performance on its vulnerability are examined and discussed (normal vs. abnormal operating conditions, continuous vs. intermittent water supply, nonseismic vs. seismic loads). Even though a number of previous studies have investigated WDN vulnerability, seldom have the system-component interaction, the abnormal operating conditions, and the seismic performance of such systems been in tandem considered in such evaluations.

Further to an overview of current methodologies for assessing the risk of failure and the vulnerability of water distribution networks under normal and abnormal operating conditions, the book discusses recent advances in seismic WDN vulnerability assessment, combining data on historical nonseismic performance of urban water distribution networks and of their components. The intent is firstly to propose a methodology for assessing the vulnerability of a WDN using available everyday measurements, and secondly to extent existing, where applicable, methodologies (such as that of the American Lifelines Alliance guidelines [5,6]) with localized knowledge on the performance and vulnerability of such networks under normal operating conditions.

Even though the work presented herein utilizes as case-study network a specific WDN topology and its historical performance for discussing WDN vulnerability, as well as the ALA guideline for the estimation of the pipe repair rates and of the fragility curves for demonstrating WDN vulnerability under seismic loads, the presented methodologies are generic and thus applicable when other WDN topologies and component fragility curves are used (e.g., when derived from seismic models, empirical data, or survival analysis).

In summary, further to providing insights into water loss management and leakage detection, a major contribution of the book is in providing a scientific approach to appraising the vulnerability of WDNs under either continuous (CWS) or intermittent water supply (IWS) operations, and under seismic and non-seismic loads, through the use of applicable models and methodologies and through the post-event or near-real-time analysis of available information.

ACKNOWLEDGMENTS

The work included herein, which amalgamates the authors' collective knowledge on water distribution networks, is partially based on actions stemming from the *Nireas-IWRC* research project. The project was cofinanced by the European Regional Development Fund and by the Republic of Cyprus through the Cyprus Research Promotion Foundation (Grant No. NEA-YPODOMI/STRAT/0308/09; 2010–2015; http://www.nireas-iwrc.org/).

Further, special thanks are extended to the Water Boards of Limassol and Nicosia (Cyprus), for providing operational data on their water distribution networks.

BOOK ORGANIZATION

The book starts with a brief description of its subject matter, the need for vulnerability assessment of WDNs, and the state of knowledge on the vulnerability assessment of WDNs (Chapter 1, p. 1).

It then discusses vulnerability assessment of WDNs under normal operating conditions and under nonseismic loads, highlighting the importance of a network's historical performance on its current state of vulnerability (Chapter 2, p. 21). The analysis presented in Chapter 2 is component-based and data-driven (based on historical performance records from a real-life WDN, spanning numerous years), and it is clustered by several classes: by WDN component (e.g., pipes vs. valves), by pipe type (e.g., water mains vs. house connections), by pipe material, pipe age, by the number of previously observed previous breaks (NOPB), etc. The real-life urban WDN (from the city of Nicosia, Cyprus) which is used as the data source for Chapter 2, is also utilized as case study for most of the succeeding chapters.

The discussion is then switched to the case of intermittent water supply (IWS) as a demonstration of how historical performance under normal operating conditions affects the network's performance under abnormal operating conditions (Chapter 3, p. 131).

Chapter 4 (p. 161) discusses topological robustness of water distribution networks, and how indices such as the betweenness centrality may be used to get an insight into the behavior of a WDN and its vulnerability under varying operating conditions.

Chapter 5 (p. 173) extends the study, to the seismic vulnerability, using the guidelines of the American Lifelines Alliance (ALA) as a starting point. The seismic vulnerability is augmented in order to include any available information on past nonseismic damage that has been compiled in the form of survival curves.

Hydraulics and spatial (topological and path) considerations are also introduced in the analysis (Chapter 6, p. 209), producing vulnerability heatmaps for the WDN in study (Chapter 8, p. 247). Real-time monitoring of WDNs is the subject of Chapter 7 (p. 227), with a brief overview of sensor-placement optimization strategies and a presentation of change-point analysis for anomaly detection (such as water loss). The book concludes with a chapter on disaster resilience of WDNs (Chapter 9, p. 269) and with a short epilogue on interconnected networks (p. 283).

An effort has been made to keep the aforementioned chapters as self-contained as possible, so that they can be reviewed independently of the other chapters.

LIST OF ACRONYMS

AC	**A**sbestos **C**ement
AI	**A**rtificial **I**nteligence
ALA	**A**merican **L**ifelines **A**lliance
ANN	**A**rtificial **N**eural **N**etworks
AMR	**A**utomatic **M**eter **R**eading
A.V.	**A**ir **V**alve
B.V.	**B**all **V**alve
C	**C**ouple
CFD	**C**omputational **F**luid **D**ynamics
CUSUM	**CU**mulative **SUM**
CyRPF	**Cy**prus's **R**esearch **P**romotion **F**oundation
C.V.	**C**heck **V**alve
DAC	**D**iscriminant **A**nalysis **C**lassification
DBMS	**D**ata**B**ase **M**anagement **S**ystem
DDM	**D**ata-**D**riven **M**odeling
D.I.	**D**uctile **I**ron
DMA	**D**istrict **M**etered **A**rea
DS-C	**D**istribution **S**ystem – **C**ouple **M**odel
DS-HC	**D**istribution **S**ystem – **H**ouple **C**onnection **M**odel
DS-WM	**D**istribution **S**ystem – **W**ater **M**ains **M**odel
DSS	**D**ecision **S**upport **S**ystem
EPA	**E**nvironmental **P**rotection **A**gency
EPANET	**E**nvironmental **P**rotetion **A**gency **NET**work modeler
F.H.	**F**ire **H**ydrant
ID	**Id**entity
IR	**I**nfrared **R**adiation
GA	**G**enetic **A**lgorithm
GIS	**G**eographical **I**nformation **S**ystem
GP	**G**aussian **P**rocess
gpd	**g**allons **p**er **d**ay
GPR	**G**round **P**enetrating **R**adar
GPS	**G**lobal **P**ositioning **S**ystem
HC	**H**ouse **C**onnections
HR	**H**azard **R**atio
H.V.	**H**ydrant **V**alve
IAWQ	**I**nternational **W**ater **Q**uality **A**ssociation
IWA	**I**nternational **W**ater **A**ssociation
IWRMT	**I**ntegrated **W**ater **R**esources **M**anagement **T**ool
IWS	**I**ntermittent **W**ater **S**upply
IWSA	**I**nternational **W**ater **S**upply **A**ssociation
MCS	**M**onte **C**arlo **S**imulation
O&M	**O**perations and **M**aintenance
OFWAT	**OF**fice of **WAT**er Services (UK government)

PGD Permanent Ground Deformation
PGV Peak Ground Velocity
PN Pipeline Network
PPFA Pipe-by-Pipe Failure Analysis
PVC Poly-Vinyl Chloride
RDBMS Relational Database Management System
RuLSIF Relative unconstrained Least-Squares Importance Fitting
SCADA Supervisory Control And Data Acquisition
Sub-DMA Sub-D Metered Area
S.V. Switch Valve
TTF Time To Failure
US-EPA United States' Environmental Protection Agency
UWDN Urban Water Distribution Networks
WB Water Board
WBL Water Board of Limassol
WBN Water Board of Nicosia
WDD Water Development Department
WDN Water Distribution Networks
WDNA Water Distribution Network Authorities
WDS Water Distribution Systems
WFA Valve-by-Valve Failure Analysis
WLTF Water Loss Task Force
WM Water Mains
WRF Water Research Foundation
WSN Wireless Sensor Network
WSRA Water Services Regulation Authority

CHAPTER 1

Introduction

Classical risk analysis is principally concerned with investigating the risks surrounding a plant (or some other object), its design and operations. Such analysis tends to focus on causes and the direct consequences for the studied object. Vulnerability analysis, on the other hand, focuses both on consequences for the object itself and on primary and secondary consequences for the surrounding environment. It also concerns itself with the possibilities of reducing such consequences and of improving the capacity to manage future incidents.

Lövkvist-Andersen, A.L., et al. (2004),
Conference of the Society for Risk Analysis.

1.1 INTRODUCTION

Each year, hundreds of kilometers of pipes across the globe are repaired or replaced in an attempt to mitigate the effects of pipe bursts and of water loss, and to maintain the uninterrupted transport of water. Existing water distribution systems are increasingly at risk due to numerous factors (both internal and external to the distribution networks) and the accidental or deterioration-based breakage of urban water distribution networks (WDN). This situation represents a significant problem for communities and authorities.

Water distribution networks deliver water from its sources to the customers of the network. Being able to assess the reliability of the network against different hazards helps water distribution agencies prioritize their interventions and ensure a minimum reliability level of the network. Therefore, water distribution agencies are required to develop and implement new methods for monitoring, repairing (or replacing) aging WDN infrastructures, as well as modeling deteriorating WDN conditions, and proactively devising strategies to keep the networks in operation. In essence, water distribution agencies are faced with the increasingly more complex

Urban Water Distribution Networks
DOI: http://dx.doi.org/10.1016/B978-0-12-813652-2.00001-3

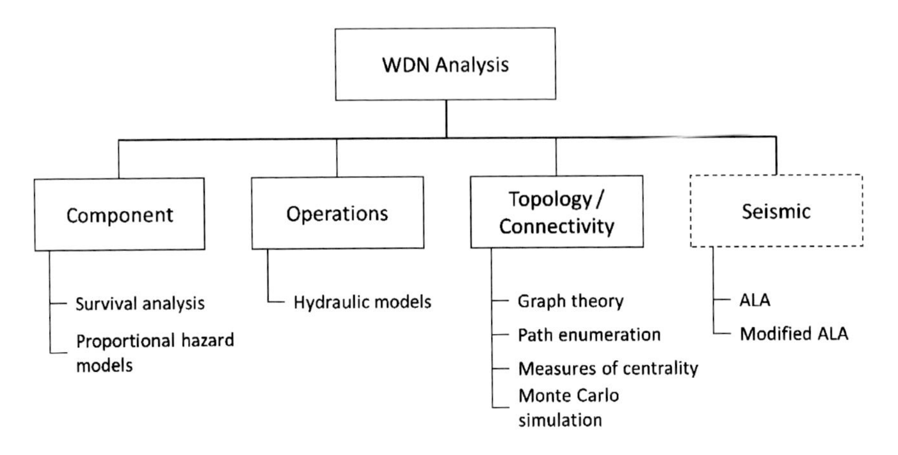

Figure 1.1 Water distribution network vulnerability analysis (schematic).

task of intelligently and efficiently assessing (or modeling) the condition of a pipe network, while managing the network in ways that maximize its reliability and minimize its operational and management costs. The question that usually arises is whether an organization should repair or replace deteriorating water mains and, in either case, what should be the sequence of any such repairs as part of a long-term network rehabilitation strategy.

Research to-date has helped identify a number of potential time-invariant and time-dependent risk factors contributing to pipe breaks and to WDN vulnerability. Among them are factors such as a pipe's age, diameter, material, and number of previous breaks, as well as the network's operating pressure and water flow. Component-based risk assessment, though, is only one of the many dimensions pertaining to the general problem of assessing the reliability of WDNs. Network topology and hydraulic operations are also of significant impact to WDNs, whilst abnormal operations (such as seismic loads) are also risk-of-failure sources that need to be considered (Fig. 1.1).

Granted that risk assessment of critical infrastructure poses a major problem for the engineering community, the problem is even more challenging when it comes to underground infrastructure and "lifelines", such as the urban water distribution networks. The challenges arise from a number of sources: (i) the system reliability depends on both the network topology and the reliability of the system's constituent components; (ii) WDNs consist of various system components of variable design parameters; (iii) there is a lack of historical data on the components' performance over time and under varying operating/loading conditions; and (iv) the networks are vastly inaccessible for visual inspections.

WDN vulnerability, though, is not static. Not only is it the product of many time-invariant and time-variant factors, but it is itself also dynamic in nature. A WDN performs one way under normal operations and another way under abnormal operating conditions. Abnormal operating conditions may be the result of seismic loadings, the implementation of intermittent water supply policies, or simply the scheduled (or unscheduled) maintenance operations in the WDN.

For example, recent cases of WDNs subjected to intermittent water supply operations have documented the adverse effects of such policies on the vulnerability of the networks, while recent earthquakes have shown that following a strong earthquake the caused damage of the various urban lifelines may cause a series of immediate, short-term, and long-term problems.

1.2 WHY THE NEED FOR VULNERABILITY ASSESSMENT OF WATER DISTRIBUTION NETWORKS?

A water distribution network is, in a way, a "living organism" with several of its components being constantly in flux, subjected to internal and external forces of time-variant and invariant parameters, and which with every component failure its survival as a whole is affected. After all, as it will be discussed in Chapter 2 (p. 21), WDNs don't differ much from living organisms for which every past failure incident ("illness") brings them closer to catastrophic failure ("death"). Hence why survival analysis (a method stemming from biological sciences and which attributes pipelines with characteristics similar to those of patients) is among the most common methods for assessing the probability of failure of water pipelines, and why one needs to constantly assess the fragility of a network's components and with that the vulnerability of the network as a whole.

The need for vulnerability analysis is further compounded by the importance of WDNs in our lives, as they constitute one of the primary "lifeline networks" of modern urban societies. Granted their importance to society, it is thus imperative to develop knowledge and to device methods for continuously and accurately assessing their state of affairs and their vulnerability under both normal and abnormal operating conditions.

Furthermore, WDN vulnerability relates to water quality and the risks to the consumers' health, as natural or malicious threats to the WDN threaten the health and well-being of the citizens that the WDN serves. Pipe deterioration leads to pipe breakages, breakages lead to water loss and to infiltration of pathogens in the network, pathogens lead to water contamination, contamination leads to health threats for the population being serviced by the WDN.

Finally, WDN vulnerability is associated with the disruption of residential, commercial and industrial activities and the cause of severe direct and indirect economic losses which, in the case of indirect losses, are higher the more developed the society is. Direct losses relate to the cost of repair, while indirect losses relate to the way the economy is affected by the disruption of the lifeline.

1.3 CURRENT OPERATIONS & MANAGEMENT PRACTICES FOR WATER DISTRIBUTION NETWORKS

The globalization of the problem of diminishing water resources has increased the level of interest in it, and inspired a lot of scientific research on the topic. The range of research topics related to potable water resources is vast, among which prominent are the topics pertaining to the sustainable management of WDNs. In its general definition, sustainability relates to the ability to endure and sustainable development refers to development that meets the needs of the present without compromising the ability of future generations to meet their own needs. Within the WDN domain, sustainable management is defined as the application of practices for the operation and management of WDNs in ways that allow the networks to thrive while benefiting both current and future generations, and while limiting the depletion of water resources.

However, the biggest problem currently faced worldwide by the managing authorities of water supply networks is that most of the pipe networks have already reached their useful life and that, as a result, most WDNs need maintenance, repair or even replacement to a great extent. In fact, the daily volume of water lost in such networks is enormous and, as a result, this reality has brought to the foreground the need for improved operations and maintenance (O&M) actions and for the development of leakage management plans/actions for numerous scenarios [55].

As aforementioned, in recent years the problem of water resources' shortage (especially the shortage of potable water) has grown and has expanded beyond the boundaries of countries where water shortage is traditionally encountered. As a result, national research has grown to international collaboration and joined research agendas, to the exchange of knowledge and to the development of good-practice guidelines in the form of manuals issued by national and international water-related organizations which, in turn, have subsequently been promoted to the various entities responsible for the management of the WDNs. The developed manuals primarily focus on the proper restructuring of WDNs for the improvement of their performance, on the efficiency of operations and the increased effectiveness in leakages detection, on fiscal improvements and on the development of decision support system tools for the sustainable management of WDNs [96,120,126,155].

1.3.1 Management Model Proposed by the International Water Association (IWA)

In 1999, the International Water Supply Association (IWSA) and the International Water Quality Association (IAWQ) consolidated in creation of the International Water Association (IWA) which immediately turned its focus on addressing the aforementioned sprawling problem of diminishing water resources. The vision of the newly formed organization was to bring together, worldwide, all experts on water issues to achieve an efficient and sustainable method of water management which would be extracted through collective knowledge. The organization has since been organized into specialty divisions. The Water Loss Task Force (WLTF) is IWA's specialty group focusing on research and knowledge development on water loss reduction and which is tasked with reaching out to all water industry governance groups, utilities and employees, associated individuals and companies, and community representatives at all levels of government, including national and international agencies, in order to save water losses and reduce future water demand.

The WLTF has issued a manual which was the result of its various research activities and work. This manual has the form of guidance and is referring to how to achieve the best possible management of water supply networks [116]. Furthermore, the WLTF had moved a step forward by issuing a second guidance manual [135] which deals with the detection and repair of leaks in water pipes of water distribution networks, while also reporting regularly on WLTF's research work in relation to the management of WDNs, through IWA's monthly "Water21" magazine [58,82,104,107] and in several scientific journals.

The strategies and WDN structures presented in the succeeding pages are those currently used by several large urban centers across the globe for the proper and efficient management of their water supply networks, based on the aforementioned instructions issued by the WLTF in the form of guidance notes. As a preamble to the methods described in subsequent sections, one should note the following key concepts:

District Metered Area (DMA)

Water distribution systems are large and complex networks that are designed and built to serve the water supply needs of an urban area. For the best and most simplified management of networks, the WLTF urges that the water supply networks should be divided into sections, called "District Metered Areas." These areas can be defined as sections of the network that comprise separate regional water meters [116] as well as one entry point and one exit point. Furthermore, the WLTF states that the water pressure within the pipelines of a DMA should be steady, at about 3 bars. Therefore, there should not be large deviations in altitude within DMAs, especially in gravity-based networks. For cases where this is unavoidable, pressure valves could be used in order to ensure constant pressure across the DMA. In the case of the three largest urban centers in Cyprus[1] (the cities of Nicosia[2], Limassol[3], and Larnaca[4]), their WDNs are divided in approximately 20 DMAs each.

The importance of DMA division is justified by:

1. Better monitoring. Smaller regions can be monitored better than a large unified network. By separating the network into zones, the manager of a water system can directly detect and localize the area which has water loss problems. Furthermore, DMAs enable agencies that are charged with the management of a water network to isolate the part of the network in which the incident has occurred so as to repair the problem without affecting the entire network. The fact that parts of the network can be isolated without interrupting the water supply is one of the biggest advantages offered by DMAs since the uninterrupted water supply to the consumers is of paramount importance to both the manager and the consumers of the water supply network.

[1] https://en.wikipedia.org/wiki/Cyprus
[2] https://en.wikipedia.org/wiki/Nicosia
[3] https://en.wikipedia.org/wiki/Limassol
[4] https://en.wikipedia.org/wiki/Larnaca

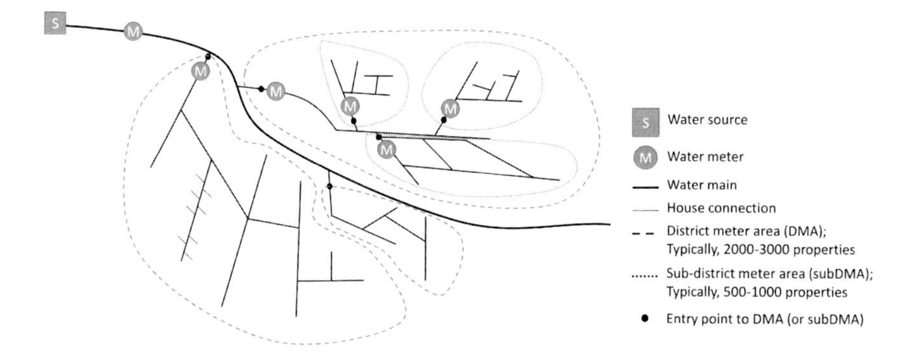

Figure 1.2 A typical arrangement of a WDN.

2. Uniform pressure across the WDN. It has been proven that pressure is one of the most critical parameters affecting pipe failures and thus water loss [158]. Unlike large-area unified WDNs where the water pressure in the network pipes is varying and thus very difficult to be controlled, monitored, and kept within uniform pressure levels, DMAs provide ease of control over pressure levels and thus lower levels of water loss across the WDN.

Subdistrict Metered Area (Sub-DMAs)

As mentioned above, the division of a WDN into DMAs is based on the network's spatial extent as well as on the variations in altitude within the area that the DMA covers. The WLTF has moved a step further, proposing the development of Subdistrict Metered Areas (Sub-DMAs). It is logical that through further separation of DMAs into subregional divisions increases the efficiency of the network administrator to manage the piping network. For oversized WDNs, a simple division into DMAs is not enough to achieve the goals sought by DMAs, because their size would be too large for their efficient monitoring, and hence better management control of these networks is sought through the subdivision of the DMAs into smaller sections. Fig. 1.2 illustrates a typical water distribution network arrangement, composed of DMAs and Sub-DMAs.

Data Loggers and Modeling

In developed networks, i.e., networks which are well-structured and organized, WDN operators make use of data loggers to monitor the behavior of their networks in real time. Data loggers are embedded on permanent basis onto critical points of the network and their purpose is the collection of useful information from the pipeline system for the continuous monitoring of the network status. Typical data collected by data loggers are the volume of water flow, the water pressure within the pipeline and the water quality within the pipe. Furthermore, there are data loggers which also act as noise correlators and they are used for localizing water leaks.

The information from the data loggers is typically received by use of mobile computer stations and by "passing over" the points where the data loggers have been installed (i.e., inspectors move to the points where the data loggers are installed and extract the data when in proximity to the loggers). The retrieval of information has been made easier and faster in recent years and for technologically-advanced WDNs, as wireless technology has increasingly been utilized in transmitting the data to remote base stations. Therefore, there is a continuous flow of information from the WDN to the WDN operator, on the key parameters of the network's operation. As a result, WDN managers are able to have live monitoring of the WDN's state, and maintenance crews can immediately respond to any challenge posed by the WDN.

Furthermore, for well-structured and monitored WDNs there exist hydraulic models of the network, as well as geographic information system (GIS) maps and decision support systems (DSS) for the automated processing of real (or near-real) time information on the performance of the network. This collected information and the DSS-deduced knowledge is constantly compared with the information calculated by use of the hydraulic simulations and of other modeling tools, and subsequently utilized in operations and in the sustainable management of the monitored WDN.

1.4 WATER LOSSES

The water volume channeled through the distribution network is separated into two main categories, namely "revenue water" and "nonrevenue water", with several subcategories under the aforementioned two [56], as depicted in Fig. 1.3.

The "revenue water" category corresponds to the authorized water consumption that brings income to the authorities. On the other hand, the "nonrevenue water" category refers to the volume of water which relates to financial losses for the authorities. It has to be clear that the "nonrevenue water" category does not refer only to the water losses due to network inef-

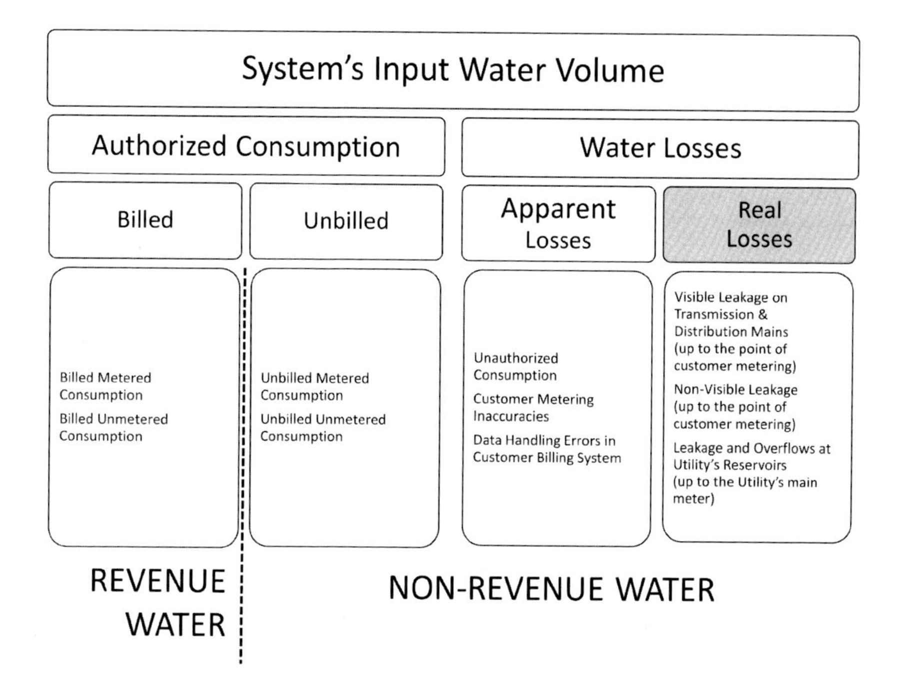

Figure 1.3 Categorization of the water volume channeled into the WDN.

ficiencies but also to a group of other parameters that cause loss of income for the water authorities.

1.4.1 Nonrevenue Water

There are three different types of water losses:

- Unbilled water
  "Unbilled water" is the authorized consumed water quantity that is not invoiced.
- Apparent losses
  "Apparent losses" are composed of the quantity of water that is stolen by unauthorized consumers, water volume which is not counted due to faulty metering and water volume that is not invoiced because of data handling errors in customer billing system.
- Real losses
  "Real losses" are water leakages due to the distribution network system's inefficiencies. Real losses are separated into the "*background leakages*" and the "*burst leakages.*" Background leakages are invisible leakages from the pipe system fittings. Individually they are too small to be de-

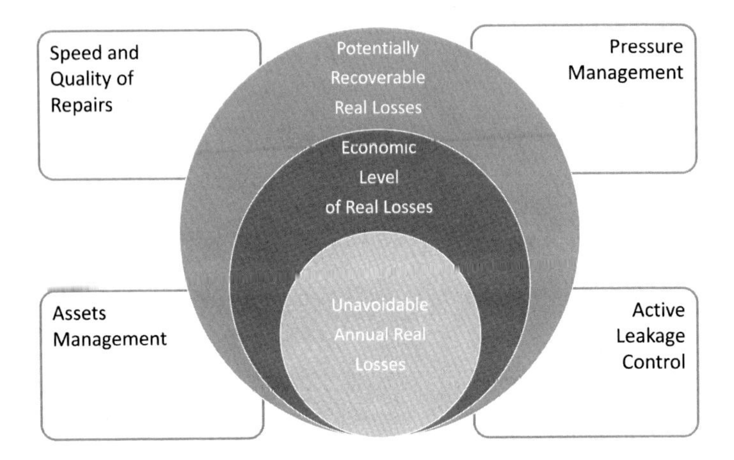

Figure 1.4 Leakage management.

tected by visual or acoustic inspection, thus pressure management serves as a very important parameter in controlling them. Burst leakage is the leakage due to the failure of the pipelines. Burst leakage could be visible or invisible, depending on the rate of the water loss [116]. The lower the rate of water loss a leak has, the more likely it is for the volume of water lost to remain in the lower ground layers and not rise to the surface.

"Real Losses" is the category of the nonrevenue water on which the WDN managers focus the most. Most research works dealing with WDN aim at optimizing the network management through actions associated with the real losses. As shown in Fig. 1.4, the optimization of real losses is based on four parameters: (1) pressure management; (2) speed and asset management; (3) pipeline and asset management; and (4) active leakage control. For calculation and comparison purposes, the real losses are always estimated based on the yearly water volume.

It is natural that the real losses cannot be reset, so there will always be a lost water volume, the "*Unavoidable Annual Real Losses.*" Furthermore, there is an additional volume of water, the "*Economic Level of Real Losses*", whose restoration is not economically profitable for the authorities. In short, there are leaks in the system whose restoration costs are more than the value of water lost due to them. Hence the "*Potentially Recoverable Real Losses*" are the leaks that exist in the system network and their rehabilitation cost is less than the cost of the water loss that would be saved by repairing the problem.

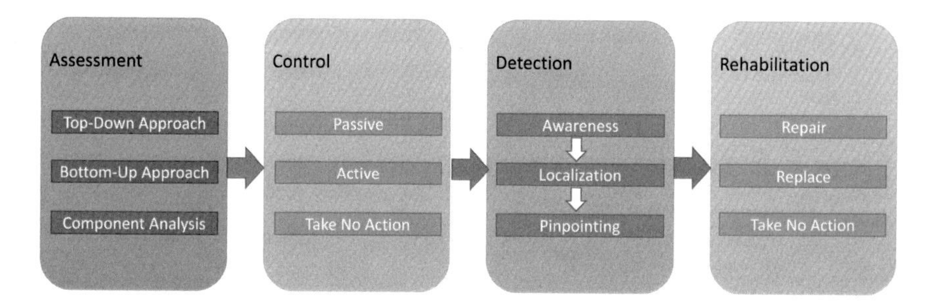

Figure 1.5 Management of real water losses.

1.5 MANAGEMENT OF REAL LOSSES

The category of water losses which is associated with the network condition and has a prominent role for the nonrevenue water is the one involving real losses, i.e., water losses due to inefficiencies of the WDN. For this reason the authorities place particular emphasis on the development of a rational network leakage management model.

As illustrated in Fig. 1.5, the management of real losses comprises three types of activity. The first activity, the "*leakage assessment*", deals with the estimation of the water losses due to leakages within the network. The second one, the "*leakage control*", is related to the strategic plans for the control of leakages and the restoration of a WDN upon fixing leakages. The last activity, termed "*leakage detection*", is related to the indication, detection, and localization of the leakages within the WDN [141].

1.5.1 Leakage Assessment

Leakage assessment deals with the estimation of water losses due to leakages within the network, and there are three different methods for the assessment of real water losses: the *top-down approach*, the *bottom-up approach,* and the *component analysis*. The assessment could also be performed using a combination of the above methods [54].

Top-Down Approach

The top-down leakage assessment method attempts the estimation of the water loss volume due to leakages through the water balance in WDN. There are two main top-down approaches that are used, the IWA approach and the one that is used by the Water Services Regulation Authority (OFWAT) in the UK [141].

The underlying approach of this type of methods relies on calculating the water volume of the real losses by subtracting the authorized consumption and the apparent losses from the system's input volume (Fig. 1.3). This type of methodology calculates the total losses due to leakages and does not provide analytical results regarding the type of real losses. Also it does not specify which areas of the network experienced the most serious leakage problem. For this reason it is used in conjunction with the one or both of the alternative methods which are presented below.

Bottom-Up Approach

This approach is used for calculating the real losses of a certain network area. This area could be a DMA, or a Sub-DMA or even a region of the network selected by the manager. The application of this approach on all areas which form the entire network can provide the means for calculation of the total real losses in the WDN.

There are two different ways of implementing the approach. The first is the "*24-Hour Zone Measurement*" [150] and the second is the "*Minimum Night Flow Analysis*" [45].

Component Analysis

Component analysis is a method used for the allocation of real losses into the different leak types, as these are indicated in Fig. 1.3 and it is usually employed for the evaluation of alternative management options for tackling leakage incidents.

In order to perform a component analysis, certain parameters of the examined WDN are needed. Such parameters are the average flow rates, the average run-times of different types of background and burst leakages, the basic infrastructure data and the infrastructure condition factor [54].

1.6 NONSEISMIC AND SEISMIC VULNERABILITY ASSESSMENT OF WDNS: STATE OF KNOWLEDGE

As discussed in Section 1.2 (p. 3), beyond the standard nonseismic risk/vulnerability assessment, there is a need for seismic risk/vulnerability assessment. Seismic risk is in essence a common abnormal operating condition for which we need to account and prepare, and thus Water Agencies need to be able to assess given the various operating parameters of their WDNs.

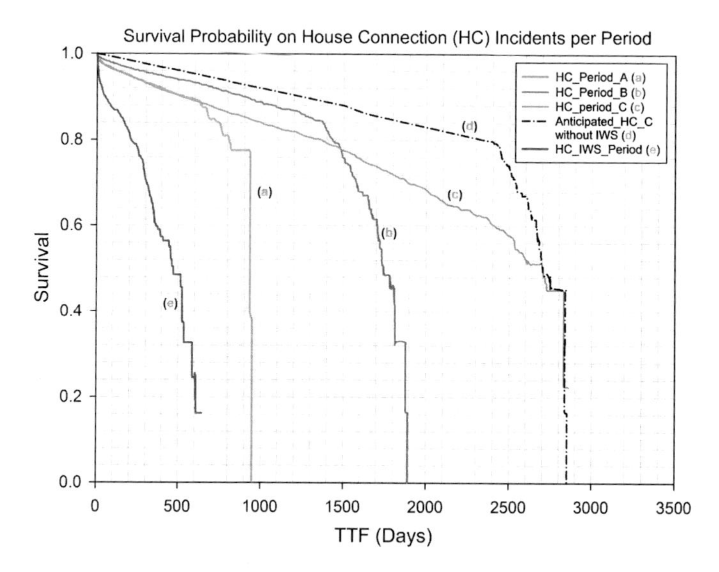

Figure 1.6 Loss of survivability of house connections (HC) due to historical performance – The case of intermittent water supply (adopted from Agathokleous, 2015).

As will be discussed in detail in subsequent sections (Chapter 3, p. 131), abnormal operating conditions greatly affect network vulnerability in normal operating conditions. Thus, historical nonseismic performance matters.

An example of this behavior, showcasing the effects of abnormal operating conditions on WDNs is the case of intermittent water supply (IWS). Fig. 1.6 depicts the behavior of a case-study WDN under IWS operations, graphing the survival probability of house connection pipes vs. their time to failure (TTF), for four time periods (A, B, C, IWS) which are discussed in detail in Chapter 3 (p. 131). As shown in Fig. 1.6, IWS reduces the survival probability of several WDN components (such as the house connection pipes), thus increasing their vulnerability and, along that, the vulnerability of water distribution networks. This increase in vulnerability depends on the severity and frequency of past break events and on the topology of the network.

1.6.1 Nonseismic Vulnerability Assessment

The studies and methods reported upon in literature with regards to nonseismic assessment of WDN are primarily on deterioration modeling, pipe-break forecasting and network monitoring. The intent has been to assist owners of water distribution networks in improving their understanding of the systems' behavior over time, their deterioration rate and their reliability with respect to presumed risk factors, so that they could intelligently arrive at "*repair-or-replace*" decisions on a more scientific basis. The studies have traditionally attempted the identification of statistical relationships between water-main break rates and influential risk factors. Most such studies show a relationship between failure rates and the time of failure (age of pipes), and some of them suggest a method to optimize the replacement time of pipes.

Shamir and Howard [152], for example, reported an exponential relationship, and Clark [44] developed a linear multivariate equation to characterize the time from pipe installation to the first break. Both studies used a multivariate exponential equation to determine the breakage rate after the first break. A study by Andreou et al. [8,9] suggested a probabilistic approach consisting of a proportional hazards model to predict failure at an early age, combined with a Poisson-type model for the later stages. They further asserted that stratification of data (based on specific parameters) would increase the accuracy of the model. A nonhomogeneous Poisson distribution model was later proposed by Goulter and Kazemi [66] to predict the probability of subsequent breaks given that at least one break had already occurred. Finally, Kleiner and Rajani [96] developed a framework to assess future rehabilitation needs using limited and incomplete data on pipe conditions.

More recent research on WDNs extended the studies on reliability with additional risk factors and analyses tools. Prasad et al. [140] outlined a multi-objective genetic algorithm and introduced a new reliability measure, called network resilience, that tried to provide surplus head above the minimum allowable head at nodes and reliable loops with practicable pipe diameters. A number of frequently-used methods for the risk analyses of water supply systems were presented by Tuhovcak et al. [161]. The methods address the identification of qualitative and quantitative risks posed by the individual system components, the evaluation methods and interpretation of results, with an emphasis on the Hazard Analysis and the Critical Control Points (HACCP) method. Park [128,129] presented a method to assess and track changes in the hazard functions between water main breaks by using the proportional hazards model.

Several findings on risk assessment and prioritization of "repair-or-replace" actions were reported by Christodoulou et al. [27,28,33,38] based on neurofuzzy systems, survival analysis, and geospatial clustering of WDNs under both normal and abnormal operating conditions. Their findings reinforced, among other, the importance of risk factors such as pipe age, pipe diameter/length, pipe material, and of the network operating conditions, and pointed out the importance of the number of previously observed breaks (NOPB) as a risk-of-failure factor.

Evaluation of the probability of failure was also addressed in the work reported by Carrion et al. [21], who analyzed failure data registered in a water supply network and then used an extended version of the Nelson–Aalen estimator, modified in order to accommodate left-truncation besides right-censoring (LTRC). Influencing factors on water pipes survival were identified and by use of a semi-parametric model based on the Cox proportional hazards model the effect of each factor over the failure risk of a pipe section was estimated. Evaluation of the residual lives of water pipes was also the focus of work by Park et al. [132] and Park [130], who used the proportional hazards model (PHM) as the basis for the calculation of the economically optimal replacement times of pipes. The survival times used in the proportional hazards modeling process were defined as the economically optimal replacement times of pipes, and the break rate of an individual pipe was estimated using the General Pipe Break Model (GPBM).

In terms of introducing a network's topology to its risk level, recent work by Pinto et al. [137] introduced several theoretical concepts of a proposed theory of vulnerability of water pipe networks (TVWPN), based on the structural vulnerability theory. Related work has also been reported by Yannopoulos and Spiliotis [177], who presented a methodology for evaluating water distribution system reliability based on the minimum cut-set approach, combining the mechanical reliability and the hydraulic reliability of a WDN.

1.6.2 Seismic Vulnerability Assessment

Seismic risk assessment of lifeline systems has also been the subject of extensive past research. For example, Nuti et al. [121] proposed a methodology for the reliability assessment of electric power grids, water and road systems, neglecting the interdependence between the networks, while Poljnašek et al. [139] proposed a method for gas and electricity transmission networks considering the increased vulnerability due to interdependency. Cimellaro et al. [42] proposed a performance index for evaluating the functionality of a road network during extreme events such as earthquakes that can be utilized within the general PEOPLES framework for measuring disaster resilience indices for a community at various scales. Further, Esposito [53] worked on the seismic performance evaluation of gas distribution networks and produced fragility curves for these systems.

In terms of water supply networks, Romero et al. [146] discussed the possible seismic hazards and presented results corresponding to a severe event in California, and Wang et al. [169] proposed a methodology for seismic risk assessment and for the identification of critical links of water supply systems. The US Federal Emergency Management Agency (FEMA) has also developed the nationally applicable standardized methodology and software program HAZUS-MH MR3 [119], which estimates potential losses from earthquakes, hurricane winds, and floods.

The research has dealt in general with all lifelines and their interdependency, and also specifically with water distribution networks. In terms of water supply networks, work on their seismic performance can be traced back in the early 1980s, with Ariman and Muleski [11] presenting an updated and detailed review of the earthquake response and seismic-resistant design of underground piping systems, noting the need for additional research work on WDN and citing as major difficulty the serious lack of field data. Moghtaderizadeh et al. [115] presented a method for seismic reliability assessment of lifeline networks (including water distribution systems), analyzing lifeline component failures resulting from ground shaking and fault differential movements. The method utilized a fault-rupture model and a polynomially-bounded algorithm for computing the lifeline network reliability.

A method for assessing the seismic performance of a water delivery system with the aid of geographic information system (GIS) technology was presented by Hwang et al. [78], in which pipe damage from ground shaking and soil liquefaction were estimated separately and two types of pipe damage (leak and break) were included. The serviceability of a water delivery system was determined from the connectivity and flow analysis of a seismically damaged water network, which was established from a Monte Carlo simulation.

Shortly after, a review of the seismic response of buried pipelines was offered by Datta [47], who considered the modeling of the soil–pipe system and the seismic excitation. Also, Isoyama et al. [79] presented a practical procedure for estimating seismic-related damage to water pipes, considering the pipe material and diameter and also the ground conditions, the liquefaction susceptibility and the seismic intensity. A standard fragility curve was proposed as a function of peak ground acceleration based on observation data from the 1995 Kobe earthquake. This standard fragility curve was modified by the aforementioned factors, and the relationship between topography and damage rate was analyzed qualitatively using GIS.

Yang et al. [174] presented a probabilistic model of seismic damage prediction of buried pipeline, in which both the seismic response and the resistance of pipe structures were treated as random variables. A disjoint algorithm for reliability analysis of a network was then presented, in which sharp-product operation is adopted to construct a disjoint minimal path set of the network.

A discussion by Laatch et al. [99] based on a study of the first four years of the American Lifelines Alliance (ALA) project concluded that the ALA projects had resulted in new national consensus guidelines and standards from ANSI-approved standards organizations.

Pitilakis et al. [138] presented the RISK-UE methodology for the seismic risk assessment of utility systems (potable water, waste-water, gas system, telecommunication, electric power) and transportation infrastructures (port, airport, road, and railway systems). The methodology provides a uniform basis for the reduction of the consequences of lifeline damages in urban areas by use of a detailed inventory for every element. This approach also includes a reliable seismic hazard assessment, appropriate selection of fragility models, estimation of the "global value" and economical impact of lifeline damages and losses.

Jacobson and Grigoriu [80] discussed how the seismic performance of a water supply system depends on the individual performance of its components and on system configuration. Complete fragility analysis is performed on pipes, including hazard due to both seismic wave and fault displacement. The influence of seismic wave hazard is assumed to be more prominent for pipes located far away from a seismic source, while fault displacement hazard is assumed to affect only pipes located close to the seismic source, i.e., faults.

The study by Unen et al. [163] is an effort to more reliably and accurately assess the seismic response of utility network systems during a major seismic event by topologically modeling the lifelines as a system of interacting networks.

Wang et al. [169] described a process for seismic risk assessment and identification of critical links of water supply systems subjected to earthquakes. The probabilistic performance of water supply systems is measured using the system serviceability index (SSI), the damage consequence index (DCI), and the upgrade benefit index (UBI). Direct Monte Carlo simulation was used for estimating the performance indices, and it was then shown that the probabilistic characteristics of SSI can be attributed to system characteristics (e.g., demand distribution pattern) of the water supply system, while UBI was shown to be the primary index in seismic mitigation, with critical links being the pipes. It was also recognized that the values of UBI corresponding to different upgrade scenarios can be estimated using conditional samples from a single run of direct Monte Carlo simulation instead of repeated runs.

Yazdani and Jeffrey [178] noted that topological aspects of system resilience, viewed as the antonym to structural vulnerability, can be assessed in connection with the network architecture, robustness, and loop redundancy. Deterministic techniques from complex networks and spectral graph theory were utilized to quantify well-connectedness and to estimate the loop redundancy. They assessed the node/link failures and isolation of the demand nodes from the source(s) by using graph connectivity and expansion properties. Among other measurements, two metrics of meshedness and algebraic connectivity were proposed for quantification of redundancy and robustness, respectively.

Romero et al. [146] discussed the possible seismic hazards and presented results corresponding to a severe event in California, while Wang et al. [169] proposed a methodology for seismic risk assessment and for the identification of critical links of water supply systems.

Rajah et al. [142] discussed how, for pipelines, fragility curves would depend on the joint types and characteristics of the pipes (especially their axial, translational, and rotational flexibility), and they then proceeded to classifying the pipes into two major groups: segmented (flexible joints) and continuous (rigid joints). They also noted that "*the most commonly available fragility formulations for pipelines are based on damage and repair data collected after past earthquakes on segmented pipelines, and are primarily empirical. Since continuous pipelines typically perform better in a seismic event, direct damage data available from past earthquakes on continuous pipelines is limited. Therefore, fragility formulations for continuous pipelines are typically developed using damage data collected on segmented pipelines and modified to reflect the seismic performance of continuous pipelines in practice. This provides a reasonable approach to estimate seismic risk and has been proposed by the American Lifeline Alliance (2001) [5] for water pipelines. A disadvantage of this approach is that it fails to provide a mechanism to account for variability in design codes, regulations, requirements, and practice.*"

Pineda-Porras and Ordaz [136] presented a comprehensive state-of-the-art revision of the seismic parameters that have been employed as damage indicators for pipelines, and the most important seismic fragility functions proposed to-date, describing two prediction models for the Mexico City water system. The results of this research reveal that a previous damage estimation study for this network, based on a PGV (peak ground velocity) fragility function, overestimated the expected number of pipe repairs caused by earthquakes with magnitudes around 8.0–8.1 and lower, and underestimated the damage for stronger earthquakes.

Fragiadakis et al. [59,61] presented a framework for the seismic risk assessment of water supply networks based on the pipe vulnerabilities that are determined considering survival curves obtained from historical data of damage. The network reliability is assessed using Graph Theory, while two alternative approaches are compared for calculating the system reliability: the path enumeration method and Monte Carlo simulation. The methodology proposed can estimate the probability that the network fails to provide the desired level of service and thus allows the prioritization of retrofit interventions and capacity-upgrade actions for the existing water pipe networks.

Esposito et al. [53] reviewed the seismic risk assessment of the L'Aquila gas distribution network in a performance-based earthquake engineering framework. The process employed probabilistic seismic input analysis, fragility models for the evaluation of gas system components. Their analysis indicated that system performance may be underestimated when spatial correlation and ground failure are ignored.

1.7 CONCLUDING REMARKS

Nowadays, with the evolving global climate changes, the higher frequency and severity of droughts (especially in arid regions), the increased urbanization and the deterioration of urban water distribution networks, the safeguarding and the sustainable utilization of such networks has become more pressing than ever. It is within these pressing, and continuously worsening, global conditions that the vulnerability assessment of water distribution networks under both normal and abnormal operating conditions should take a prominent role in the operations and maintenance of water distribution networks. An assessment that should aim at not only the identification of the sources of vulnerability for the WDNs and their components, but also at the analysis of and knowledge-extraction from WDN performance datasets, and at the mitigation of these sources of vulnerability.

In the fight for reduced WDN vulnerability, WDN agencies have at their disposal tools such as the implementation of DMAs, supervisory control and data acquisition (SCADA) systems, water pressure management methods, water loss detection and reduction methods, and WDN modeling. Despite, though, that all these tools are made available to the scientific community, to practitioners and to WDN agencies, and that the related knowledge-base is continually updated, the value in using the aforementioned tools and methods lies in the customization and localization of them to fit the specific needs of a WDN.

Subsequent chapters will discuss related general concepts on the vulnerability assessment of water distribution networks, as well as present a number of case studies showcasing how such localization can be achieved and how WDN modeling and analysis can generate knowledge, which in turn can be transfused to operations and maintenance practice.

CHAPTER 2

Vulnerability Assessment of Water Distribution Networks Under Normal (Continuous Water Supply, CWS) Operating Conditions and Nonseismic Loads

Vulnerability refers to the inability to withstand the effects of a hostile environment.

Wikipedia, The Online Encyclopedia.

2.1 INTRODUCTION

A deterioration, vulnerability, and time-to-failure mathematical model is at the core of an integrated tool for the monitoring and management of WDNs. A modern and realistic mathematical model is structured in two parts. The first one is the data repository (database) of historical information on the performance of the network, and the second is a data analysis component. Chapter 2 presents both a simple, and minimal, database architecture applied to a case-study WDN (that of Nicosia's Water Board, WBN), as well as the mathematical models used in processing the aforementioned database of water loss incidents and the deduced knowledge extracted from such an analysis.

Even though the design and development of the aforementioned database follows the needs and requirements of the WBN, the database was structured and is presented in a manner that is transferable and applicable to any other agency which is related to the operations and maintenance of urban WDNs. The developed mathematical model can be used in forecasting water loss incidents and in extracting useful information and conclusions on the WDN in study. Through the mathematical model developed in this research work, the network owners may evaluate the condition of their network, assess historical incident and risk-of-failure data, and prioritize O&M work based on the inherent risk and cost of selected actions. The

Urban Water Distribution Networks
DOI: https://doi.org/10.1016/B978-0-12-813652-2.00002-5

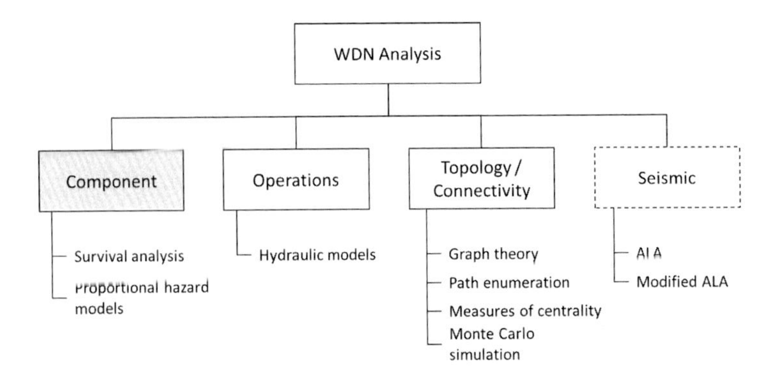

Figure 2.1 WDN component analysis (schematic).

goal is to enable field engineers and maintenance technicians of water utilities to better manage collected condition assessment information about their networks, to process historical records with a number of analytical and numerical models, to identify underlying data patterns with artificial intelligence techniques and eventually to assess the corresponding risk of failure of each WDN element.

The analysis included in Chapter 2 on continuous water supply operations (and subsequently in Chapter 3, p. 131, on intermittent water supply operations) focuses on a WDN's components (Fig. 2.1) and their vulnerability over time, with the developed data models stratified and clustered by several sources of vulnerability.

2.2 VULNERABILITY-RELATED DATABASE MANAGEMENT SYSTEM FOR WATER DISTRIBUTION NETWORKS

2.2.1 Overview

The first of two objectives of Chapter 2 is the presentation of a simplified architecture of a database management system (DBMS) used in assembling the knowledge related to the operational behavior of water distribution networks, as applied to the Water Board of Nicosia (WBN), Cyprus. The development of the database was achieved upon a review of relevant literature and based on knowledge gained from the authors' participation in relevant research programs, as well as from the exchange of views and experiences with the technical staff of several water agencies.

Even though WDNs differ from one to another, since each WDN has specific characteristics that set it apart from other WDNs, there exist specific elements that are common to all WDNs. Thus, apart from data related to the specific characteristics of a WDN, the underlying DBMS consists of several standard information fields which are transferable to other WDNs. Such information is, for example, an inventory of the WDN pipes and valves (location, physical properties), historical records of leakage incidents (date/time, location, type, cost, etc.) and water flow data (pressure, flows, etc.). In essence, it is necessary for a WDN DBMS to include information on the factors that affect the lifetime of the WDN in observation in ways that such data can be analyzed, used in modeling the behavior of the WDN over time, and give valuable insights into a WDN's past, current, and future vulnerability.

2.2.2 State of Knowledge

Research related to WDN modeling has concluded that a number of risk factors contributes to pipe failures, some of them being time-dependent and the rest being time-invariant. The center of attention for the most of such research activities was placed on expressing the relation between pipe failures and the factors related to it. Among the many risk factors contributing to pipe failure, as reported by the various research efforts to date, the main ones are reported to be the age, diameter, and material of the WDN pipes, the historical data of previous breaks ("*number of observed previous breaks*"; NOPB), the network's operating pressure, the temperature fluctuations in the soil housing the water pipes (typically over a span of 24 hours), and the external loadings the WDN is subjected to (e.g., traffic, construction work, and seismic loads).

The studies to date pertaining to the management of water distribution networks have shown a strong relation between the failure rate and the age of WDN pipes (and by extension the age of a WDN). Shamir and Howard [152] referred to an exponential relationship between the failure rate and the age of the pipes, while Walski and Pelliccia [168] applied such an exponential relationship to WDN data stratified according to the material of pipes. Also, Clark, Stafford and Goodrich [44] referred to the development of a linear multivariate equation that characterizes the time of the placement/installation of the pipeline until its first failure. They also developed an exponential multivariate equation for estimating the rate by which the pipeline exhibits breaks (leaks) after its first break. A thorough literature review on the topic by O'Day [122] reinforced the conclusion that a pipeline's age is a strong evidence for possible future failures. Further, O'Day argued that the location of a pipeline is a risk factor. In their work, Kettler and Goulter [90] presented an inverse linear correlation between the failure rate and the pipe's diameter, as well as a linear relationship between a failure and the pipe's age. In the work by Park et al. [128], the hazard characteristics with regard to the factors causing breaks to several of pipe materials were identified. Their results confirmed that pipes whose material type and/or NOPB differ have different hazard rates. A similar analysis by Christodoulou [27] further asserted the findings, producing survival curves for several NOPB and material-type classes of WDN pipes.

A summary of presumed risk-of-failure factors and thus of related data requirements for a WDN DBMS was listed by the US Environmental Protection Agency (EPA) [155], in which a comparison was also provided of the eight main rehabilitation models for WDNs used in the USA and in Europe. On the basis of these eight models, the data requirements for the development of a WDN rehabilitation model are as presented in Table 2.1.

The findings of the aforementioned research work were also reinforced by the research activities of [30,33,37,167], who have developed and documented the main sources of uncertainty and the risk-of-failure factors for water leakage in WDN. The researchers presented several alternative numerical and artificial intelligence (AI) methods which were developed for, and tested on, the improvement of risk analysis models (primarily statistical) for the WDNs of Limassol and New York City. Since the AI methodology allows for the study of large datasets without a priori knowledge of the interrelationships and distributions of the input parameters, its use enabled the researchers the study of relationships and interactions within a huge and seemingly random database and of relationships which are difficult to detect and study statistically. AI also helps in deducing conclusions on the relationships between various risk factors through pattern recognition. Hence, relationships hidden in the historical network data were at first detected and then extracted and turned into knowledge and decision rules. Further, the interaction and correlation of various uncertainties and their sensitivity on network changes was studied by the use of correlation and sensitivity analysis. The aforementioned research works converged on the following factors as key contributors to pipe failures, including them as input parameters in their proposed DBMS: a pipe's location, "birth" date, diameter, material, thickness, flow pressure, and depth of installation (in relation to the surrounding groundwater level), as well as the pipe's number of previously observed breaks, incident date, observed leakage volume,

Table 2.1 Data requirements for the pipe rehabilitation model of a WDN.

Data description	Importance
Pipe Material	Required
Pipe Age	Required
Pipe Length	Required
Pipe Diameter	Required
Number of Observed Previous Breaks (NOPB)	Required
Soil Data	Highly significant
Traffic Data	Highly significant
Pipe Location	Highly significant
Water Pressure	Highly significant
Failure/Defect Type	Highly significant
Pipe Condition	Highly significant
Type of Corrective Action	Highly significant
Type of Joints	Usefull
Leakage Rates	Usefull
Incident Date	Usefull
Date Pipe Repaired	Usefull
Date Pipe Visually Inspected	Usefull
Economic Data	Usefull
Rehabilitation Cost	Usefull
Utility Locations	Usefull
Tree Locations	Usefull
Elevation Contours	Usefull

"repair or replace" action, rehabilitation cost, and time – all in relation to a pipe's ID which serves as a unique identifier for each pipe's historical performance.

2.2.3 DBMS Architecture

As aforementioned, and based on related literature, the parameters which are considered as necessary for the development of the DBMS are: the data referring to the pipe ("*ID*", "*Location*", "*Birth Date*", "*Diameter*", "*Material*", "*Thickness*", "*Flow Pressure*", and "*Depth*"), as well as data on the specific incidents affecting the vulnerability of the WDN in the study (i.e., "*Incident Date*", "*Date of Rehabilitation*", "*Rehabilitation Action*" (Repair or Replace), "*Rehabilitation Cost*", as well as "*Rehabilitation Time*", "*Groundwater Level*", and "*Observed Failure Type*"). These data fields are

Table 2.2 Vital pillars and key parameters contained in a DBMS for WDNs.

Pillar	Important input data
Pipes, Inventory Of	Pipe ID/Location/Birth Date
Pipes, Physical Properties Of	Pipe ID/Diameter/Material/Thickness/Depth of pipeline
Leakage Incidents, History Of	Pipe ID/Incident Date/Observed Leakage/ Repair Cost/Repair Time
Water-Flow Data	Pipe ID/Pipe Flow Pressure/Groundwater Level

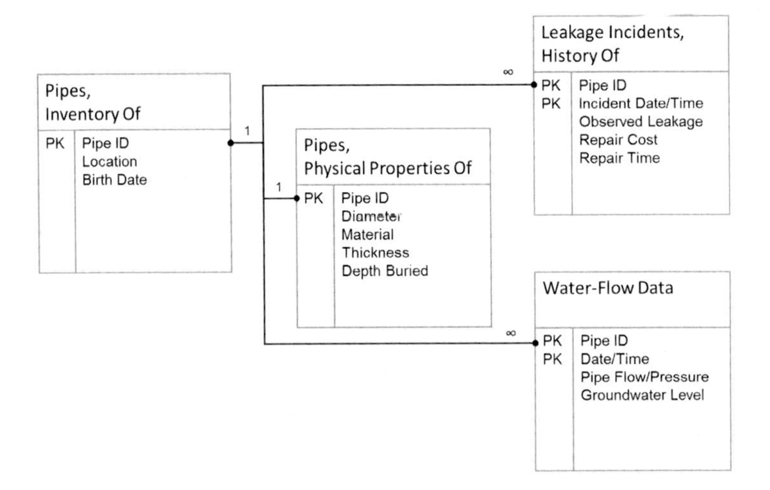

Figure 2.2 DBMS architecture (schematic).

grouped into four clusters (listed in Table 2.2) and treated as the proposed DBMS pillars.

The proposed simplified DBMS architecture (shown schematically in Fig. 2.2) provides the relational links between the aforementioned DBMS tables and their data fields, also showing the fields acting as primary keys (PK) for each of the DBMS tables. A primary key is a special relational database table column (or combination of columns) designated to uniquely identify all table records, and as such a unique identifier a primary key must contain a unique value for each row of data (null values are not allowed for primary keys).

The proposed DBMS architecture and related data (Fig. 2.3) also serve as the basis of the mathematical analysis for the vulnerability assessment of WDNs.

Street Name	DMA	Sub-DMA	Findings		Action		Rehabilitation's Details				Reason
							Diameter		Material	Location	
List of street names	List of DMA names	List of Sub-DMA names	A.V.	Nipple	Cleaning	Plug	3/8"	50mm	A/C	Close	Damage
			Area	No Problem	Closing	Reconnection	1/2"	63mm	D.I.	Close (using T)	Material/Hardware Failure
			B. V.	Not in the WDN Jurisdiction	Connection	Recording	3/4"	75mm	Galvanized	Crossing	Material/Hardware Failure (Deterioration)
			Bush	Other	Conversion	Reinstallment	1"	80mm	P.V.C.	Crossing (using T)	Material/Hardware Failure (Material Quality)
			C.V.	Pump	Disconnection	Removal	1 1/4"	90mm	Plastic		Material/Hardware Failure (Network Load)
			Clamp	Ring-Main	Expansion	Removal	1 1/2"	100mm			Tree Roots
			Closed Area	S.V.	Finding	Repair	1 3/4"	110mm			
			Closed B.V.	Saddle	Indictment	Replacement	2"	150mm			
			Closed Ferrul	Saddleferrul	Inspection	Soil Redeposition	3"	200mm			
			Closed S.V.	Seal	Installment	Telemetry	4"	250mm			
			Coupling	Sieve	Lifting	Tightening	6"	300mm			
			F.H.	Stealing Water	Lowering	Unblockage	8"	350mm			
			Ferrul	Stuffy House Connection	Maintenance	Warning	10"	400mm			
			House Connection	Tank	Opening	Wash out	12"	450mm			
			Low Area Pressure	Temporary Pipe	Other	Water Meter Sealing	16mm	500mm			
			Main Pipe	Unplugged House Connection			20mm	550mm			
			Manifold	Wasting Water			25mm	600mm			
			Maxifit	Water Meter			32mm	700mm			
			Maxistep	Water Meter Connector			40mm	800mm			
			Misc	Zone							

Figure 2.3 The fields used in the mathematical models.

Important Input Data

The main parameters needed for the proposed DBMS are presented and explained below.

- **Pipe ID** For each series of the mathematical analysis, each pipe should have a unique identification number which is subsequently used as primary key in the various DBMS data tables. One should note, however, that even though the "*Pipe ID*" is needed to identify a pipe and to track its performance over time, it is practically impossible for WDN managers to know the exact start and end point of each 6 m pipe within the network and to perfectly identify a pipe. As a side note, Water Boards (WBs) generally use 6 m-long pipes of hard plastic for the water mains, and one-piece soft plastic pipes for the house connections, all without spatially knowing where they are laid. Further, it is very complicated to define each house connection because two or three houses may be connected to the main pipeline through the same saddle. In order to address these shortfalls, the following are assumed within the context of this chapter:
  - *Street Level.* In order to simplify the problem and make the mathematical modeling of the distribution network feasible, the discretization of the main pipes and the house connections was made at "street level" (i.e., each street segment was considered as a single pipe). Consequently, the unique ID for each pipeline is set as a combination of the street name and the street's post code.
  - *Segment Level.* Street level analysis is not the most precise method though. Pipes do not run an entire street in a single segment, but rather a street consists of many pipe segments (typically of 5–6 m in length, each). Thus, preliminary vulnerability results are then drilled-in and the mathematical models are extended so as to deduce knowledge at segment level. The segments for each pipeline correspond to the street numbers where a pipe is located. For this analysis the unique ID for each pipeline is set to the street name and number, as well as the post code of the pipe's location.
- **Pipe location** For both the street and segment level analysis, the location of a pipe is the DMA name/number. In cases where the DMA is divided into smaller sections, the pipe's location is the name of the corresponding sub-DMA.
- **Pipe birth date** The pipe's birth date is the date on which the pipe in study was installed in the network. For the pipes that were replaced at one point in time, the birth date was taken as the pipe's replacement

date. For all pipes which are studied and there is no information about their installment day, this is assumed to be the time period in which water was officially provided by the WB authorities to the pipe's DMA.

- **Pipe diameter, material, and thickness** The physical properties of the network's pipes, as given by the WB authorities. These properties are updated after every replacement of a pipe.
- **Pipe flow/pressure and depth and groundwater level** The network's operational properties are also given by the WB authorities.
- **Incident date, observed leakage, "repair or replace" action, rehabilitation cost and time** This information is related to each leakage incident, as reported upon by the WB repair crews and technical staff.
  - *Incident Date.* The important dates for a leakage incident are the date that the leakage is reported and the date on which the leakage repair is completed. The date that is taken into consideration for the mathematical analysis is the date that the leakage incident comes to the attention of the WB.
  - *Observed Failure Type.* This relates to the exact location of the leakage, the failure type, and the reasons which caused the leakage.
  - *Repair or Replace.* A vital parameter for the mathematical analysis is the action that was taken by the repair crew, for it indicates the reset of the time counter (a repair action continues the age of the pipe, whereas a replace action assumes a new pipe is installed and thus resets the age counter of the pipe).
  - *Rehabilitation Cost and Rehabilitation Time.* Finally, the number of fittings and the length of pipe that were used, as well as the time that was needed for completing a "repair or replace" action and the number of labor-hours employed on the incident are recorded. The data is used in the estimation of the rehabilitation cost and duration of each leakage incident.

2.3 MATHEMATICAL MODELING

The second of two parts of this chapter relates to a proposed methodology and mathematical model for the processing of the WDN vulnerability DBMS. A mathematical vulnerability model is a proactive management strategy that assists network owners in evaluating the condition of their network, assessing historical incident and risk of failure data and in prioritizing the work based on the inherent risks and the cost of action.

The proposed model is based on two different mathematical approaches, statistical analysis and survival analysis, which are used for knowledge extraction from the previously developed DBMS and for drawing useful conclusions about the network's historical performance. The intent is for these mathematical models, coupled with the DBMS, to provide a useful tool for: (1) the analysis of WDN vulnerability data (typically manifested as water loss incidents); (2) the development of action priority lists for the maintenance of WDN pipelines (the "repair or replace" dilemma faced by WDN Agencies); and (3) the arrival at a more economical and efficient management of WDNs.

An originality of the developed mathematical model is the subsequent correlation of intermittent water supply periods (Chapter 2, p. 21) with the increased WDN vulnerability and with the frequency of future leak incidents. The intent is the study, through the statistical correlations, of the effects of the intermittent water supply (IWS) periods on the vulnerability of water distribution networks, and on the increase of the number of leakage incidents. Also by use of cost analysis, useful conclusions can be extracted as to whether WDNs are burdened by IWS operations, and whether the number of pipe breaks and of water leakages increases during and following IWS periods. In summary, the result of this analysis will ascertain whether IWS operations are an advised, good, and sustainable measure (or not).

2.3.1 State of Knowledge

The management of UWDNs is an always contemporary and in-demand research topic for the water industry. The worldwide reduction of potable water reserves has led researchers to UWDN research in the hope of finding ways to save water. There is, actually, an extended work in the research literature related to water loss and sustainable UWDN management, especially as applied to developing countries and countries with arid climatic conditions.

Defining the Problem

Water distribution network authorities (WDNA) are, on a daily basis, striving to upgrade their networks and repair (or replace) damaged pipes, in an attempt to mitigate the effects of pipe bursts and water loss and to maintain the uninterrupted transport of water to their consumers. In doing so, WDNAs have in mind four guiding principles:

- Keep the quality of the water at the highest possible standards,
- Improve their service to consumers,
- Operate networks cost-efficiently,
- Maintain networks cost-efficiently.

The use of new technologies and methods for monitoring, repairing and/or replacing aging infrastructure for the sustainable management of UWDN is not enough. Nowadays, as technology is rapidly evolving, dynamic modeling of the UWDN behavior is increasingly gaining more importance, and the modeling of a network's deteriorating infrastructure conditions should be center-staged. In fact, managers of water systems are looking for methodologies which are dynamically modeling a network's deterioration over time and proactively devising "replace or repair" strategies. The ultimate goal of water authorities is the minimization of their network's vulnerability, the maximization of the network's reliability, and the minimization of the operational and management costs, through an intelligent and efficient assessment of their network and by use of mathematical and/or numerical models.

Thus, UWDN agencies are in need of a mathematical tool which would model and simulate a WDN behavior, and by use of such mathematical models the agencies would be able to not only monitor their networks in real time but also be provided with a decision support tool for taking maintenance actions and reducing the network vulnerability.

Relevant Studies in the Literature

The uninterrupted supply of water to consumers, the reduction of network damages, the efficient use of water resources, and the effective management of a WDN are the ultimate goals for every organization which is in charge of such networks. Currently, a variety of WDN management techniques is available for use by the Water Boards. These techniques may either be in the form of guidance notes by the water associations [55,76,100,117], or of service packages by consulting firms, or of autonomous tools emanating from research products [75,103,135]. The majority of such management techniques are related to the measures taken by the network managers to expeditiously detect network leakages, and to their actions for restoring the networks to the previous operating condition.

The two main disadvantages of the techniques mentioned above are that (1) they do not prioritize simultaneous leakage incidents, and (2) the resulting precaution measures are not based on each specific network's topology but, rather, they are activated periodically across the network. This situation

Table 2.3 Classification of mathematical models according to Kleiner and Rajani (2001).

Class	**Model**
Deterministic	Time-exponential Time-linear
Probabilistic multivariate	Proportional hazard Time-dependent Poisson Accelerated lifetime
Probabilistic univariate group-processing	Cohort survival Bayesian diagnostic Semi-Markov chain Break clustering Data filtering

has prompted researchers to develop mathematical models which simulate the operating mode and condition of the WDN, so that they could enable WDN managers to prepare maintenance strategies as well as prioritization plans for the cases where multiple leakage incidents may be encountered.

Initially, researchers tried to develop a single-objective mathematical model expressing the failure pattern or system reliability, while subsequent studies expanded the work by presenting multiobjective failure models. Through the years, researchers have moved a step further by developing models which simulate UWDN behavior and include decision support systems (DSS) for the network managers. Finally, some recent research activities focused on abnormal operating conditions due to exogenous factors that affect the condition of a network. An overview of the mathematical models related to structural deterioration of the WDNs was provided by Kleiner and Rajani [97], with a summary of these different models, grouped into classes by article, presented in Table 2.3 and briefly discussed in succeeding pages.

Shamir and Howard [152] and Clark et al. [44], using different approaches, developed mathematical relationships which express the time from pipe installation until the first failure incident, and the failure frequencies after the first failure occurred. Walski and Pelliccia [168] investigated the correct sequence of steps needed for choosing the repair or replace action for failing pipes or for problematic network segments. Andreou et al. [8] used proportional hazards models to estimate failure at an early pipe age, and a Poisson-type model for the later stages. Also Goulter and Kazemi [66] presented a nonhomogeneous Poisson distribution model which predicts the probability of successive breaks after the first failure occurred.

The core equations of the mathematical models [96] mentioned above are presented in Table 2.4.

Rehabilitation-Oriented Models

Kleiner et al. [95] and Kleiner and Rajani [96] proposed a framework to assess future rehabilitation needs using limited and incomplete data with regard to pipe conditions, while a mathematical model for the optimal scheduling of pipe replacement was presented in Park [127]. The model develops an economically sustainable threshold break rate for the replacement of water mains within a WDN by minimizing the total cost resulting from replacement and repairs. The output is a design chart that displays the optimal threshold break rate as a function of the pipe diameter and length. The optimum number of repairs that a pipeline should have before its replacement can be calculated using the chart.

The approach presented by Mailhot et al. [109] identifies the number of breaks that a pipe should have in order to be replaced. The model considers the time to failure between two successive pipe breaks as random variables. The time to failure for the first two breaks of a pipe, is described by probability density functions with nonconstant hazard functions while the rest of time intervals between two successive breaks is described by probability density functions with nonconstant hazard functions. Expressions for the time evolution of the average number of pipe breaks per unit time have been derived and an optimal replacement criterion is defined based on a cost function using conditional probabilities for estimation of the expected future costs. The critical pipe break order, at which the replacement of a pipe should be made, is identified by the minimization of the cost function.

A strategy for the optimal replacement of water mains by use of a Bayesian approach, *EPAnet* and genetic algorithms (GA) is presented by Dridi et al. [50]. Two key parameters of the WDN's deterioration, the structural integrity and the hydraulic capacity, are used by the model which in turn combines hydraulic and mathematical models. The rehabilitation of the network is carried out in five-year plans, and at the beginning of each planning phase the information related to the status of the WDN pipes is updated so that each successive plan is scheduled under the new network conditions.

An approach for optimizing the replacement and rehabilitation activities for WDN pipes, based on the assumption that the occurrence of breaks in

Table 2.4 Mathematical models of structural deterioration of WDNs.

Mathematical model	Notation	
Shamir and Howard (1979)	t	Time elapsed (from present; in years)
$N(t) = N(t_0) \times e^{A(t+g)}$	$N(t)$	Number of breaks pre unit length per year
	$N(t_0) = N(t)$	At the year of installation of the pipe
	g	Age of the pipe at the present time
	A	Coefficient of breakage rate growth
Clark et al. (1982)	x_i	Regression parameters
$NY = x_1 + x_2 \times D + x_3 \times P + x_4 \times I + x_5 \times RES + x_6 \times LH + x_7 \times T$	NY	Number of years from installation to first repair
	D	Diameter of pipe
	P	Absolute pressure within a pipe
	I	% of pipe overlain by industrial development
	RES	% of pipe overlain by residential development
	LH	Length of pipe in highly corrosive soil
	T	Pipe type (1 = metallic, 0 = reinforced concrete)
Walski and Pellicia (1982)	t	Time elapsed (from present - in years)
$N(t) = C_1 \times C_2 \times N(t_0) \times e^{A(t+g)}$	$N(t)$	Number of breaks pre unit length per year
	$N(t_0) = N(t)$	At the year of installation of the pipe
	g	Age of the pipe at the present time
	A	Coefficient of breakage rate growth
	C_1	[Break frequency for (pit/sandspun) cast iron with (no/one or more) previous breaks] over [overall break frequency for (pit/sandspun) cast iron]
	C_2	[Break frequency for pit cast pipes 500 mm diameter] over [overall break frequency for pit cast pipes]
Andreou et al. (1987) – Early Stage	T	Time to next break
$h(t, Z) = h_0(t) \times e^{(b^T)} \times Z$	$h(t, Z)$	Hazard function
	$h_0(t)$	Baseline hazard function
Andreou et al. (1987) – Late Stage	Z	Vector of covariates
$h = \lambda = e^{(b^T)} \times Z$	b	Vector of coefficients to be estimated by maximum likelihood
	h	Hazard (constant at the late stage)
Goulter and Kazemi (1988)	P	Probability of failure
$P(x) = (m^x \times e^{-m})/x!$	m	Mean number of subsequent failures occurring in the cluster domain
	x	Number of subsequent failures occurring in the cluster domain

a pipeline segment follows a nonhomogeneous Poisson process, was presented by Hong et al. [72]. In their study, the optimal replacement time was extracted through the minimization of the expected annual average cost during the service period of the pipeline segment.

Christodoulou et al. [32] presented an integrated decision support system (DSS) for arriving at "repair-or-replace" decisions, as part of a long-term pipe system asset management program that could be used by the authorities so as to improve the reliability of their WDNs.

Xu et al. [172] presented a pipe break prediction model for the WDN of Beijing City. The mathematical model uses two data-driven techniques: genetic programming and evolutionary polynomial regression. The model uses data concerning the recorded pipe break incidents for the pipelines of the specific network during the period from 1987 to 2005. The results showed that the model has a great capability to obtain reliable predictions and hence it can be used to prioritize the pipes rehabilitation. Subsequently, Xu et al. 2013 [172] presented the development of a DSS which was used in the deployment of cost-effective pipe maintenance plans for WDNs. The model comprises a prediction tool which was developed using genetic programming and an economically optimal pipe replacement tool. As a case study, the optimal pipe replacement time for a specific part of Beijing's WDN was determined, using pipe properties and pipe breakage data from 2008 to 2011.

A research report published by Grigg et al. [67] of the Water Research Foundation (WRF) addressed the need for the rehabilitation of WDNs and presented a tool for assessing risk and organizing data to aid in capital investment planning for pipelines. In addition, methodologies to assess whether and how to rehabilitate or replace a pipeline by integrating information about cost of failure into asset management decisions were offered.

Decision Support Models Targeting the Condition Level

The research works by Vanrenterghem et al. [167], Aslani [12], and Christodoulou et al. [30,31] dealt with New York City's WDN, modeling its degradation and providing a risk analysis framework for evaluating the structural degradation of the water mains using statistical modeling techniques and neurofuzzy systems. Subsequently, Christodoulou et al. [26, 38,29,31,33,37] developed a framework for integrated GIS-based management, risk assessment, and prioritization of water leakage actions. Similar to those was research work by Vairavamoorthy et al. [165].

Park [128] proposed a proportional hazard model which estimates and then tracks changes of the hazard function for the various pipe types in an UWDN, between the observed water main failures. Park et al. [131] presents two failure rate models, for the pipes failure rate and the estimation of the optimal replacement time of the pipes in a WDN. The research effort examines the performance of the log-linear rate of occurrence of failure and the power law process by the use of maximized log-likelihoods for different modeling approaches, in which the method of observing failures differs (failure-time based data or failure-number based). It was proved that the "failure-time based" method is better compared to the "failure-number based." Also, the log-linear rate of occurrence of failure modeling has better performance than the law process when the "failure-time based" method is used. Subsequently, Park et al. [134] presented more sophisticated proportional hazard models for the time period within successive failures of water mains, which included the time-dependent effects of covariates. Their case study includes 150 mm individual cast iron pipes of a WDN which are categorized into seven groups according to past break history. A proportional hazard model is developed for each one of the seven groups. This methodology provides more information on the status of the pipe breaks than the one in Park et al. [131]. This extra information pertains to the main effects of the factors of failure (including the effects of time to the failure), to the changes in the condition of the pipes as more breaks occur and to the survival probabilities of the pipes for each type of break. The hazard analysis was also applied by Tuhovcak et al. [162] who combined it with Critical Control Points methods for the risk analysis of UWDN.

Kanakoudis and Tolikas [87] suggested a management system that estimates the performance level of a distribution network if a new action is being taken on it. The proposed model takes into consideration the repair and replacement costs of failing pipes. Further, Kanakoudis [86] presented a methodology for the hierarchical analysis of the preventive maintenance policy of WDN that uses performance indices which are estimated by an economic analysis of the cost of actions taken for the network rehabilitation. The goal for Kanakoudis and Tsitsifli [88], Tsitsifli and Kanakoudis [159], Kanakoudis and Tsitsifli [84] and Tsitsifli et al. [160] was the development of a model that could correctly classify water distribution pipes, to define the pipe characteristics responsible for the behavior of the pipes and to predict whether a pipe would fail or not. The model uses discriminant analysis classification (DAC) to categorize the pipes in failure or success.

The results show that DAC can be a good method for predicting the reliability of the WDN when the available information by the authorities is of good quality.

A method for risk analysis of WDN pipes, which can be used as a DSS for asset renewal plans by the water authorities, was suggested by Salehpour [149]. The model analyzes the water-quantity risks associated with the failure of the pipelines of a WDN. The model uses as a case study one of the DMAs of the WDN in Goteborg city. It combines the fault tree analysis with a software tool (CARE-W) to analyze the risk of pipe failure. The fault tree model provides probability estimations, whereas the software enables estimations of the consequences of the pipe failure. A dynamic fault tree method is also presented by Lindhe [105]. The model evaluates the uncertainties and provides information on risk levels, failure probabilities, failure rates and downtimes of the system. It also identifies where risk-reduction measures are needed most. Furthermore, the model comprises an economic analysis for the identification of the most cost-effective risk-reduction alternative. In addition, two alternative models for risk-based, multicriteria decision analysis for evaluating and comparing risk-reduction measures have been developed which are based on risk ranking. They can consider uncertainty in estimates and include criteria related to different risk types and economic aspects.

Tabesh et al. [156] worked on data-driven modeling techniques, such as artificial neural networks (ANN) and neurofuzzy systems. They utilize more comprehensive and accurate prediction tools for the pipe failure rate and for better depicting the pipe network. These data-driven modeling (DDM) techniques allow the inclusion of multiple risk factors within the analysis, which make it more complicated but at the same time more accurate and realistic. This analysis method is also used by Christodoulou et al. [31,33,37,39].

Fattahi and Fayyaz [57] developed a multiobjective model having as variables the water distribution cost, the volume of leakage, and the social satisfaction level. The mathematical model was used for optimizing urban water management.

A project that involves the development of a complete risk-based asset management model was presented by Park et al. [133]. It combines a repair/replacement prioritization tool for the assets of WDNs and a comprehensive strategy for the maintenance of existing infrastructure. The model comprises of geographical information systems (GIS), risk-based asset prioritization, and hydraulic modeling. The project was implemented in Florida (USA) to enable the Tampa Water Department to assess the condition of its WDN, to assign risk ranking, and to prioritize repair/replacement actions.

Bentes et al. [16] presented a method for assessing the vulnerability of WDNs. Their suggestion is based on three water pipe network examples, focusing on tracing the weakest parts of the distribution system and giving guidance for improving the condition of the network.

The work by Christodoulou et al. [40] presents a spatio-temporal analysis model that can be used as a DSS for increasing the efficiency of maintenance strategies related to UWDN. The suggested model utilizes classical statistical tools, neurofuzzy systems and GIS-based spatio-temporal clustering and visualization techniques. The spatio-temporal analysis allows for spatio-temporal clustering and pattern recognition, it helps devise repair-or-replace strategies and it reinforces the belief that intermittent supply increases the vulnerability of UWDN.

WDN Monitoring Models

An automated model for WDN monitoring that consists of a low-cost sensor technology and an Artificial Neural Network (ANN) model is presented in Khan et al. [91]. The sensor, which provides input to the mathematical model, measures the opacity or cloudiness of the water flow within a WDN. The ANN model uses the time series data produced by sensors to construct an empirical model for time series prediction and classification of events. The field results show that the proposed model can be a monitoring tool for WDN and it can be used for leak detection.

"*Project Neptune*" [151] is a research work that targeted the development of a complete management system for the WDNs. The model is based on three pillars: the enhancement and upgrade of the WSN monitoring system, the development of tools for the management of pressure and energy (consumed by the system), and the development of an integrated DSS for the evaluation of rehabilitation strategies.

In Eliades and Polycarpou [52] the problem of leakage detection (of small magnitude) within a WDN is formulated and a mathematical model for leakage detection is presented. Algorithms that analyze the discrete inflow signal, which are recorded by standard flow sensors, are designed so the leakage incidents of the WDN be detected early. The model is based on multiple components: a yearly seasonal, a weekly periodic, and an uncertainty component, along with the leakage fault signal. The results proved the efficiency of the proposed mathematical to detect leakages of smaller magnitudes, in comparison to the night-flow technique that is the frequently used method by the water authorities for this purpose.

Models Targeting Specific Parts of WDNs

A model that identifies pipes and valves of high importance within WDNs was presented by Jun et al. [83]. The model consists of two failure analysis methodologies, Pipe-by-Pipe (PPFA) and Valve-by-Valve (VVFA), which were developed to prioritize the importance of pipes and valves in a water distribution system. The model is based on a matrix algorithm that can be easily implemented to any WDN, and analyses can be performed very efficiently regardless of the size of the water distribution system by coupling with the *EPAnet*. Results of PPFA and VVFA can be used to prioritize the order of maintenance of a water distribution system.

Models Studying the Behavior of WDN Under Abnormal Conditions

Andey and Kelkar [7] compared continuous water supply with intermittent water supply in four Indian cities. Christodoulou and Agathokleous [28] reported on the performance of UWDNs under intermittent water supply (IWS), citing that of the 12,000 water loss incidents in a two-year period the majority were related to house connections and small-diameter pipes, with an increase in incidents during the intermittent supply period of about 28% compared to the normal operating conditions period (uninterrupted supply). This finding is in agreement with the fourth finding of the analysis by Christodoulou et al. [38] on the increased risk of failure of house-connection pipes.

A framework for the seismic risk assessment of aging water supply networks, based on the American Lifelines Alliance (ALA) and coupled with historical performance data and pipe vulnerabilities that are determined considering survival curves, is presented by Fragiadakis et al. [60,61]. The network reliability was assessed using Graph Theory. Also two alternative approaches, the path enumeration method and a Monte Carlo simulation, were compared for calculating the system reliability. Further, Fragiadakis and Christodoulou [59] presented an application of the aforementioned model on a small-scale DMA in the WDN of the city of Limassol (Cyprus). Based on the above, Christodoulou and Fragiadakis [35] discuss how the ALA guidelines' repair rate metric underestimates the seismic effects on the vulnerability of a network, and recommend how the damaged and undamaged network states should be included in the calculation of a pipe's probability of failure.

Discussion on Literature Review

The literature review discloses a range of mathematical models which simulate the behavior of the pipes in WDNs. The models primarily use data associated with the water mains and not data on the house connections and on fittings, which are also part of the underground/buried pipe network. As a result, existing mathematical models can not accurately model the full (realistic) behavior and operating conditions of the entire pipeline system in a WDN. Additionally, the analysis of existing mathematical models is performed at street level. This has a negative effect on drawing useful conclusions due to the large disparity in the lengths of the roads.

For these reasons, the chapter proposes a methodology for the mathematical modeling of a WDN's behavior through time that is comprised of all the elements of which a WDN consists of, as well as segment-level analysis for the deduction of more qualitative conclusions regarding a pipeline network's condition and vulnerability.

One issue that is discussed more and more recently, is associated with the impact and the residues of the IWS policies in the network condition. It is observed that the implementation of this policy causes additional fatigue to the networks, but there has not been carried out yet a particular scientific study to prove it. A part of this chapter deals extensively with this issue, through the analysis of the data of the Nicosia's WDN, in a given period where the IWS policy was implemented. The ultimate goal is the study of whether the specific period is associated with any additional deterioration of the WDN's condition and, thus, of whether IWS increases a WDN's vulnerability in the short, medium, and long term.

2.3.2 Selection of Mathematical Methodology

Development of the proposed mathematical model will include two levels. For the first level, which is related to the quantitative analysis of the available data, statistical analysis is a suitable methodology. For a more sophisticated and qualitative data analysis (second part of the mathematical model), it was decided that survival analysis (Section 2.3.2, p. 41) is the best option.

There are two main reasons for the selection of survival analysis as the main tool of the mathematical model. The first has to do with the nature of the available data. The deployment of Nicosia's pipeline network (WBN's WDN) began in the 1960s. Digitization of the records on leakage incidents in WBN's WDN was initiated in 2003, and until then there are no available data records. One of the most important features of the survival analysis method is that existence of a completed dataset is not a prerequisite for the analysis. In fact, the method has the ability to result in a reliable prediction by simply making assumptions about the missing data and by forecasting the underlying values, forecasts of which are improved on each new data series import. The second reason for choosing survival analysis relates to the behavior of networks during aging. The initial application of survival analysis was in biology for simulating the effect of the different diseases on the human organism through the passage of time. WDNs behave the same way as the human body. Over time, and with the increasing number of failure incidents (NOPB) in a pipe system, the condition of a WDN worsens. Successive repairs may prolong the time-to-failure of a WDN's components but eventually, if these components are not replaced, failure ("death") occurs. Thus, survival analysis techniques were applied to the study of WDNs, for they successfully simulate the behavior of a WDN's condition as it ages and experiences failures of varying causes over time.

Statistical Analysis

The first level of the mathematical model uses, for the purpose of quantitative analysis, the common statistical measures of central tendency as applied to the complete dataset and to the individual subsets. These common measures are the mean value (the sum of all measurements divided by the number of observations in a dataset), the median (the middle value of a dataset that separates the higher half from the lower half) and the mode (the most frequent value in a dataset). Furthermore, the calculation of the standard deviation for the dataset will depict the spread of data as to the central tendency.

Survival Analysis

As aforementioned, survival analysis was chosen as the modeling method to describe and simulate the time-related risk-of-failure relationships in UWDNs. Survival analysis is a branch of statistics dealing with deterioration and failure over time. It involves the modeling of the elapsed time between an initiating event and a terminal event [71,93]. In the case of WDNs, an initiating event is represented by the date of pipe installation, a near-terminal event is represented by the date of pipe repair and a terminal event is represented by the date of a pipe replacement. The mathematical model estimates the reliability of a system and its lifetime, subject to multiple risk factors. It aims to analyze the effect of these risk factors on the system's lifetime and the probability of survival, and on the expected mean time to failure of each individual part of the system [71,102]. The data values used in the analysis are a mixture of complete and censored observations, where "censored" means data with unknown initiating or terminal events.

The basic mathematical terms used in survival analysis are: the cumulative distribution function, $F(t)$, that denotes the probability that an individual survives until time t, the survival function, $S(t)$, that denotes the probability that an individual survives beyond time t, the hazard rate, $h(T)$, that denotes the probability that an individual at time T experiences the event in the next instant, and the cumulative hazard function, $H(T)$, that is the integral of $h(t)$ from 0 to T [26]. The correlation of the terms is shown by the equations listed below (see Eqs. (2.1)–(2.5)):

$$F(t) = \int_0^t f(x)dx, \tag{2.1}$$

$$S(t) = \int_T^\infty f(x)dx = 1 - F(T), \tag{2.2}$$

$$S(t) = \exp\left[-\int_0^T h(x)dx\right] = \exp[-H(T)], \tag{2.3}$$

$$h(x) = \frac{f(T)}{S(T)}, \tag{2.4}$$

$$H(T) = \int_0^T h(x)dx = -\ln[S(T)]. \tag{2.5}$$

Kaplan–Meier Estimator

The proposed mathematical model estimates the survival function by use of the nonparametric Kaplan–Meier estimator [89] (Eq. (2.6)),

$$\hat{S}(t) = \begin{cases} 1 & \text{if } T_{min} > T, \\ \prod_{T_{min} \le T_i \le T} \left[1 - \frac{d_i}{r_i}\right] & \text{if } T_{min} \le T, \end{cases} \tag{2.6}$$

where T is the elapsed time until the occurrence of a specified event; i is an index; D_t is the set of all failures (deaths) that occur at time T_t and d_t is the number of deaths in this set; R_t is the set of all individuals that are at risk immediately before time T_t and r_t is the number of individuals in this risk set. T_{min} refers to a minimum time below which failures are not considered (often used when data are left truncated).

A Kaplan–Meier plot/estimate of the survival function is a series of horizontal steps of declining magnitude which approaches the true survival function for that population (a large enough sample is needed). The value of the survival function between successive different sampled observations is assumed to be constant. One of the advantages of the Kaplan–Meier estimator [89] is that it can take into account missing data, thus being able to approximate the underlying probability distribution function even in the presence of incomplete datasets [93].

Hazard Ratio

The hazard rate (or conditional failure rate) is a metric which is usually used for identifying the appropriate probability distribution of a particular mechanism [71]. During survival analysis it is very useful to compare the hazard rates of two groups of similar attributes within the examined dataset, by employing the hazard ratio (HR). HR is a metric that estimates the relative risk of an event.

A regression model for the hazard function of two variables is given by [73,94]:

$$h(t,x,\beta) = h_0(t) \times r(x,b) \tag{2.7}$$

where h_0 is the baseline hazard function (when the $r(x,\beta)=1$) and $r(x,\beta)$ denotes how the hazard changes as a function of subject covariance. The hazard ratio of two datasets with covariate values of x_0 and x_1 is given by

$$HR(t,x_1,x_0) = \frac{h(t,x_1,\beta)}{h(t,x_0,\beta)} = \frac{h_0(t) \times r(x_1,\beta)}{h_0(t) \times r(x_0,\beta)} = \frac{r(x_1,\beta)}{r(x_0,\beta)}. \tag{2.8}$$

Eq. (2.8) shows that the hazard ratio depends only on the function used to model failure. This property allows the comparative examination of the risk factors and their effects on the survivability of the subject in examination. In the case of WDNs, one may compare the risk factors that influence the failure of the water pipes within a distribution network and their relative importance to the risk-of-failure metric.

The proposed mathematical model estimates the hazard ratio by use of the Cox semiparametric proportional hazards model. Cox [46] suggested that $r(x,\beta)=\exp(x\beta)$ and therefore the hazard function is transformed to

$$h(t,x,\beta) = h_0(t) \times e^{x\beta}. \tag{2.9}$$

Hence, the hazard function becomes

$$HR(t,x,\beta) = e^{\beta(x_1 - x_0)}. \tag{2.10}$$

2.3.3 Novelty of the Proposed Mathematical Model

As seen from the literature review, the modeling of the behavior of WDNs has been widely studied, with all documented mathematical models exclusively studying the behavior of the pipe elements through the passage of time and because of aging (as either individual elements, or collectively at a street level).

In developing the proposed mathematical model, effort was exerted for the elimination of some of the drawbacks of existing WDN modeling methods, primarily with regard to the class of data and the level of analysis. Also, an attempt was made to uncover possible correlations between the frequency and number of leakage incidents within a WDN and the abnormal/IWS operations the WDN is periodically subjected to.

Type of Data Used for the Development of the Mathematical Model

The majority of past research efforts related to WDNs deal with water mains. A small part of the research studies has added house connections in the analysis, while very few researchers have examined how the failure of specific network parts, such as the valves, affect the network performance.

As previously noted, WDNs behave like the human organism. Rehabilitation actions in the WDN related to any part of it, whether it is a pipe or a fitting, have a direct effect on the condition of the whole system. Moreover, in the overall cost of repairing a leakage, the cost of the failed unit is very small compared to the cost of the work to detect, localize, and restore the problem. Thus, the effect of the fittings in the network's rehabilitation cost is equally as important as that of the pipelines. Therefore, the inclusion in the mathematical analysis of data pertaining to not only pipes but also of failure incidents for the individual fittings and components of the network, enhances any utilized mathematical model simulating the overall WDN condition, bringing such condition appraisals closer to reflecting the actual condition of the network.

For this reason, the presented model is not limited to traditional data records (i.e., pipe-related leakage records) but it goes a step further by introducing in the analysis additional incident and rehabilitation data (repair or replacement actions) of additional network components (e.g., house connections, valves, and fittings). Furthermore, the chapter includes comparisons of the calculated vulnerability results of the all-component network analysis against the pipes-only network analysis.

Level of Analysis

A second point that has to be studied further is the analysis level for pipe elements, with regard to the data resolution used in the mathematical model for the calculation of the WDN vulnerability. The ideal approach for the development of a WDN model is the inclusion of all the units that constitute the system in the analysis. This is impossible, though, for two reasons. The first reason relates to the size of the network and the near impossibility of correctly mapping each individual pipe, and the second with the nature of the repairs.

A WDN consists of hundreds of thousands of pipes. For example Nicosia's WDN, whose main pipelines have a total length of 1200 km, consists of about 200,000 individual pipes (each 6 m long). Furthermore, the majority of WDNs were developed decades ago and thus there are no exact maps of where the pipes are laid. Thus, the identification of all individual pipes is almost impossible.

Even if a record of the individual pipes was possible (and/or was available), the repair of problematic pipelines would cause data conflicts for the identified elements. During the rehabilitation of a failure, if the damage is of limited extent then the part of the pipe which has been worn is cut off and replaced by a new one using connection fittings. These cases create an issue for the mathematical model, as the repaired pipe should be replaced by two parts of which one has the same characteristics with pipe, while the second is brand new. For this reason the mathematical models that examine the individual pipes of the water mains take into consideration only the replacement actions. These models are lacking in accuracy, as they do not take into account the repairs carried out for the rehabilitation of failures.

To avoid this painfully time-consuming detailed recording of breakage incidents, which in essence does not offer any specific important information on the overall condition of the network (6 m of pipes in a 1200 km network), the mathematical analysis of the pipelines is usually carried at a street level (i.e., incidents are recorded by street, and not by 6 m-long pipe segments). Such models are deficient in two aspects: (1) they lack specific assessment information for each individual pipeline and thus it is impossible to compare pipelines, and (2) the spatial resolution of the WDN condition is poor (usually we are able to produce risk heatmaps by district metered area, DMA, and/or street, but not by 6 m-long segments within the WDN).

In the case of street-level analysis, the length of the pipes is considered to be equal to that of the street. So when comparing results across the population of pipes, a pipe may be several tens of meters in length while another may be a few kilometers. Also the results for the hazard rate of the sustainability for a long pipe may not be representative, as the problem may concern a short part of the pipe.

In the mathematical model developed and utilized in this chapter, an effort was made to perform water loss analysis of water mains at a segment level and to compare the results with those of a street-level analysis. These comparisons will indicate the size of the deviation between the analysis results obtained from the two approaches. For streets of short length the difference between the two approaches may not distinct. But for streets of medium or big length, the difference in the results may be substantial.

Intermittent Water Supply (IWS)

Countries with arid weather conditions, particularly those located in the Middle East, have issues with the uninterrupted supply of water. The island of Cyprus faces the problem of poor or even null rainfall approximately every five years. In recent decades, Cyprus experienced four-year periods of adequate rainfall, followed by two years of drought. As a result of this situation, in every fifth year a policy of intermittent water supply would be enacted for the purpose of "saving" water.

Interestingly enough, data published by the water authorities regarding the annual volume of nonrevenue water indicates a significant increase during a time period that follows the termination of an intermittent supply period. This fact creates the need to investigate whether this increase in the volume of nonrevenue water corresponds to an increase in the number and/or severity of leakage incidents in the networks, and whether this relates to the implementation of the intermittent supply policy.

2.3.4 Dataset of the WBN

The data analysis presented in the following sections was developed using the WBN data for the period from 01/01/2003 to 31/12/2010. As previously mentioned, a part of the dataset is missing (01/08/2009–31/12/2009). The total number of incidents of the dataset included in the analysis is 67,730.

Pipeline Class					
Water Main		House Connection		Water Main + House Connection	
Main Pipe	Pipe	House Connection	Pipe	*Water Main Class Findings*	
Clamp	Fitting	No Problem	Pipe	*House Connection Class Findings*	
Maxifit	Fitting	Stuffy House Connection	Pipe	Misc	Misc
Maxistep	Fitting	Unplugged House Connection	Pipe	Seal	Misc
Ring-Main	Fitting	Ferrul	Fitting		
Saddle	Fitting	Closed Ferrul	Fitting		
Saddleferrul	Fitting	Saddleferrul	Fitting		
A.V.	Component	B. V.	Component		
C.V.	Component	Bush	Component		
F.H.	Component	Coupling	Component		
S.V.	Component	Manifold	Component		
		Nipple	Component		
		Sieve	Component		
		Temporary Pipe	Component		
		Water Meter	Component		
		Water Meter Connector	Component		

Figure 2.4 Split of the database's *"Findings"* data field based on the part type and the class of pipe.

Classes of Pipes and Categories of Analysis

The first step in designing the categories of analysis (Fig. 2.3) was the assignment of the database's "*Findings*" field (Section 2.2.3, p. 25) based on their part type and the pipe class.

The mathematical analysis consists of the following three pipe classes: (WM) "water main"; (HC) "house connection", and (C) "couple" (which is the coupling of the first two classes). The different part types found in a WDN are: pipes, fittings, components, and misc. A classification/categorization of the findings is shown in Fig. 2.4.

The findings regarding "*Area*", "*Closed Area*", "*Closed BV*", "*Closed SV*", "*Low Area Pressure*", "*Not Our Problem*", "*Other*", "*Pump*", "*Stealing Water*", "*Tank*", "*Wasting Water*", and "*Zone*" were not assigned to any of the pipeline classes because these incidents were not related to problems within the distribution system. The data field "*Saddleferrul*" was assigned to both classes (WM and HC) as it is a component which relates to both the water mains as well as the house connections. For the coupling class,

Table 2.5 Categories and classes used in the analysis.

Category	Class	Reference
Pipeline (P)	Water Main (WM) House Connection (HC) Couple (C)	*Analysis of pipe data*
Pipeline Network (PN)	Water Main (WM) House Connection (HC) Couple (C)	*Analysis of pipe and fitting data*
Distribution System (DS)	Water Main (WM) House Connection (HC) Couple (C)	*Analysis of pipe, fitting and component data*

the "*Saddleferrul*" was inserted in the analysis of the record set only once. In essence, the number of records in the Couple class is the sum of the WM incidents plus the HC incidents, minus the saddleferrul incidents. The number of incidents (data records) applicable to this mathematical analysis is 58,015.

Further to the various analyses stemming from the mathematical model in use, in relation to the condition of WBN's network, the chapter also includes a comparison of the changes in the results when records related to the fittings and other parts of the distribution system are added to the dataset. Actually, the data analysis included herein is divided into three categories, each one consisting of the three pipeline classes mentioned in Section 2.3.4 (p. 47). The different categories, as well as a reference for what is included in each one of these, are presented in Table 2.5.

- **Pipeline** The "*Pipeline*" category analysis includes only the data associated with the water pipes. It is the data currently used for modeling the condition of WDNs.
- **Pipeline network** The "*Pipeline Network*" category analysis goes one step further by introducing data associated with the fittings, which are used to connect together the pipes of the network. This analysis category is considerably more accurate in analyzing the condition of the WDNs in relation to the "Pipeline" analysis.
- **Distribution system** The difference of the "*Distribution System*" compared to the "*Pipeline Network*" analysis is that it introduces additional data associated with the components of a WDN. The components are all the parts of a WDM associated with the monitoring and measurement of various parameters within the network. They are parts

that are easily installed in, or uninstalled from, the network such as the water flow meters, air valves etc. The analysis of the entire dataset is expected to give a considerably more accurate picture of the condition of the network in study.

Data Stratification for Analysis at the Street Level and at the Segment Level

The WBN's dataset will be analyzed on two levels, as per Section 2.3 (p. 28). The first level is the street level while the second is the segment level.

- **Street Level** The mathematical analysis at the street level, considers that the pipes installed along a street are a single (unified) piece and the length of a pipe is equal to that of the street it is located at. Thus, all the incidents regarding a particular street are considered to be related to the same pipe. By extent, the number of pipes in a WDN is equal to the number of the streets in the network. For the case study examined in this chapter (i.e., the network of the city of Nicosia, Cyprus), the number of incident records that are applicable to the street-level analysis is 58,015.
- **Segment Level** For the analysis at the segment level, the length of the segment (corresponding to the length of the pipe unit) should be decided upon. The determination of the segment was decided on the basis of three parameters. The first parameter was that the length of segments should be as uniform as possible (i.e., the assumed segments should be of approximately equal lengths across the WDN). The second parameter is associated with the ease of identification/assignment of the pipe sections, so that it is easy for the workmen to report on-site the pipeline segment for which a rehabilitation action was taken. The last parameter relates to the dataset and the fact that records on leakage incidents had already been taken prior to this research effort and thus the assignment is retroactive. For this reason it was decided that the segment length would be equal to the length of the facade of the land plot outside of which a pipeline passes. This assumption meets all of the aforementioned three parameters. The length of the facade of the plots is similar, it is easy for workers to indicate the segment id where the rehabilitation is carried out on the basis of the street number, and finally the majority of the dataset records already on file (years 2003–2010) has the street number information specified for each recorded leakage incident. As a further facilitating action, it was decided

that the segments would be defined with respect to a street address's odd numbering scheme. Of all the available 58,015 incident records, 46,041 include street numbering.

Data Clustering

The fields "*Action*" and "*Diameter*" of the dataset (Fig. 2.3) and the calculated number of observed previous breaks (NOPB) for each pipe element contain too many selections/classifications, which makes it difficult to extract useful and reliable conclusions by use of the survival analysis. For this reason, the selections were grouped in clusters in order to facilitate the analysis and the extraction of useful and generalized results. These analysis clusters are presented in Fig. 2.5.

Assumptions in the Analysis

The assumptions made for the mathematical (survival) analysis are as listed below.

- The initial date of birth of each pipe was considered to be the installation date of the part of network to which the pipe belongs,
- The NOPB is reset when there is an action of convention, expansion, installment or replacement (i.e., the pipe's birth date is then updated),
- The censor variable, D, reflects the change in condition of the part whose survival is examined. A value of 1 indicates a change in condition, while a value of 0 indicates an unchanged condition status. The data field "*Action – Cluster*" is the one used for changing the value of D. A "Replacement" action sets D equal to 1, while an "Inspection" or a "Repair" action sets D equal to 0. In practice, the value of $D = 1$ represents the rebirth of the part element and a reset of its NOPB, while 0 indicates that there is no change in the date of birth and of the NOPB for the particular part.

2.4 RESULTS

The presentation of results is evolutionary and is divided into four sections, as shown in Fig. 2.6 and briefly described below. Fig. 2.6 should serve as a reference to the progression in the analysis process and to the sections in the chapter that these analyses can be found in.

- **Statistical Analysis**

  The first section presents the results of the statistical analysis. It begins with the information related with the dataset used in the analysis and

Action				Diameter		NOPB	
Name	Cluster	Name	Cluster	Size	Cluster	NOPB Range	Cluster
Closing	Visual Inspection	Lifting	Inspection	3/8" - 1 3/4"	House Connection	1 - 4	Small
Finding	Visual Inspection	Lowering	Inspection	16mm - 32mm		5 - 8	Small - Medium
Inspection	Visual Inspection	Maintenance	Inspection	2" - 4"	Small Main	9 - 12	Medium
Opening	Visual Inspection	Plug	Inspection	40mm - 100mm		13 - 16	Medium - Large
Other	Visual Inspection	Reconnection	Inspection	6" - 8"	Main	> 16	Large
Soil Redeposition	Visual Inspection	Reinstallment	Inspection	110mm - 250mm			
Recording	Visual Inspection	Tightening	Inspection	> 8"	Ring Main		
Telemetry	Visual Inspection	Wash out	Inspection	> 250mm			
Water Meter Sealing	Visual Inspection	Conversion	Repair				
Cleaning	Inspection	Expansion	Repair				
Connection	Inspection	Repair	Repair				
Disconnection	Inspection	Removal	Replacement				
Installment	Inspection	Replacement	Replacement				

Figure 2.5 Data clustering for the performed survival analysis.

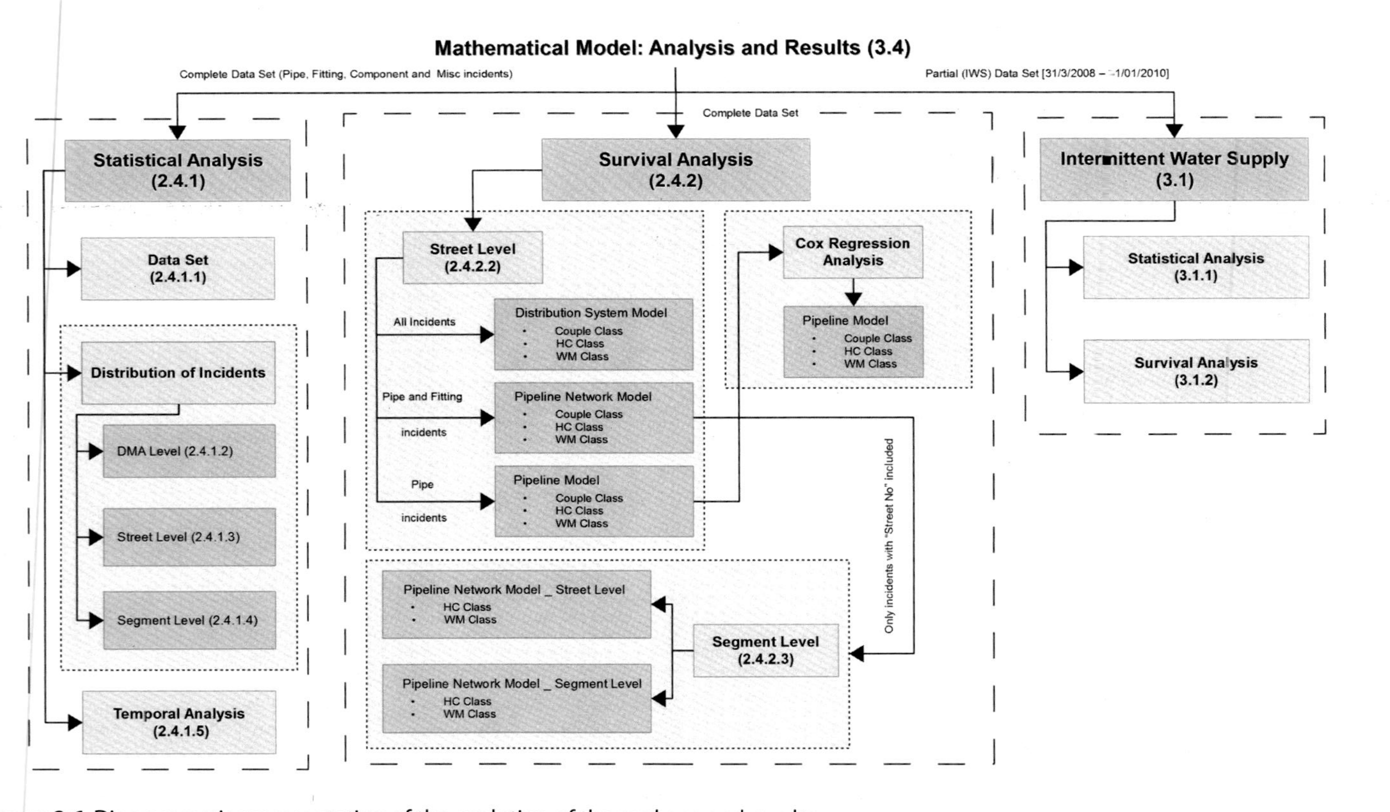

Figure 2.6 Diagrammatic representation of the evolution of the analyses and results.

continues with the data distribution at DMA level, street level and segment level. The section concludes with the results of the time series analysis.

- **Survival Analysis**
  The second section is related with the results of the survival analysis. Part "A" of the results begins with the outcome of the survival analysis models at street level. The first analysis series at street level is the "Distribution System" model that uses the complete dataset for the three model classes (*Coupled*, *House Connection,* and *Water Main*). The second one is the "Pipeline Network" model that uses for the analysis the data related to pipes and fittings. The last analysis at street level is associated with the "Pipeline" model, and uses only pipe data. Part "B" of the survival analysis results is about the segment level analysis. This part compares the changes of the "Pipeline Network" model's survival analysis results at segment level compared with that at street level. The comparison involves the House Connection and Water Main model classes.
- **Intermittent Water Supply Period**
  The third section is about the specific analyses associated with the period that the intermittent water supply policy was applied.
- **Water Board of Limassol**
  The last section compares the analysis outcomes for the WBN with those of the WBL dataset (based on knowledge from previous research activities).

2.4.1 Statistical Analysis

WBN's digital dataset of network incidents comprises 67,730 records and relates to the time period from 01/01/2003 to 31/12/2010 (the part of data between 01/08/2009 and 31/12/2009 is missing). The set of the database that is directly related with the WDN contains 58,015 incidents. From these data, only for 46,041 cases the incident's street number had been assigned and hence only those records could be used for the segment-level analysis.

All information related with the dataset used in the statistical analysis and the results of the analysis of the data distribution at DMA level, street level and segment level as well as the time-series analysis are presented below.

Dataset

The complete dataset, which consisted of 58,015 incidents originating from the "*Finding*" field of the database, is presented in Table 2.6. Fig. 2.7 shows

Table 2.6 Distribution of the dataset's incidents into the *"Finding"* field.

Findings	No. of incidents
A.V.	320
B.V.	6904
Bush	8
C.V.	9
Clamp	19
Closed B.V.	49
Closed Ferrul	2
Coupling	232
F.H.	312
Ferrul	757
House Connection	32,982
No Problem	9
Stuffy House Connection	6
Unplugged House Connection	17
Main Pipe	3807
Manifold	1323
Maxifit	7
Maxistep	1
Misc	925
Nipple	111
Ring-Main	29
S.V.	782
Saddle	571
Saddleferrul	139
Seal	427
Sieve	14
Temporary Pipe	276
Water Meter	5244
Water Meter Connector	2733
Total	58,015

the division of the "*Finding*" field into the "*Part Type*" field while Fig. 2.8 illustrates the grouping of the incidents in categories according to the "*Action Cluster*", "*Pipeline Class*", "*Diameter Cluster*", "*Material*", "*Reason*", and "*Location*".

It is clear that the incidents related to the House Connections constitute the majority of the dataset. This has a direct impact on the "*Pipeline Class*",

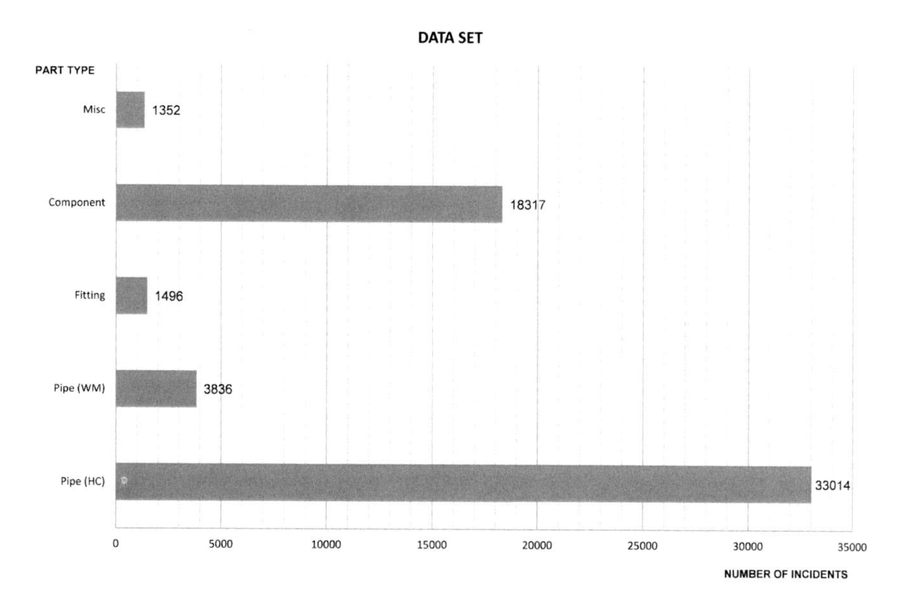

Figure 2.7 The categorization of incidents into "*Part Type*."

"*Diameter Cluster*", and "*Material*" fields. Furthermore, in the "*Action Cluster*" field, the repairs are much more than the replacements, while 11% of the incidents do not involve leakage.

Regarding the "*Diameter Cluster*" of the Water Mains, the majority of the incidents involve "*Small Main*" while those of "*Ring Main*" are very few. This, probably, has to do with the population of the pipes of various diameters that are installed in the network. Also the incidents related to Visual Inspection ("*Action Cluster*") and part of the incidents related to Components ("*Part Type*") and Misc ("*Part Type*") should not be classified in this field. The volume of missing data (11%) compared to the data with incomplete (not properly recorded) information (16.1%) is at an acceptable level and thus the sample space utilized in the analysis (89% of the dataset) is reliable.

The majority of older pipes, which are galvanized or made of either asbestos cement (AC) or ductile iron (DI), have been replaced. That is the reason why the proportion of these types of pipes does not exceed 15% of the classified incidents. Furthermore, the incidents related to Visual Inspection ("*Action Cluster*"), Components ("*Part Type*") and Misc. ("*Part Type*") should not be classified in this field. As with the previous case, the volume of missing data (11%) compared to the data with incomplete (not

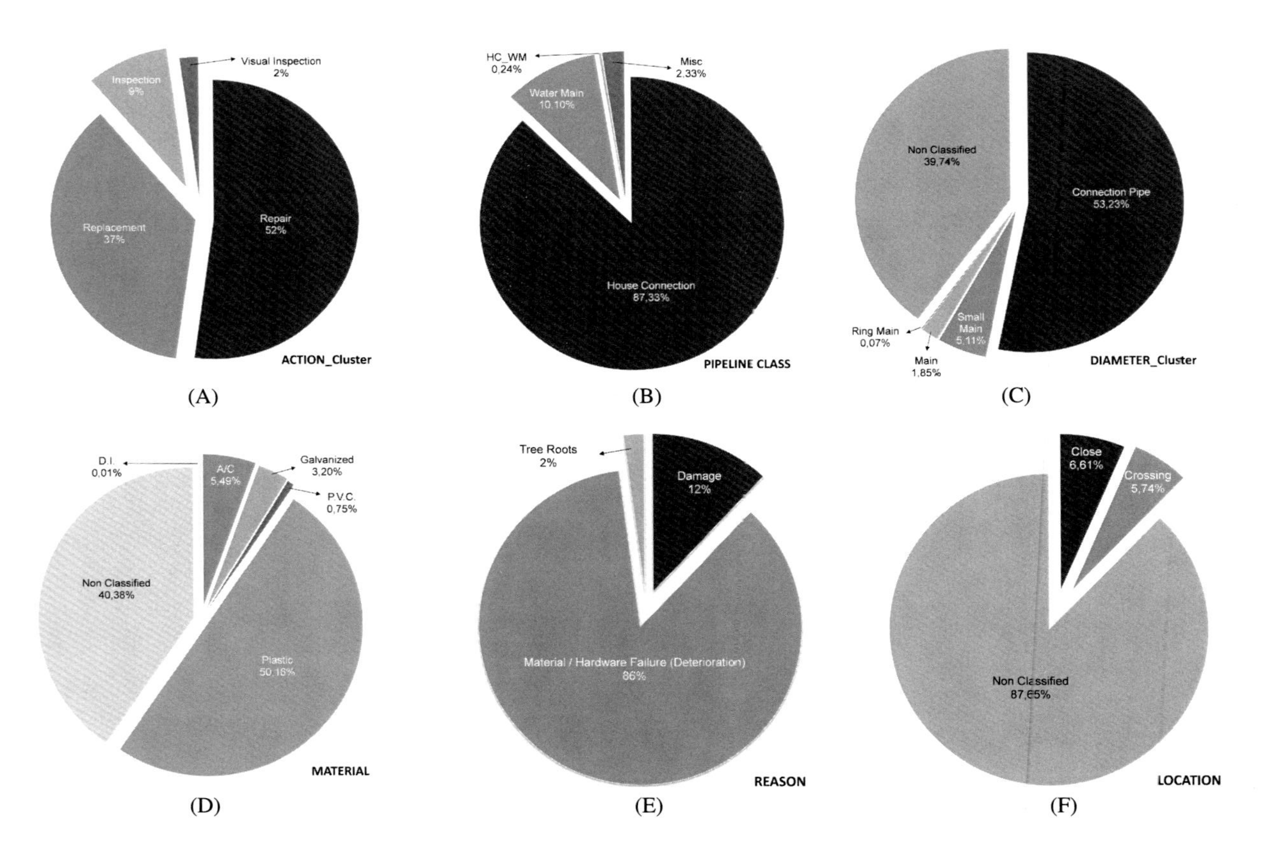

Figure 2.8 Categorization of incidents into various fields.

properly recorded) information (16.1%) is at an acceptable level and thus the sample space utilized in the analysis (89% of the dataset) is reliable.

The field "*Reason*" of the dataset is not satisfactorily filled. Only 16.1% of the cases have been classified with regards to the reason for failure (i.e., 83.9% missing), while in the case of the Inspection and Visual Inspection (of the "*Action Cluster*" field), the level of missing data was 11% of the total number of incidents. The main volume of the data with complete information on them (16.1%) is related to the material's failure due to deterioration and a small percentage is associated with damage. There are some incidents due to the roots of the trees that are planted on the sidewalks.

"*Location*" is related only to the HC pipes. As seen from Fig. 2.8F, the incidents of HC which are installed at the side of the street where the WM is located ("*Close*") are slightly more than those that are placed on the opposite side of the street. Logically, it should be the other way because the connection that crosses the road is exposed to more risk factors. The incidents that are assigned to Visual Inspection ("*Action Cluster*"), Component ("*Part Type*"), and Misc. ("*Part Type*") should not be classified in this field. However, the "Nonclassified" incidents are much more common.

Although the dataset consists of a massive volume of information, its quality is not the best since much information associated to specific database fields is not classified. This is expected to have an impact on the accuracy of the individual analyses associated with the missing data.

Data Distribution per DMA

This section deals with the distribution of the dataset's incidents per DMA. It also examines the relationship between the different fields that are distributed to the DMAs of Nicosia's WDN.

Table 2.7 presents the distribution of the dataset's incidents per DMA. Of the 24 DMAs the WBN is divided into, DMA 17 has the most incidents and is followed by DMA 7. DMAs 21, 15, and 16 are the next in descending order with regards to the number of incidents in the study period (2003–2010).

Apart from the known reasons that affect the condition of a WDN, the number of incidents in each DMA is also dependent on the size of the DMA (as the total pipeline length within a DMA increases, the number of incidents in that DMA is also expected to increase) and the year in which a DMA joined the WBN (younger areas are expected to have a smaller number of incidents; in fact, some areas joined WBN after 2003). One should also note that in Table 2.7 the areas that have almost zero incidents

Table 2.7 Distribution of the dataset's incidents per DMA.

DMA	No. of incidents
1	3507
2	2954
3	3084
4	6
5	1581
6	2382
7	5806
8	2864
9	3861
10	523
11	2
12	3
13	1056
14	2138
15	4409
16	4305
17	7340
18	1366
19	1433
20	1427
21	4800
22	2076
23	860
24	232
Total	**58,015**

do not actually belong to Nicosia's WDN, but they are supported by the WBN when incidents requiring specialized labor action occur.

Figs. 2.9 and 2.10 present the incident distribution per DMA based on the fields of the dataset. The DMA with the highest number of incidents related to pipes is DMA 17, followed by DMA 7 and DMA 21 (Fig. 2.9A). Conversely, DMA 7 has the highest number of incidents related to fittings, components, and misc., and is followed by DMA 17 (Fig. 2.9B). This shows that, based on the "*Part Type*" field, each DMA has different characteristics from the others, and there is no correlation between the classification and the number of incidents that occur within a DMA.

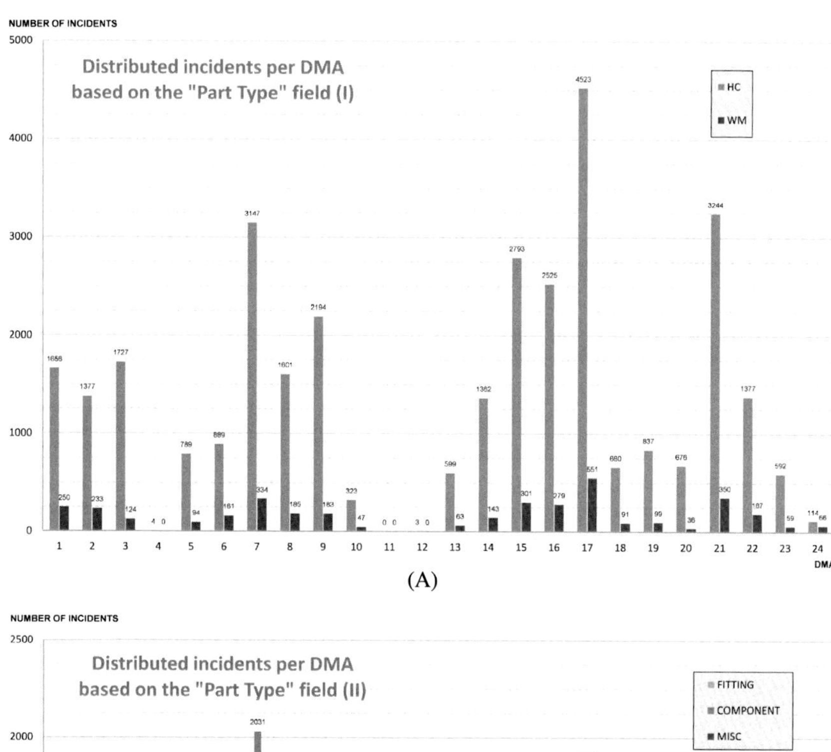

(A)

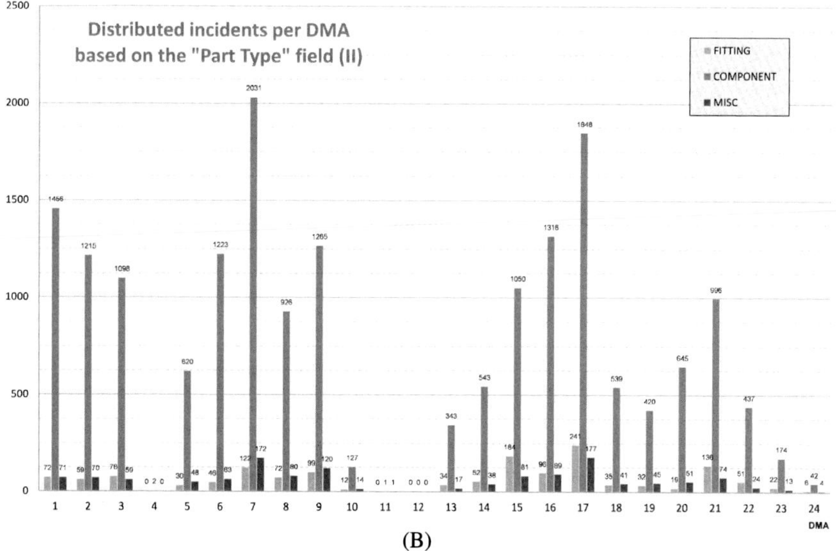

(B)

Figure 2.9 Distributed incidents per DMA based on the *"Part Type"* data field.

In contrast to the field "*Part Type*", the fields "*Diameter Cluster*", "*Material*", "*Reason*", and "*Location*" present a correlation between the number of incidents per DMA and the quota of the different categories

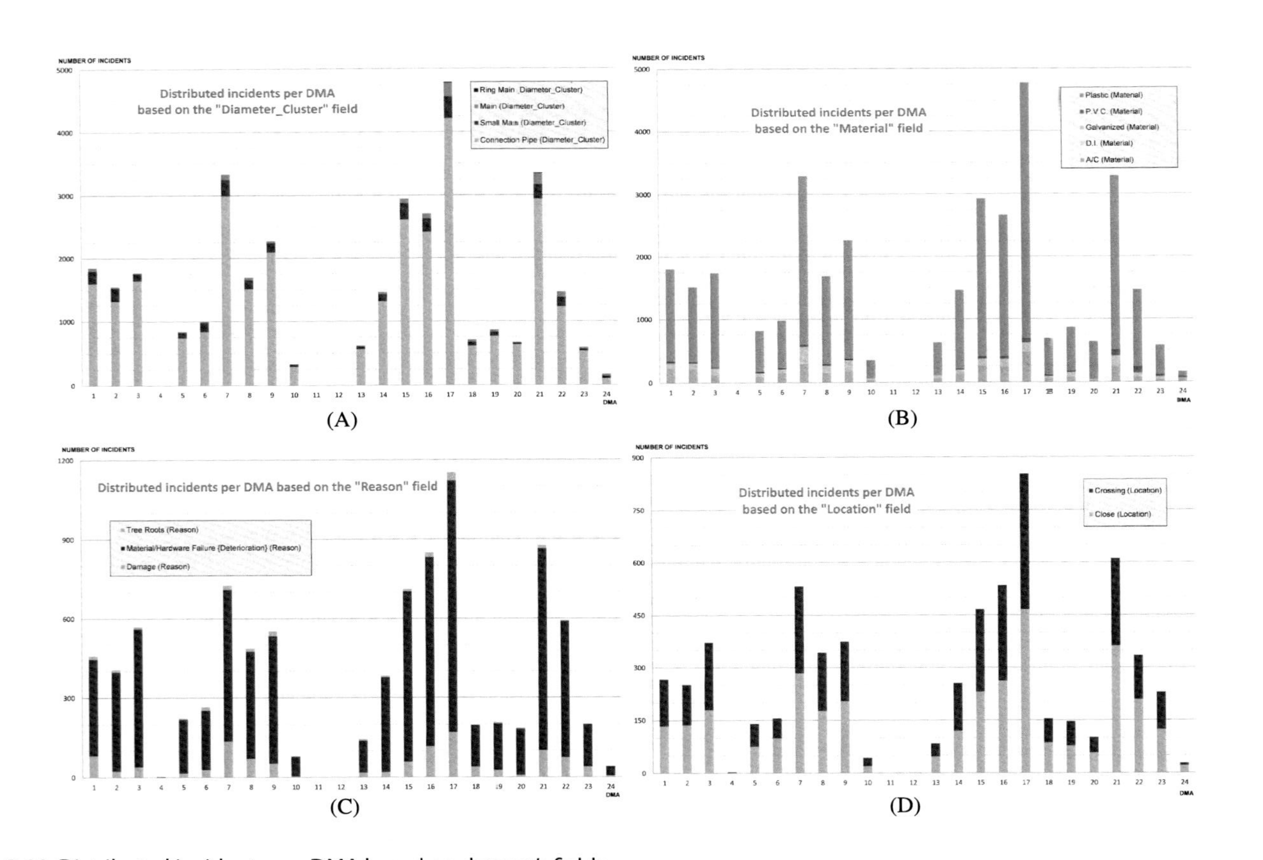

Figure 2.10 Distributed incidents per DMA based on dataset's fields.

of each field. Fig. 2.10 shows that the various types of incidents for each field are distributed proportionally based on the total incidents population of the field and the total number of incidents corresponding to each DMA.

Data Distribution per Street

The next analysis level is that of the street. Figs. 2.11 and 2.12 present the incident distribution of the different fields, for the top 15 streets in descending order for the number of incidents. The street with the highest number of incidents is the "*Strovolou Leoforos*"[1] of DMA 7 (259), followed by the "*Keryneias Leoforos*" of DMA 16 (205) and "*Athinon*" of DMA 7 (192). Even if DMA 17 has the highest incident number, none of the top 15 streets with respect to incident number belongs to it. This shows that the DMA level analysis certainly provides reliable and accurate results.

The street with the highest number of pipe incidents is the "*Strovolou Leoforos*" of DMA 7, followed by the "*Larnakos Leoforos*" of DMA 15, and "*Athinon*" of DMA 7 (Fig. 2.11A). "*Strovolou Leoforos*" of DMA 7 has also the highest number of incidents related to fittings, components, and misc., and is followed by "*Keryneias Leoforos*" of DMA 16 and "*Tseriou Leoforos*" of DMA 7 (Fig. 2.11B). Apart from the "*Strovolou Leoforos*" of DMA 7, the results show that based on the "*Part Type*" field, each street has different characteristics from the others and that there is no correlation between the classification and the number of incidents that occur in a street. This is confirmed by Table 2.8 that presents the first eight streets based on the number of WM incidents. Only two of the eight are included in the first 15 streets in descending order of the number of incidents.

As shown in Fig. 2.12, the distribution of the dataset's fields is not affected by the total number of incidents in a street, as much as it is by the classification of the "*Part Type*" field and the pipeline's age.

"*Aglantzias Leoforos*" of DMA 15 has the highest number of incidents related to the "*Ring Main*" of the "*Diameter Cluster*", and "*Keryneias Leoforos*" of DMA 16 has the highest incident number related to "*Small Main.*" "*Strovolou Leoforos*" of DMA 7 has the most incidents of the "*Diameter*

[1] The naming convention for street addresses follows the official naming protocol (as used by the Cyprus Postal Service).

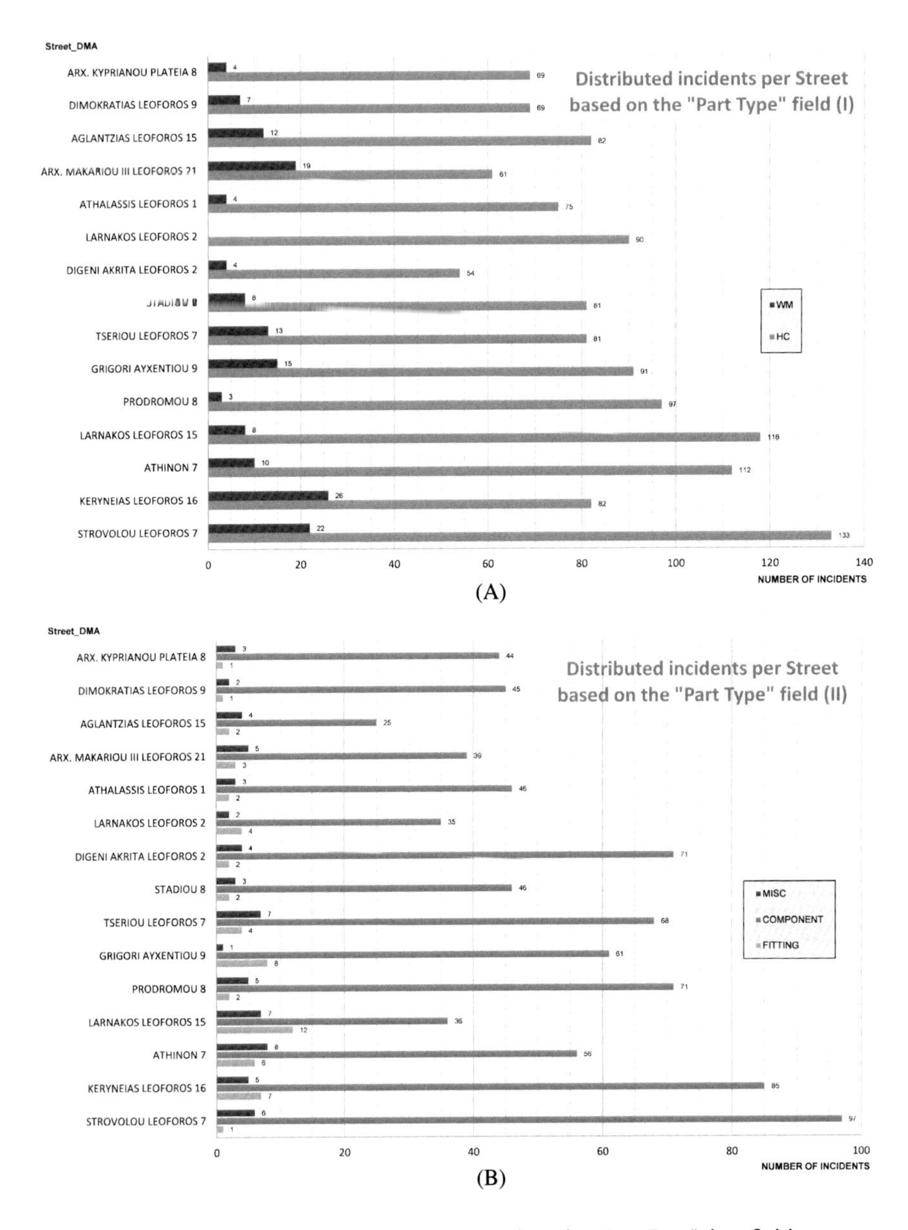

Figure 2.11 Distributed incidents per street based on the "*Part Type*" data field.

Cluster" data field, yet it has fewer incidents of the "*Ring Main*" and "*Small Main*" fields compared to the two aforementioned streets. Similarly, "*Larnakos Leoforos*" of DMA 2 has only "*Connection Pipe*" incidents.

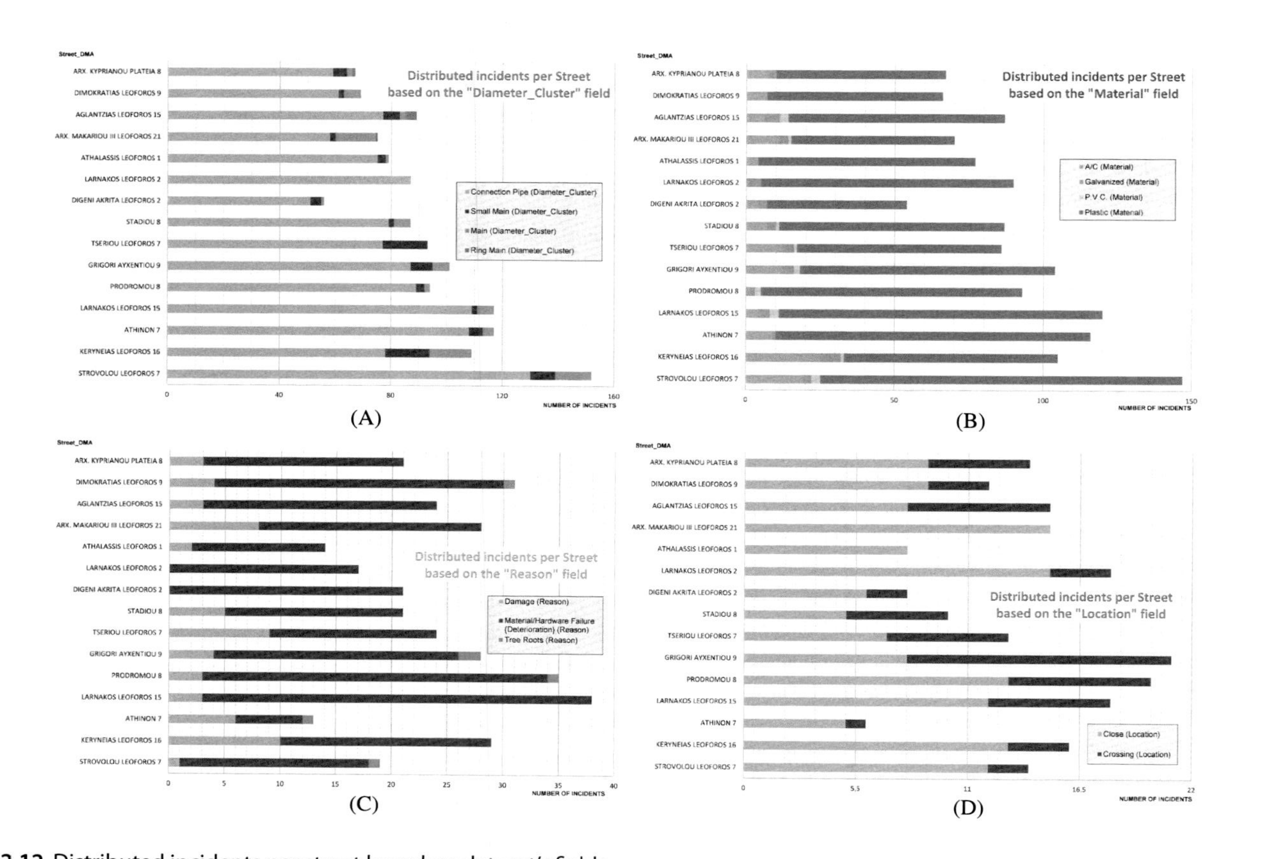

Figure 2.12 Distributed incidents per street based on dataset's fields.

Table 2.8 List of streets where most WM incidents occur.

Street	**Part type**					**Sum**
STREET NAME; DMA	*Pipe*		*Fitting*	*Component*	*Misc.*	
	WM	*HC*				
KERYNEIAS LEOFOROS; 16	26	82	7	85	5	205
LEMESOU LEOFOROS; 16	25	35	1	20	1	82
28 OKTOVRIOU LEOFOROS; 17	23	35	6	47	4	115
DIMOSTHENI SEVERI LEOFOROS; 6	22	15	2	35		74
MODESTOU PANTELI; 21	22	47		2	2	73
STROVOLOU LEOFOROS; 7	22	133	1	97	6	259
LEMESOU LEOFOROS; 1	20	15	1	20	2	58
TZON KENNENTY LEOFOROS; 14	20	36	1	20		77
Total	180	398	19	326	20	943

It can thus be deduced that the distribution of the overall "*Diameter Cluster*" field is not the same as the distribution of its constituent classifications.

"*Keryneias Leoforos*" of DMA 16 has more incidents related to the A/C Material compared to "*Strovolou Leoforos*" of DMA 7, "*Larnakos Leoforos*" of DMA 15 and "*Athinon*" of DMA 7 that have more incidents associated to the "*Material*" field.

The Damage incidents of the "*Reason*" field are more than double in "*Keryneias Leoforos*" of DMA 16 compared to those of "*Larnakos Leoforos*" of DMA 15, despite the fact that "*Larnakos Leoforos*" of DMA 15 has more than 20% volume of incidents associated to the "*Reason*" field. Also "*Larnakos Leoforos*" of DMA 15 has no incidents related to the "*Tree Roots*" classification.

Further, "*Grigori Ayxentiou*" of DMA 9, which has the most incidents associated to the "*Location*" field, has more incidents that are classified as Crossing compared to those of classification Close. On the other hand, "*Athalassis Leoforos*" of DMA 1, which has fewer incidents, has only incidents that are classified as Close.

Data Distribution per Segment

The analysis below concerns the segment level. Figs. 2.13 and 2.14 present the distribution incidents of the different fields, for the first 15 segments ordered by the number of incidents. The segment with the highest number of incidents is "*Poseidonos 16*" of DMA 15 (38 incidents), followed by "*Prodromou 65*" of DMA 8 (22 incidents) and "*Mnasiadou 21*" of DMA 6 (20 incidents). Although DMA 17 (at DMA level) and "*Strovolou Leoforos*" of DMA 7 (at street level) have the highest number of incidents, none of the two is included in the "top 10" list of incidents at the segment level. This proves that the segment level analysis could provide interesting and useful information, in addition to that obtained from DMA or street level analysis.

The segment with the highest number of pipe incidents is "*Poseidonos 16*" of DMA 15, followed by "*Metoxiou 1*" of DMA 5 (Fig. 2.13A). "*Prodromou 65*" of DMA 8 has the highest number of incidents related to fittings, components, and misc., and it is followed by "*Mnasiadou 21*" of DMA 6 (Fig. 2.11B). The results show that based on the "*Part Type*" field, each segment has different characteristics from the others and that there is no correlation between the classification and the number of incidents that occur in a street. This is confirmed by Table 2.9 that presents the first eight segments in number of WM incidents. None of them is included in the first 15 segments in order of the number of incidents.

As shown in Fig. 2.14, the distribution of the dataset's fields is not affected by the total number of incidents in a segment, as much as it is affected by the classification of the "*Part Type*" field and the pipeline's age.

"*Poseidonos 16*" of DMA 15 has the highest number of incidents related to the Connection Pipe of the Diameter Cluster, followed by "*Metoxiou 1*" of DMA 5. The graph of Fig. 2.14A contains only Connection Pipe incidents because the pipe classification of the "*Part Type*" field for the first 15 segments, ordered by incident number, consists only of HC.

The bar chart in Fig. 2.14B is of the same shape (i.e., follows the same distribution) as the "*Diameter Cluster*" chart, for the same reason as stated above (i.e., because the pipe classification of the "*Part Type*" field for the first 15 segments, ordered by incident number, consists only of HC pipes). All the incidents have been associated with the plastic material except the one that belongs to "*Themistokli Tsatsou 2*" of DMA 21, which is made of galvanized steel.

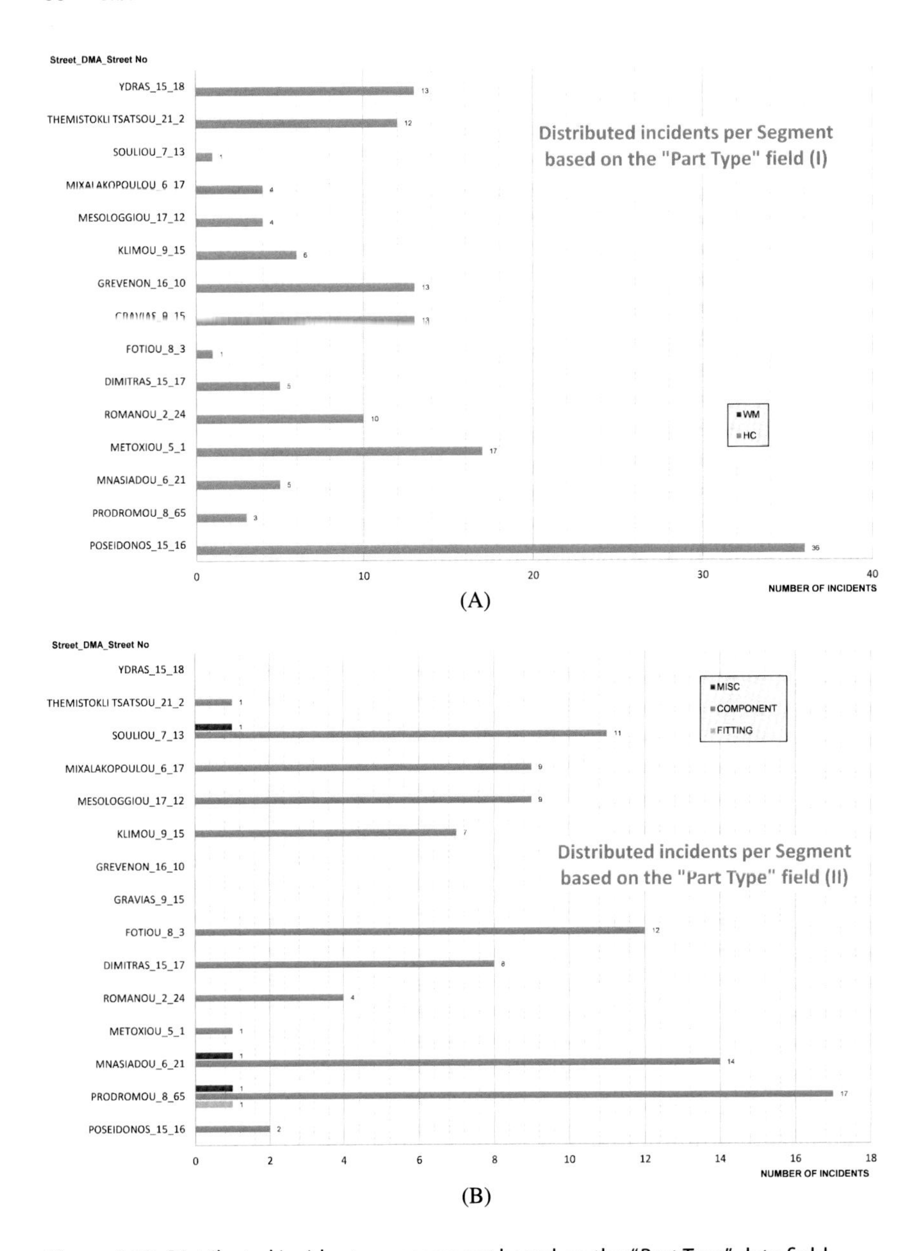

Figure 2.13 Distributed incidents per segment based on the *"Part Type"* data field.

The classified incidents for the "*Reason*" field are only of "Material/Hardware failure due to deterioration." "*Klimou 15*" of DMA 9 has the highest number of incidents and it is followed by "*Gravias 15*" of DMA 9 and "*Poseidonos 16*" of DMA 15.

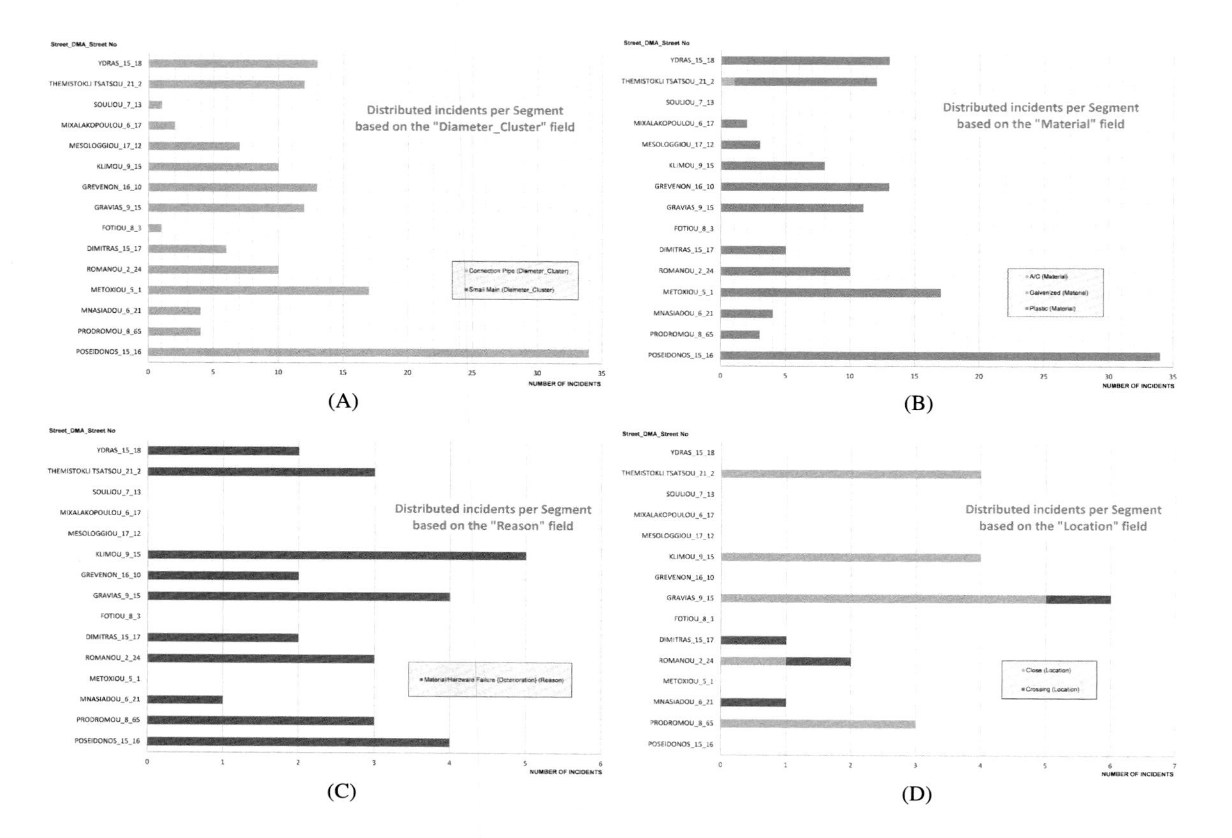

Figure 2.14 Distributed incidents per segment based on dataset's fields.

Table 2.9 List of segments where most WM incidents occur.

Segment	**Part type**					**Sum**
Street Name; DMA; Street No.	*Pipe*		*Fitting*	*Component*	*Misc.*	
	WM	*HC*				
KONSTANTINOUPOLEOS; 17; 5	5		0	0	0	5
MIXAIL KARAOLI; 10; 43	5	1	0	1	0	7
AGIOU NIKOLAOU; 2; 3	4	2	0	1	0	7
AIANTOS; 6; 25	4		0	0	0	4
SMYRNIS; 13; 79	4		0	0	0	4
THESSALIAS; 2; 39	4		0	0	0	4
Total	26	3	0	2	0	31

"*Gravias 15*" of DMA 9 has the most incidents associated to the "*Location*" field. It has 5 incidents, one of which is classified as Crossing. It is followed by "*Themistokli Tsatsou 2*" of DMA 21 and "*Klimou 15*" of DMA 9 which have four incidents each, classified as Close.

Temporal Resolution of the Dataset

Finally, we look at a temporal analysis of the dataset. The first part of the analysis deals with the yearly and seasonal distribution of the dataset's incidents per DMA. The graphs of the obtained results are illustrated in Fig. 2.15. As can be deduced from Fig. 2.15A, there is no relation between the annual distribution of incidents and the DMAs (the shape of the historical annual distribution is irregular and independent of a DMA). For example, in 2005 DMA 7 had more incidents than DMA 16, while in 2009 it had fewer. Noteworthy is the fact that DMA 17 had, every year, the highest number of incidents among all DMAs. Fig. 2.15B shows that the distribution of the seasonal incidents for all DMAs has the same pattern, even though their yearly distribution of the number of incidents is not the same. The pattern has the form of probabilistic distribution in which the incidents are increasing as summer approaches while they are reduced towards the end of the year. A small deviation is observed in the months of April, August, and December, probably due to the fact that these months coincide with holiday periods (Easter, Assumption of Mary, and Christmas periods, respectively). Because of the holiday seasons, the detection, localization, and rehabilitation for a fraction of the incidents that occur during these periods is being deferred to the following months (January, May, and September, respectively), showing an increase in incidents in those months.

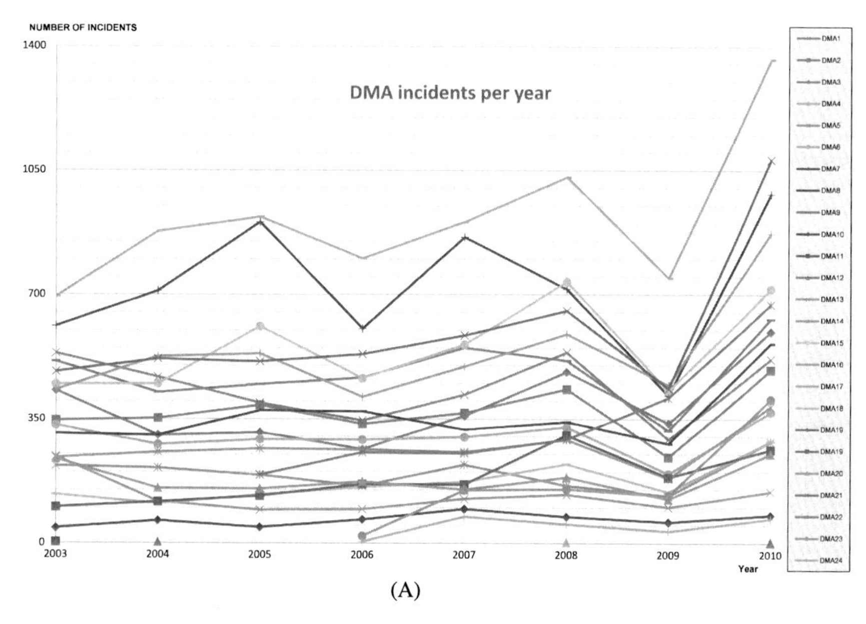

(A)

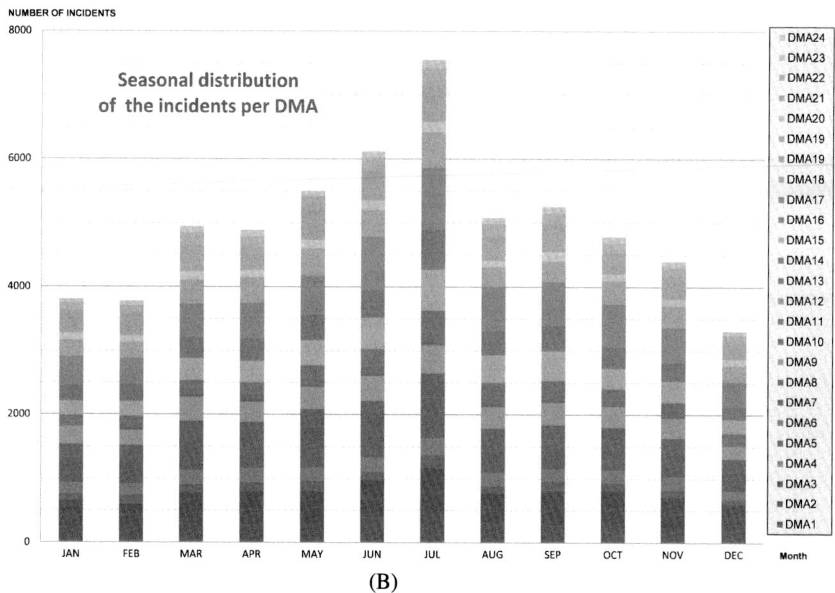

(B)

Figure 2.15 Time series analysis per DMA.

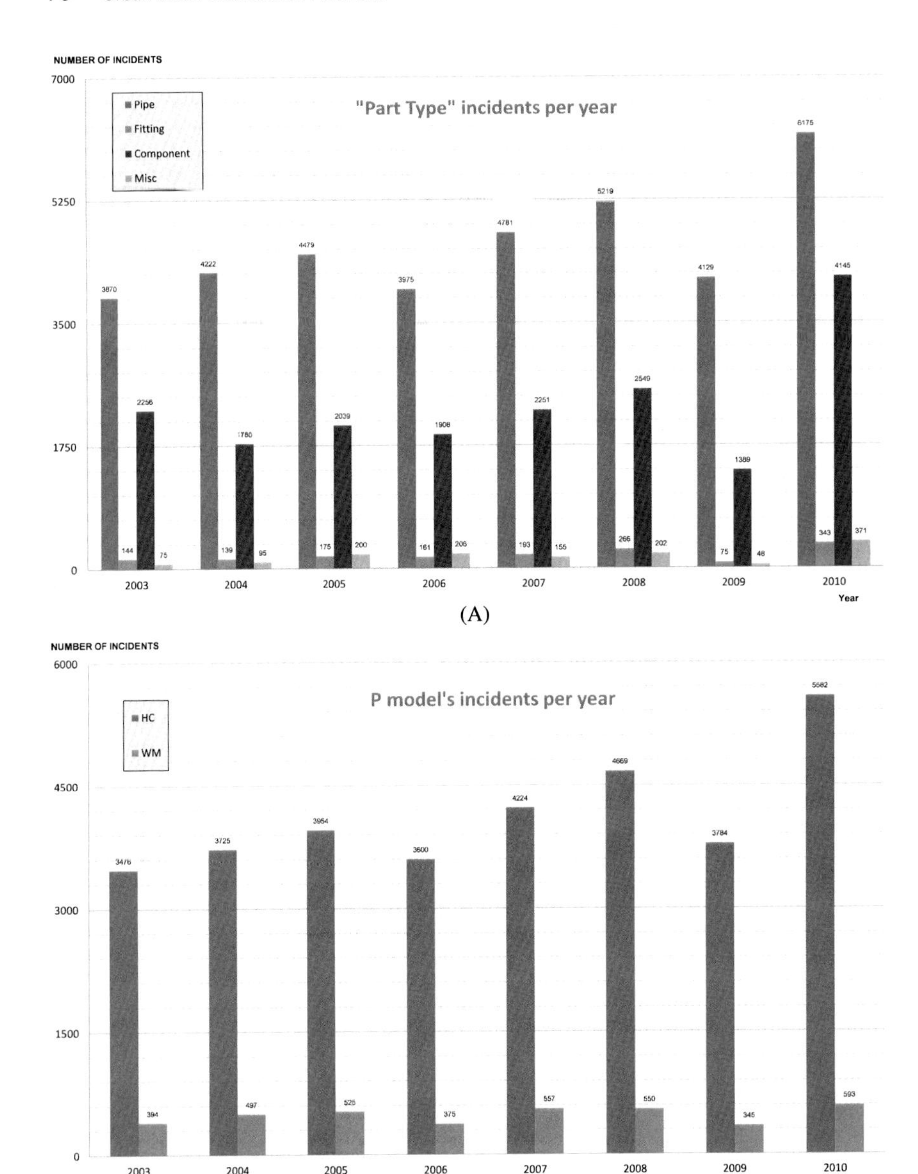

Figure 2.16 Time series analysis based on the "*Part Type*" data field.

Fig. 2.16A presents the distribution of incidents per year for the different classifications of the "*Part Type*" field, and Fig. 2.16B presents the division of the pipe incidents in those of HC and WM (i.e., the incident distribution for the "*Pipeline Class*" of the Pipeline model). The Pipe incidents are almost 50% of the dataset, and the incidents regarding Fittings and Misc. are a small part of the total. The reduced number of incidences for 2009 is due to the lack of computerized data from August to December of that year. As the figures indicate, there does not seem to be any correlation in the fluctuation in the per-year incident number for the different classifications. For example, a comparison of the distribution of the number of incidents for 2003 to that of 2006 shows that the incidents concerning pipes increased by 2.71%, those related to fittings increased by 11.81%, the component incidents decreased by 15.43%, and those of misc. increased by 174.67%. Fig. 2.16B also shows that there is no correlation in the fluctuation in the per-year incident number for the different pipeline classes. For example in 2006, compared to 2003, the number of HC incidents increased by 3.57% while those related to WM decreased by 4.82%. A more extensive analysis related with Fig. 2.16 is presented in Section 3.1 (p. 131) on the intermittent water supply period.

As shown from the above analyses, there is a pattern in the seasonal distribution of the incidents. It was thus considered beneficial to extend the study with an emphasis on the leakage incidents.

The last part of the temporal analysis is about the seasonal time series analysis for the dataset fraction associated with leakage incidents.

Fig. 2.17 illustrates the average seasonal distribution of the leakage incidents for the three "*Part Type*" categories of the Pipeline Network model. The best-fit polynomial trendline for the three categories showed a 75% accuracy for the HC incidents, as well as 11% for the WM and 50% for the Fitting categories, respectively. The shape of the three scatter charts shows that the seasonal parameter does not correlate much with the incidents distribution for the WM and Fitting categories, but it does with the distribution of the HC leakage incidents.

Fig. 2.18A presents the average distribution of the leakage incidents related to the PN model, illustrating separately the incidents distribution associated with repair and replacement as well as their coupling (the coupling data are also equal to the summation of all data from the categories presented in Fig. 2.17). The existence of a correlation between seasonality and the distribution of the repair incidents for a WDN is clear, and it has the form of a stochastic distribution. For the case of replacement actions, the correlation is not as clear as in the case of repair actions, but it is still evident. The seasonal distribution of the incidents seems to have also the form of a stochastic distribution. Regarding the distribution of their cou-

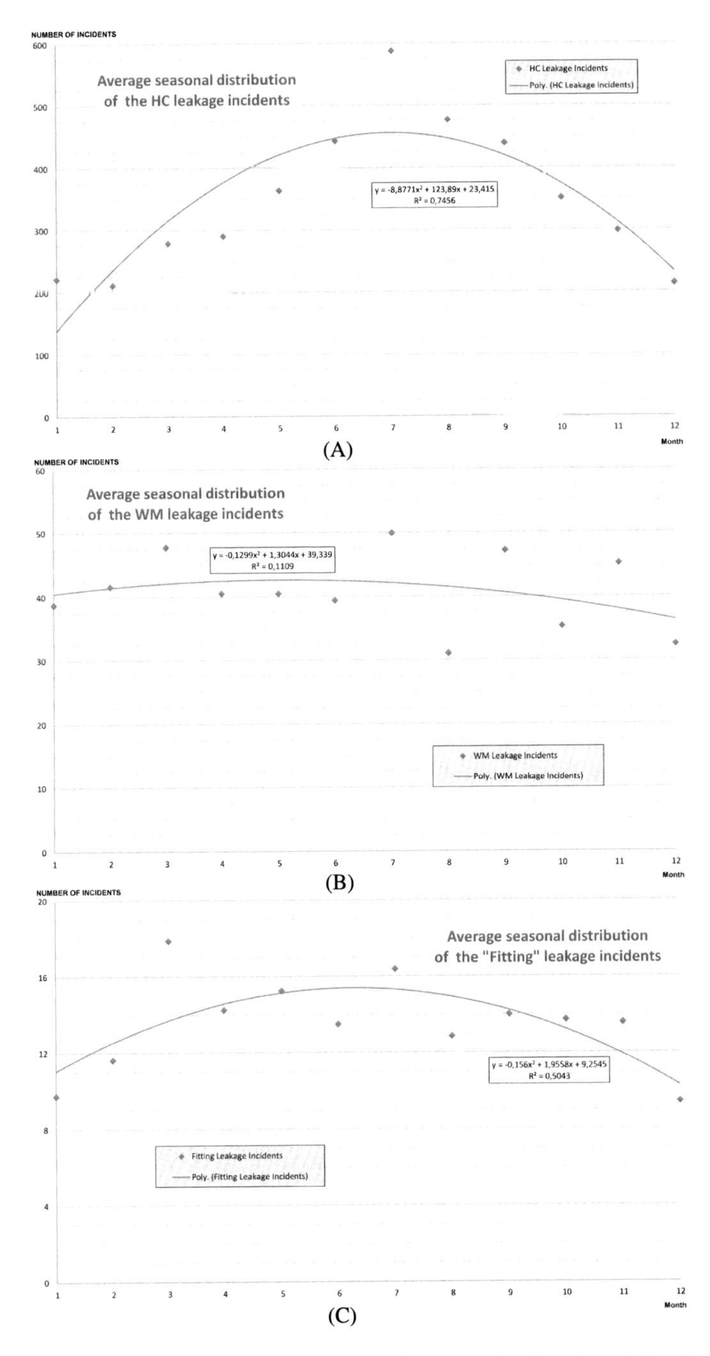

Figure 2.17 Seasonal time series analysis based on the classification of the "*Part Type*" field.

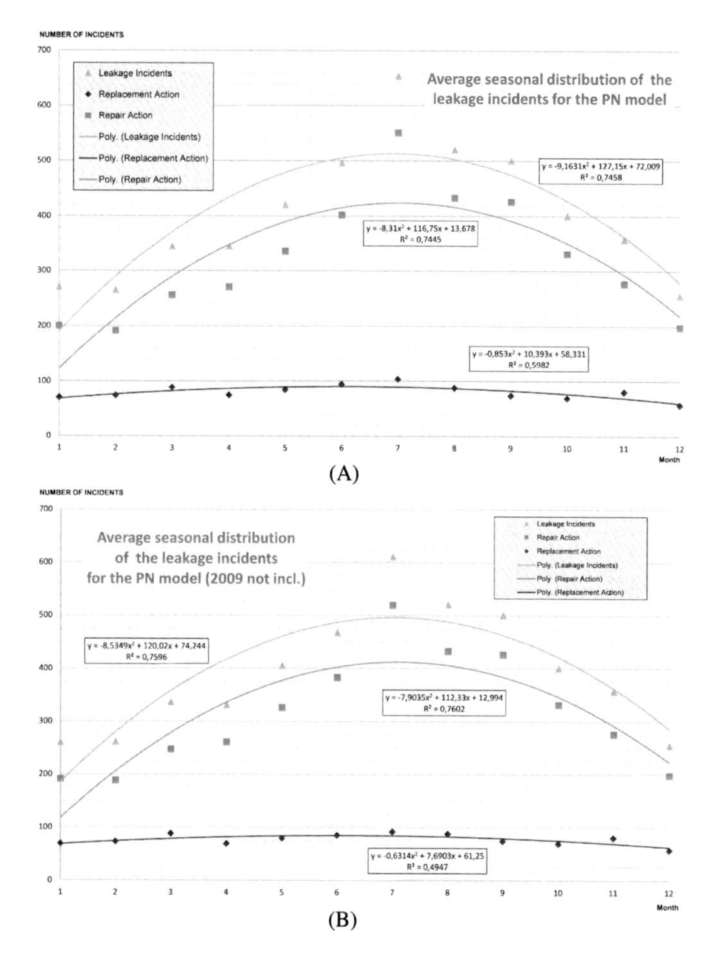

Figure 2.18 Seasonal time series analysis based on the classification of the "*Action Cluster*" field.

pling incidents, it is also apparent that the number of incidents correlates with seasonality since the repair actions are much more common compared to the replacement actions. The best fit trendline for the three distributions is that of a polynomial equation. The equation for the replacement incidents shows a 60% fit, while for the repairs and coupling we have 74% and 75% fits, respectively.

Fig. 2.18B depicts the same set of incidents as Fig. 2.18A, except that the data of 2009 were excluded from the calculation of the average distribution of leakage incidents. The success in capturing the three distributions by use of polynomial equations corresponds to 49% for the replacement action, 76% for the repair action, and 76% for their coupling.

2.4.2 Survival Analysis

The statistical analysis (Section 2.3.2, p. 41) was utilized in extracting useful preliminary conclusions regarding the quantification of the dataset and the distribution of the incidents at the DMA, street, and segment levels, as well as with respect to the different grouped variable factors. This level of analysis is not enough as it does not yield in-depth information about the relationships and correlations between the different variables and the risk-of-failure factors. To achieve this objective, the mathematical model uses survival analysis.

Survival analysis simulates the deterioration and failure in time between the initiation and a termination of an event. In the case of piping networks, an initiating event may be the installation date of the pipe while a terminating event may be a water leak initiation, a water leak observation, the replacement of a pipe, or a repair action. The method uses probabilistic models to study the variation in a pipe network's condition from the effect of several variables, and to investigate their impact on and correlation with the result, either as individuals or as a whole. Survival analysis can manage both complete and censored observations. A complete observation is a terminal event (like replacement of an element), while a censored observation is the cases for which the terminal (or initiating) event is known (or hasn't occurred yet). The terminal event occurs just once for every element.

This part of the chapter presents the results of survival analysis for WBN's dataset. First, a hypothetical example of a dataset consisting of 20 pipe incidents is presented to illustrate the methodology for survival analysis, followed by the different mathematical models used in this research and the results obtained.

Survival Analysis Example

The hypothetical set (Table 2.10) consists of pipes of two different materials and of varying age and NOPB. There are two types of incident outcome (failure or survival) which are represented by a censoring variable (binary in nature). Failure is denoted as "1" and survival as "0." The NOPB variable is divided into two classes because of the small ratio between the different NOPB and the number of incidents, so that meaningful conclusions could be extracted.

Table 2.10 Dataset of hypothetical example for demonstrating survival analysis.

Pipe ID	Age (years)	Survival time (months)	Censor variable	NOPB	NOPB cluster	Material
1	35	3	1	0	No breaks	Plastic
2	32	5	1	1	NOPB (1–4)	Plastic
3	27	6	1	1	NOPB (1–4)	Galvanized
4	29	6	0	0	No breaks	Galvanized
5	33	6	0	2	NOPB (1–4)	Plastic
6	36	8	1	3	NOPB (1–4)	Galvanized
7	21	10	0	0	No breaks	Plastic
8	40	12	1	1	NOPB (1–4)	Plastic
9	37	7	1	2	NOPB (1–4)	Plastic
10	36	9	1	0	No breaks	Galvanized
11	36	13	0	1	NOPB (1–4)	Plastic
12	35	13	0	0	No breaks	Plastic
13	38	14	1	0	No breaks	Plastic
14	38	9	0	1	NOPB (1–4)	Plastic
15	25	10	1	3	NOPB (1–4)	Galvanized
16	27	11	1	1	NOPB (1–4)	Plastic
17	33	20	0	4	NOPB (1–4)	Galvanized
18	37	15	0	2	NOPB (1–4)	Galvanized
19	34	17	0	0	No breaks	Galvanized
20	38	19	1	0	No breaks	Plastic

To determine the survival time of the pipes relative to the material type, the incidents should be separated according to the material of pipes and be ordered according to the survival time, starting with the smallest values. The first step is the calculation of the proportion of surviving pipes. The proportion for each survival time is equal to the ratio of the number of the known elements that survived after this point in time and those that survived up to that moment. The proportion of surviving pipes is estimated only for survival times involving complete events, and at survival time 0 it is equal to 1. The next step is the calculation of the cumulative proportion surviving for each survival time. This is carried out by multiplying the proportion surviving of a specific survival time, with the previous one. At survival time 0, the cumulative proportion surviving is equal to 1.

Table 2.11 presents a detailed calculation of the cumulative proportion of pipes surviving at each time step. Pipes related with censored data are

Table 2.11 Calculation of the cumulative proportion surviving, for the survival analysis example dataset.

Material	Pipe ID	Survival time (months)	Number known to have survived	Failures	Censored	Proportion surviving	Cumulative proportion surviving
Plastic	–	0	12	–	–	1.000	1.000
	1	3	12	1	0	$(12-1)/12=0.917$	$1.000 \times 0.917=0.917$
	2	5	11	1	0	$(11-1)/11=0.909$	$0.917 \times 0.909=0.833$
	5+	6	10	0	1	–	–
	9	7	9	1	0	$(9-1)/9=0.890$	$0.834 \times 0.890=0.741$
	14+	9	8	0	1	–	–
	7+	10	7	0	1	–	–
	16	11	6	1	0	$(6-1)/6=0.833$	$0.834 \times 0.890=0.617$
	8	12	5	1	0	$(5-1)/5=0.800$	$0.617 \times 0.800=0.494$
	11+, 12+	13	4	0	2	–	–
	13	14	2	1	0	$(2-1)/2=0.500$	$0.494 \times 0.500=0.247$
	20	19	1	1	0	$(1-1)/1=0.000$	$0.247 \times 0.000=0.000$
Galvanized	–	0	8	–	–	–	1.000
	3, 4+	6	8	1	1	$(8-1)/8=0.875$	$1.000 \times 0.875=0.875$
	6	8	6	1	0	$(6-1)/6=0.833$	$0.875 \times 0.833=0.729$
	10	9	5	1	0	$(5-1)/5=0.800$	$0.729 \times 0.800=0.583$
	15	10	4	1	0	$(4-1)/4=0.750$	$0.583 \times 0.750=0.438$
	18+	15	3	0	1	–	–
	19+	17	2	0	1	–	–
	17+	20	1	0	1	–	–

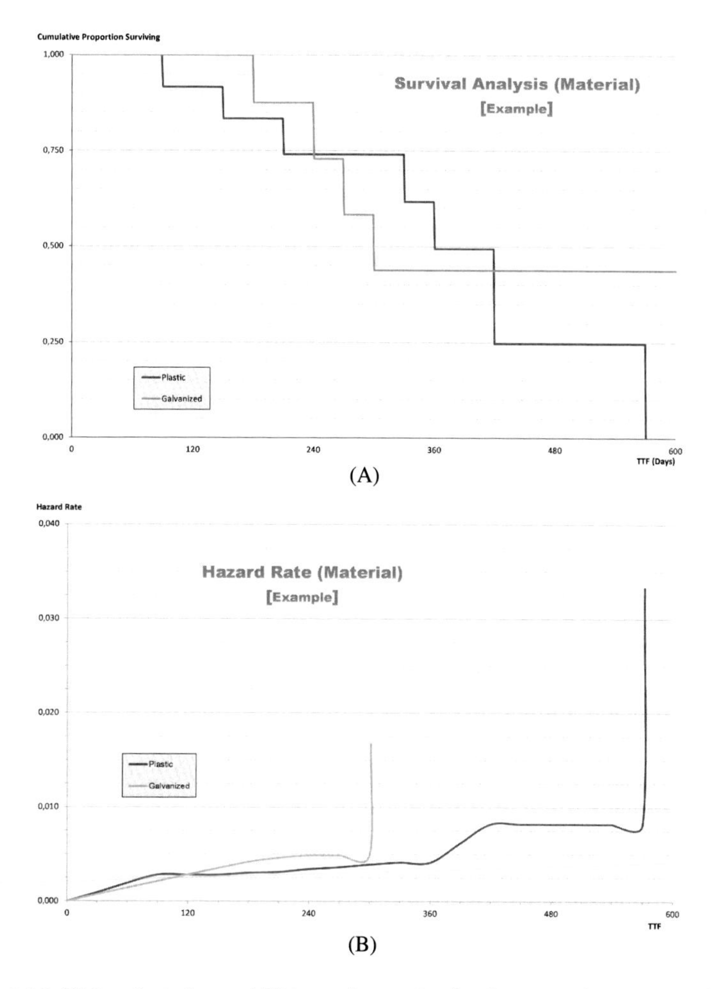

Figure 2.19 (A) Survival plot and (B) hazard rate plot for the example on survival analysis.

indicated with a "+" sign. Fig. 2.19A shows the survival plot for the two types of materials.

The hazard rate of a variable can be calculated by using the survival plot. It is equal to the drop in the cumulative proportion surviving divided by the time corresponding to that range. For plotting the hazard rate diagram, the time to failure (TTF) of the survival curve of a variable is divided at equal time intervals and the hazard rate for each of them is calculated. For this example, time intervals of 30 days were used for the calculation of the hazard rate. Selection of the time interval was based on the number of

Table 2.12 Calculation of the hazard rate for the survival analysis example dataset.

Material	Survival time (days)	Cumulative proportion surviving	Difference in cumulative proportion surviving	Hazard rate
Plastic	0	1.000	0.0000	0.0000
	90	0.917	0.0833	0.0028
	150	0.833	0.0833	0.0028
	210	0.741	0.0926	0.0031
	330	0.617	0.1235	0.0041
	360	0.494	0.1235	0.0041
	420	0.247	0.2469	0.0082
	570	0.000	0.2469	0.0082
Galvanized	0	1.000	0.0000	0.0000
	180	0.875	0.1250	0.0042
	240	0.729	0.1458	0.0049
	270	0.583	0.1458	0.0049
	300	0.438	0.1458	0.0049

incidents and the total survival time. Table 2.12 presents the calculation of hazard rate for the presented example while Fig. 2.19B illustrates the hazard rate plot.

The manual calculations were also performed with the help of a commercial statistical software (NCSS) for verification purposes. For comparison reasons, an additional analysis was made using the same WBN dataset, but with the inclusion of the pipes' age. The resulting analysis plots are presented in Fig. 2.20.

Further, if in Eq. (2.9) we let $x = X_i$ denote a collection of p risk parameters $(X_1, X_2, \ldots, X_p)$, with $h_0(t)$ being the baseline hazard function for a subject with covariates 0 and β_i being the regression coefficients, then we get the Cox regression model represented by Eq. (2.11). Table 2.13 presents the computed Cox regression model as estimated by using NCSS

$$H(t, X) = h_0(t) \times \exp\{\beta_1 \times X_1 + \beta_2 \times X_2 + \cdots + \beta_p \times X_p\}. \tag{2.11}$$

This model is of high interest since it adds information about the variables' proportional hazard rate to the survival analysis model.

Comments on the Results From the Survival Analysis Example

- The survival plots relative to the material of the pipes, which do not take into account the pipe age, are the same (Fig. 2.19A and

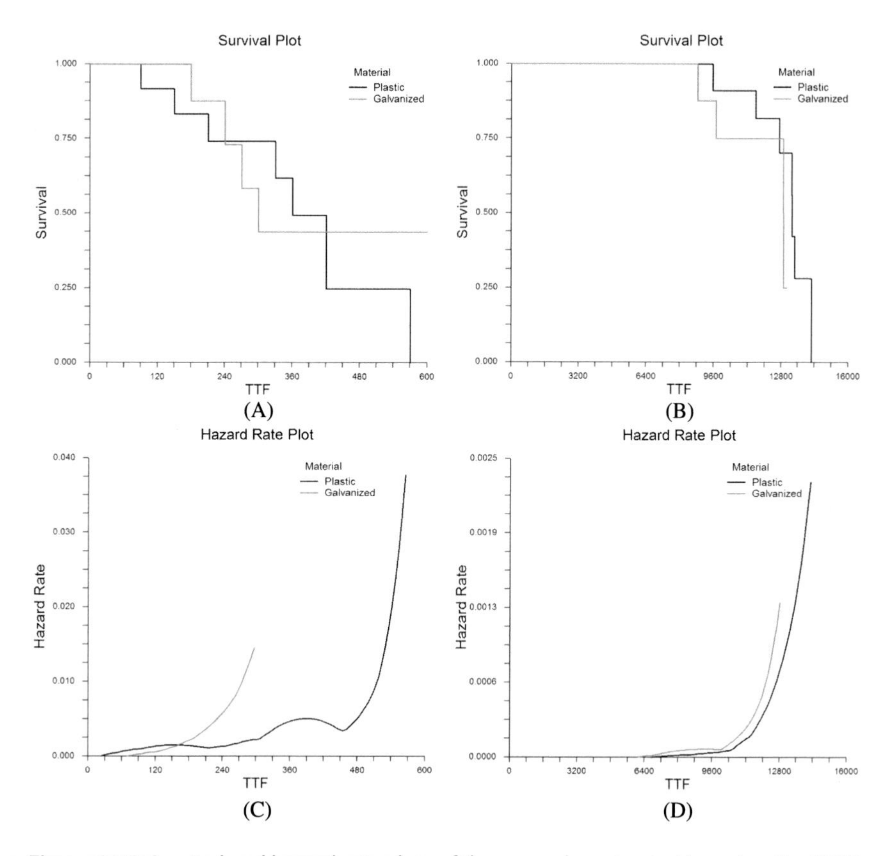

Figure 2.20 Survival and hazard rate plots of the example, extracted by using the NCSS software.

Table 2.13 Cox regression model for the dataset in the survival analysis example.

Model	**Variable categories**	**Regression coefficient (β_i)**	**Standard error of (β_i)**	**Risk ratio (exp[β_i])**
No Pipe Age	Material="PLASTIC"	0.253374	0.633147	1.2884
	Estimated Cox Regression Model: Exp [+ 0.253374*(Material="PLASTIC")]			
Pipe Age	Material="PLASTIC"	−0.908707	0.769497	0.4030
	Estimated Cox Regression Model: Exp [– 0.908707*(Material="PLASTIC")]			

Fig. 2.20A). They show that initially galvanized pipes have a higher proportion of surviving, but after 240 life-days their survival rate be-

comes worse than that of plastic pipes. The population of plastic pipes dies at 570 days, while for galvanized pipes the lowest cumulative proportion surviving is reached in 300 days and remains constant for 300 days. This is because the last incident for the plastic class is a complete event (so the cumulative proportion surviving becomes zero), while the last complete event for the galvanized pipes happens at day 300. So, *accurate conclusions about the survival of a variable can only be exported up to the time when the last complete incident is observed.*

- The more incidents there are, the more time intervals with survival data are available, and thus the survival curve has more data points. Thus, *the drop of the survival curve becomes smoother with the increase in the number of events over time.*
- Fig. 2.20B presents the survival plot for the dataset based on the material variable taking into account the pipe age. The only common point compared to the other two survival plots is the number of drops in the cumulative proportion surviving, which is seven for the plastic and four for the galvanized pipes (corresponding to the number of complete incidents). The introduction of the information related to the pipe age into the analysis affects the time range between the first and the last complete incident as well as the size of reduction in a variable's survival, because of the appearance of a complete incident. Based on the graph, the rate of reduction in the survival plot of the plastic pipes is smaller than that of the galvanized pipes throughout their lifespan. Therefore, *an irrational choice of the pipes' age (for cases where there are no available data) will lead to the extraction of inaccurate results.*
- A comparison of Fig. 2.19B (the hazard rate was exported from the survival plot) with Fig. 2.20C (the hazard rate was calculated by use of the Nelson–Aalen estimator as implemented in the NCSS software) shows that they follow the same pattern, differing at the point where the last complete incident occurs for both material types. In this case, the difference in hazard rate values between the two plots is not so important (it is caused by the small number of complete incidents and the methodology of calculating the hazard rate from the survival plot). For example, the hazard rate for plastic pipes at the time of the first incident, using a time interval of 30 days, was found to be 0.0028 (= 0.083/30). If the chosen time interval were equal to 10 days then

the hazard rate for the same point would be 0.0083. So, *what is important in a hazard rate plot is the proportion change in hazard rate per time interval, as well as the time period at which the slope of the hazard rate line starts to increase dramatically.*

- *The line that depicts the hazard rate of a variable, whose last event is a complete incident, is curved and tends towards infinity at the point that corresponds to the time of the incident. The hazard rate line of the plot for variables whose last event is not a complete incident stops at the time when the last complete incident is observed. At that point the shape of the line becomes curved while the final hazard rate value can be equivalent to the proportion surviving at this time divided by the time interval.* For example, the proportion of the galvanized pipes surviving when the last complete incident was observed is 0.438, while the used time interval is 30 days. Hence the hazard rate at that time is equal to 0.0146, which is very close to that of the NCSS software (Fig. 2.20B).
- The hazard rate reflects the deterioration rate over time. Thus, *the hazard rate plot cannot decrease – it only increases or remains constant over time.* Any reductions in the hazard rate plots seen in this example are because the incidents do not correlate to a realistic scenario and their number is very small compared to the duration of the example's survival time.
- As in the case of the survival plots, the inclusion in the hazard rate analysis of the information related to the pipe age affects the time range between the first and the last complete incident as well as the size of the variable's hazard rate (Fig. 2.20D). Hence, *an irrational choice of the pipes' age (for cases where there are no available data) will lead to inaccurate results.*
- Table 2.13 presents the Cox regression models for the material variable (extracted by using the NCSS software) for the two versions of the example analysis (one related with the survival time of the study's pipes and one incorporating pipe age). *The Cox regression is an important analysis tool since the "Risk Ratio" of a variable indicates the variation of the dataset's hazard rate if that is increased by one.* In both case studies only plastic material is referred, since the data related to the hazard rate of the galvanized material is not of sufficient quality. In the first case, the risk ratio for the plastic pipes is equal to 1.2884. This means that if the next incident to be observed is associated with plastic material

then the hazard rate is increased by 28.84%. If it is increased by 3, then the hazard rate will be multiplied by 1.28843. In the second case, the risk ratio of plastic is much lower because the example takes into account the pipe's age and the levels of hazard rate values are much lower. Thus, the effect of plastic to the total hazard rate is less important.

Street Level Analysis

The first part of the survival analysis relates to the analysis of the incidents at the street level. In this part all the nine models, originally presented in Table 2.5 (p. 48), are analyzed. The presentation of results begins with the three classes of the Distribution System model, followed by those of the Pipeline Network, and Pipeline models.

Survival Analysis of the Distribution System Model

The analysis series related to the Distribution System model uses the whole dataset. This analysis is not expected to lead to accurate results because part of the input information is not related with the underground part of the case-study WDN. The analysis was partly carried out for comparison reasons with the results from the other two models (as also presented below).

Fig. 2.21 presents the hazard rate plots of the model's classes for NOPB and DMAs.

- The first observation is that the hazard rate plots of Fig. 2.21 do not start from 0. This is because the data contains several complete incidents of young elements, mainly due to the fact that the analysis is at the street level. For example, when a 25-year old pipe fails, a complete incident which is aged 25 years is added into the dataset. The subsequent failure of a pipe that is located in the same street will add to the data a complete incident whose age is equal to the difference in time between the two successive complete incidents. Thus, as the number of element failure incidents in a street that have short time intervals between them increases, the phenomenon in which the hazard rate does not start from zero becomes more likely. This phenomenon is eliminated when the analysis is performed at the segment level, as the elements' short length minimizes the probability of existence of several complete incidents in the same segment within a short time period.

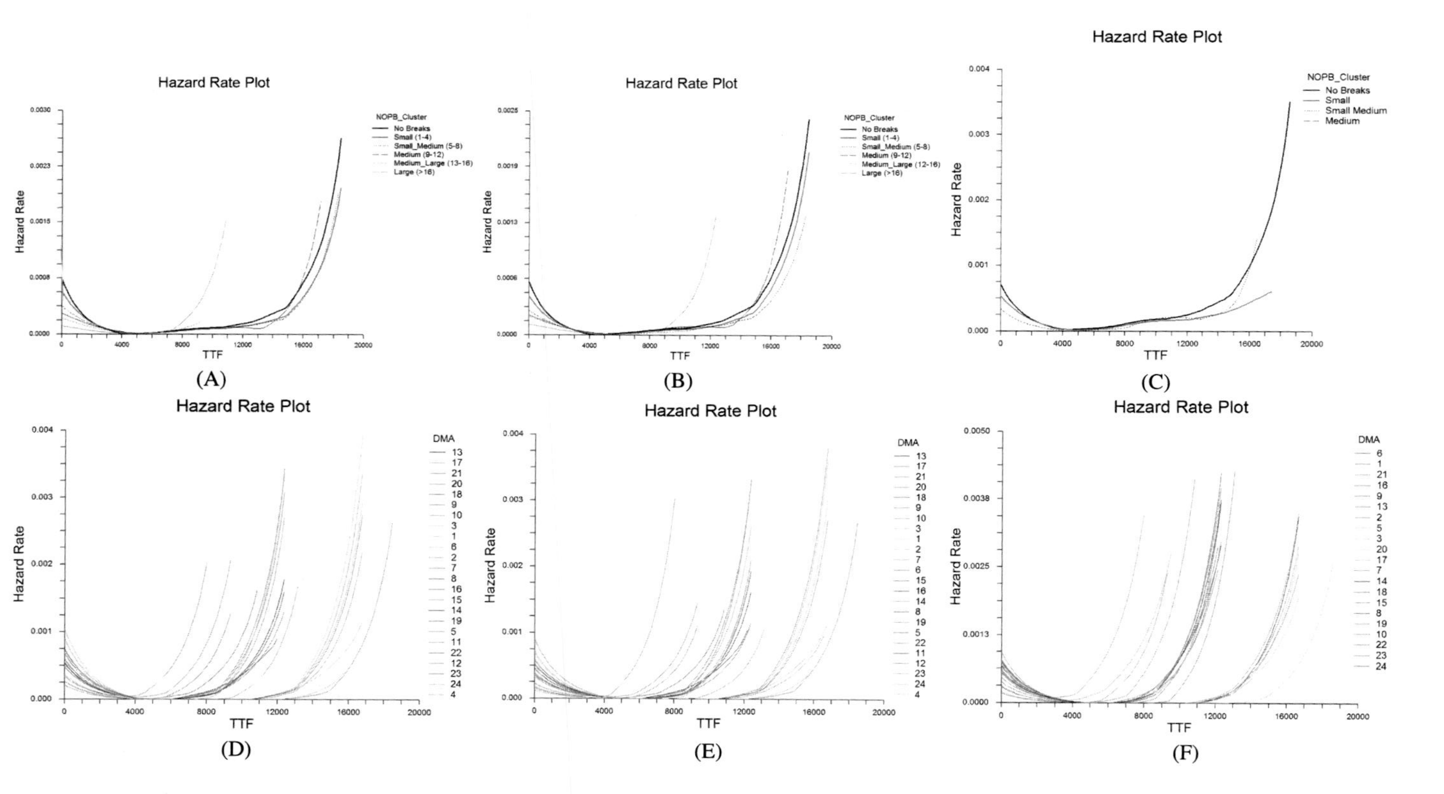

Figure 2.21 Hazard rate plots of the NOPB and DMA for the DS model (Street Level/All Incidents).

- The second important observation is related with the hazard rate plots concerning the "*NOPB Cluster.*" The hazard rate should be increasing with respect to NOPB. Unlike this, the cluster of "No Breaks" (NOPB = 0) has a higher hazard rate compared to the categories "Small" and "Small–Medium." This is because the data concerning the NOPB variable is biased. This plot examines the hazard rate according to the number of repairs that a pipe had before it fails. Based on this, the dataset should comply with two conditions. Firstly, it should only include repair and replacement incidents, and secondly, it should include only successive incidents whose last event is a complete incident. For example, the incidents regarding the cluster of "No Breaks" should include all events whose NOPB is zero and they are complete incidents, or they are censored incidents which belong to series of consecutive censored incidents whose terminal event is related to replacement action.
- The above observation requires a time-consuming data processing, especially for a growing data volume. The main objective of the hazard rate plot is not the choice of the cluster which has the highest durability (this is known), but the identification of the point in time after which the hazard rate dramatically increases and how it varies with the increase of an element's repair incidences. The hazard rate plot in Fig. 2.21 concerning the NOPB is quite realistic to achieve this objective. Therefore the calculation of hazard rate by use of the whole dataset is considered satisfactory.
- The hazard rate plot of C class (Fig. 2.21A) showed that after the 14th repair of an element, the condition deteriorates dramatically and the selection of replacement instead of repair would probably be a better decision. The same is true for the hazard rate of the HC class while for WM class there are no data related to the Medium, Medium–Large, and Large clusters (Figs. 2.21B–C, respectively). The first four clusters do not seem to affect the useful life of the pipes for the C class model. The condition of the elements is considered very good during the first 28.5 years of their life. For the next 4.5 years, the condition is satisfactory, and thereafter the pipes enter into the time period of significant deterioration. The corresponding time intervals for the other two models are 30 and 5.5 years for the HC class, and 24.5 and 5.5

for the WM class, respectively. The results coincide with the age expectancy specified by the manufacturers of pipes, which reaches close to 30 years.

- It has also been observed that the range in time between the hazard rate lines of the different DMAs is relatively large. As shown by the plots, the range in time between lines of HF is relatively large. This directly correlates with the age of the networks. As stated above, commenting the example of survival analysis, the filling of the pipe's age is very important for the survival analysis. The given age for each pipe was associated with the time period that water was supplied to the area where it belongs. Based on the year (2003) during which the incident records initiated, the pipes' ages of the DMAs range from 40 up to 15 years. There is a group of 8 DMAs whose age is approximately 40 years and another group of 12 DMAs whose age is 30 years. The age of the remaining four ranges from 25 to 15 years. Even though the missing data (before 2003) are censored, the variance between the three age groups is noticeable since the difference in the groups' age compared to the age of the DMAs as well as the time that the recorded incidents extend is significant.
- The DMA which has the lowest hazard rate is DMA 20 while the one with the highest hazard rate is DMA 21, for all three classes of the Distribution System model (Figs. 2.21D–F). These results match reality. DMA 20 is the area within the walls of the old city of Nicosia. Because of the area's archaeological importance, it is highly protected from interventions on its infrastructure, hence the WBN does not have free access to intervene on the WDN wherever and whenever WBN desires. For this reason the monitoring and management of the particular DMA is at the highest level it can be, which results in the excellent condition of this DMA. Apart from the factor related with age, there is another one for which the DMA 21 has the highest hazard rate. This area has joined the WBN during the last decade. The municipality where it belongs was the responsible authority for its management during the first 20 years of its life. Because municipality authorities do not have the proper equipment and expertise to manage a WDN (unlike the Water Board), the condition of the DMA when it first joined WBN was not adequate. As a result, the DMA was in need of several rehabilitation actions which burdened its hazard rate performance.

(a) Couple Class

This part of the analysis relates to the Couple class of the Distribution System model. Fig. 2.22 presents the survival plots of the class for the different variables. The analysis used the incidents of the Distribution System model.

- Fig. 2.22A corresponds to the survival plot for the "*Part Type*" variable of the C class model (Distribution System). The plot indicates that the phenomenon observed in Fig. 2.21 (p. 83) and relating to the fact that the hazard rate plots do not start from 0 (due to the increase in a pipe's failure incidents on a street that have short time intervals between them) affects mostly the factors of Fitting, Component, and Misc. The survival probability for the Pipes is the highest while that of Fittings is the lowest. The rate of the survival line's decline for Misc., Components, and Fittings is similar. The decline rate of the survival probability for all the factors is low up to the 35th year of their lifespan while after that point it suddenly increases. The pipes' lifetime has a high probability to exceed 40 years.
- Fig. 2.22B corresponds to the survival plot for the Pipeline class variable of the C class model (Distribution System). The factors of interest in this plot are those of HC and WM since the Misc. is also presented in the previous plot (Fig. 2.22C) and the HC-WM contain few incidents that belong to Fitting factor and are associated with HC as well as WM. The WM presents an increase in the pipe's failure incidents on a street that have short time intervals between them compared to HC. The survival probability for the HC pipeline class is higher than that of WM. The shape of survival line for the two parameters is similar. The decline rate of the survival probability of both parameters is low up to the 38th year of their lifespan while after that point it suddenly increases. The HC elements have a high probability to exceed this span of life.
- Fig. 2.22C corresponds to the survival plot for the Distribution System model. The survival probability of the model's elements is reduced by 44% during the first 5 years of their life (for the reason explained above). The shape of the model's survival line is similar to that of the HC (Pipeline class) due to the large amount of incidents related to HC in the dataset. The decline rate of the survival probability is low up to the 38th year of the model's lifespan while after that point it suddenly increases. The element's lifetime has a probability of 34% to exceed this age.

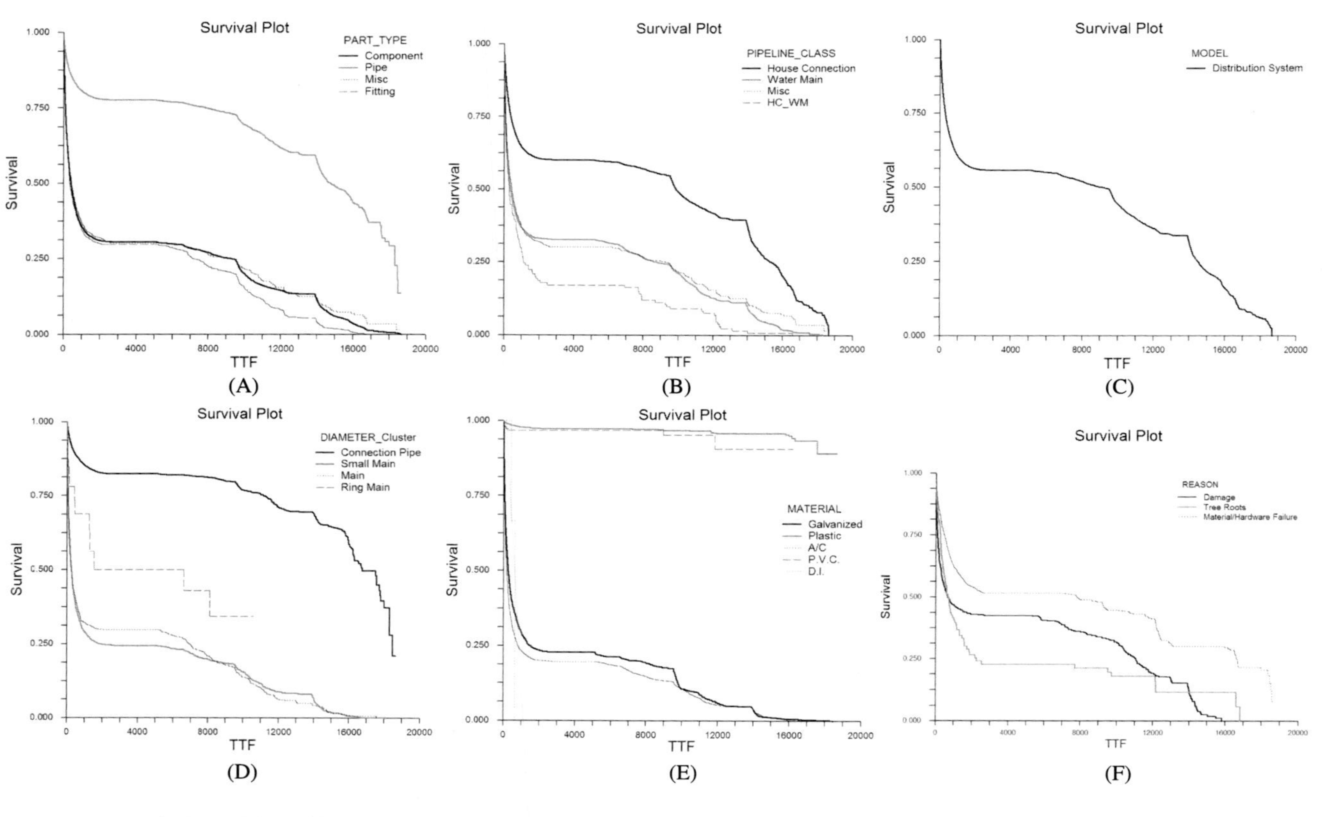

Figure 2.22 Survival plots of the different variables for the DS-C model (Street Level/All Incidents).

- Fig. 2.22D corresponds to the survival plot for the Diameter Cluster variable of the C class model (Distribution System). The survival probability for the Connection Pipes is the highest. The decline rate of the survival probability for this factor is low up to the 48th year of its lifespan when the probability to survive is 48%. The incidents involving the Ring Main are very few in the total dataset (It is seen from the stepped shape of the line survival.) hence conclusions drawn from it will not be reliable. During the first 28 years of the element's life, those correlated with the factor Main have a greater chance to survive than those of Small Main. From that point onwards, it happens the other way around.
- Fig. 2.22E corresponds to the survival plot for the "*Material*" variable of the C class model (Distribution System). The incidents involving the P.V.C. and D.I. categories are very few compared to the total dataset, and for that reason the survival curves for these two material types cannot accurately depict reality. The survival probability for the Plastic category is much higher than that of Galvanized and A/C, which have similar survival values over time. The results are consistent with what happens in reality. Plastic pipes have a longer lifespan and they have actually been put into service later than pipes of other materials. Further, the A/C and Galvanized pipes are replaced when an incident occurs even if the damage can be repaired.
- Fig. 2.22F corresponds to the survival plot for the "*Reason*" variable of the C class model (Distribution System). As shown through the statistical analysis, the incidents involving the "*Reason*" variable are very few compared the total dataset. Hence the survival lines for the three factors cannot accurately depict reality. Based on the existing incidents, the survival probability for the Material/Hardware Failure is the highest. Furthermore, during the first 38 years of the elements' life, those correlated with the Damage factor have a greater chance to survive than those related to Tree Roots. From that point onwards, the opposite occurs.

(b) House Connection Class

This analysis section relates to the HC class model (Distribution System), as per Fig. 2.6 (p. 52). The analysis is based on the incidents of the Distribution System model that concern the HC class, and Fig. 2.23 presents the survival plots of the class for the different variables.

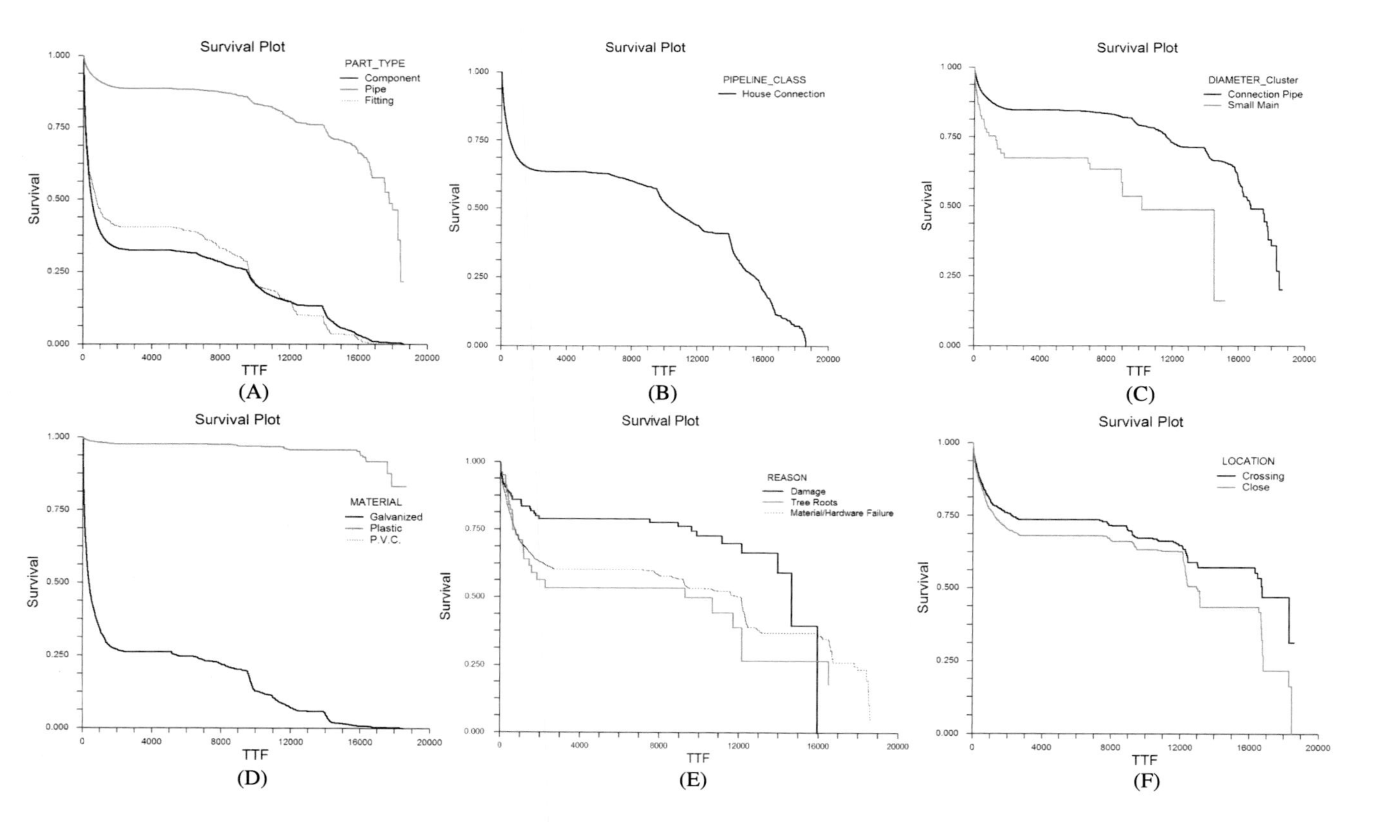

Figure 2.23 Survival plots of the different variables for the DS-HC model (Street Level/All Incidents).

- Fig. 2.23A corresponds to the survival plot for the "*Part Type*" variable of the HC class model (Distribution System). The plot indicates that the HC pipes are not subjected to a lot of failure incidents with short time intervals between them. The survival probability of the Pipe factor is the highest. The decline rate of the survival probability for this factor is low up to the 48th year of its lifespan when the probability to survive is 60%. During the first 32 years of the element's life, those correlated with the Fitting factor have a greater chance to survive than those related to Component. From that point onwards, the opposite is true.
- Fig. 2.23B corresponds to the survival plot for the "*Pipeline Class*" variable of the HC class model (Distribution System). The survival probability of the model's elements is reduced by 35% during the first 5 years of their life. The decline rate of the survival probability is low up to the 38th year of the model's lifespan while after that point it suddenly increases. The elements lifetime has a probability of 42.5% to exceed this age. The shape of the model's survival line is similar to that of the C model (It is shifted by about 9% upwards). Hence the dataset is biased with respect to the HC incidents, due to their large number in relation to the dataset.
- Fig. 2.23C corresponds to the survival plot for the "*Diameter Cluster*" variable of the HC class model (Distribution System). The survival probability for the Connection Pipes is the highest. It is almost identical with the corresponding of the C model. A very slight deviation downwards (it does not actually affect the survival probability) is due to the fact that very few data concerning both types of pipe (HC and WM) have been used as independent factor in the C model, while for the HC class model they are added into the HC incidents. The decline rate of the survival probability for the Small main factor is low up to the 37th year of its lifespan when the probability to survive is 48%. The linear shape of the plot for the survival probability reveals that the facts related to this factor are few.
- Fig. 2.23D corresponds to the survival plot for the "*Material*" variable of the HC class model (Distribution System). The incidents involving the P.V.C. factor are negligible. The survival probability for the Plastic factor is much higher than that of Galvanized. The reason for this is the same as that for Fig. 2.22E (C model).
- Fig. 2.23E corresponds to the survival plot for the "*Reason*" variable of the HC class model (Distribution System). The survival probability

plot reveals that a small proportion of the dataset incidents which are related to the HC class are correlated with the "*Reason*" factor. Hence the survival lines for the three factors cannot accurately depict reality. Based on the existing incidents, up to the point just before the last complete incident occurs the survival probability for the "*Damage*" factor is the highest. Furthermore, the elements that are correlated with the Material/Hardware Failure factor have a greater chance to survive than those related to the Tree Roots. The results of this plot do not keep up with the corresponding results of the C model.

- Fig. 2.23F corresponds to the survival plot for the "*Location*" variable of the HC class model (Distribution System). This factor is only related to HC model. The survival probability for the elements of the Crossing factor is better than that of the Close factor. The decline rate of the survival probability for both factors is low up to the 45th year of their lifespan when the probability to survive is 53% for the Crossing and 43% for the Close factor, respectively.

(c) Water Main Class

This part of analysis is related to the WM class model (Distribution System), as per Fig. 2.6 (p. 52). The analysis is based on the incidents of the Distribution System model that concern the WM class, and Fig. 2.24 presents the Survival plots of the class for the different variables.

- Fig. 2.24A corresponds to the survival plot for the "*Part Type*" variable of the WM class model (Distribution System). In contrasting to the corresponding survival plots of the C and HC models, in this case the survival probability of the Component factor is the highest. The rate of decline of the survival probability for this factor is low up to the 42nd year of its lifespan when the probability to survive is 55%. The survival probability of the Pipe factor is slightly higher than that of Fitting.
- Fig. 2.24B corresponds to the survival plot for the "*Pipeline Class*" variable of the WM class model (Distribution System). The survival probability of the model's elements is reduced by 45% during the first 5 years of their life. The decline rate of the survival probability is low up to the 38th year of the model's lifespan while after that point it suddenly increases. The elements lifetime has a probability of 20% to exceed this age. The shape of the model's survival line shows that the dramatic reduction in the model's survival probability during the first years of its lifespan affects the corresponding plot of the C model.

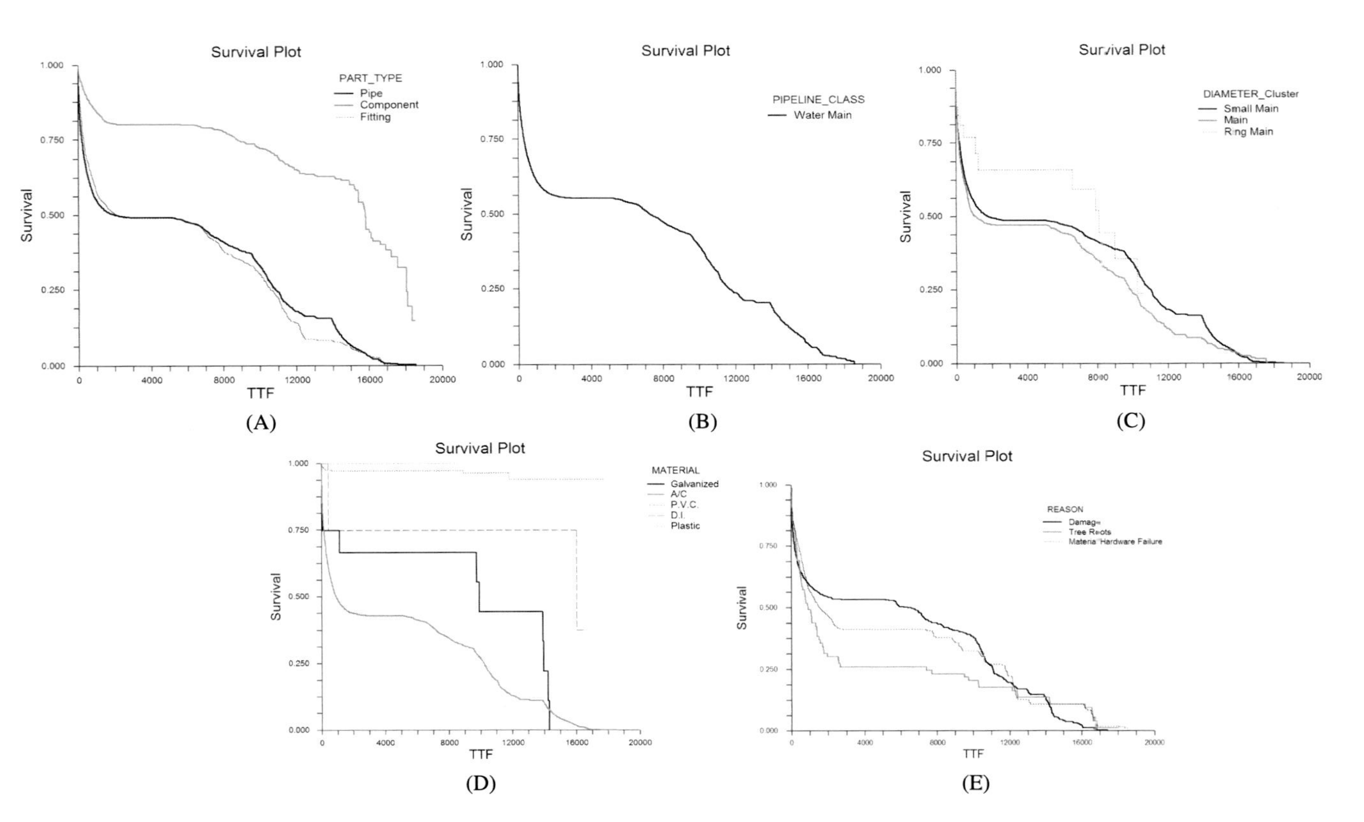

Figure 2.24 Survival plots of the different variables for the DS-WM model (Street Level/All Incidents).

- Fig. 2.24C corresponds to the survival plot for the "*Diameter Cluster*" variable of the WM class model (Distribution System). As in Fig. 2.22D (C model), the incidents involving the Ring Main are very few in the complete dataset, hence conclusions drawn about this class are not very reliable. The survival probability for the Small Main factor is higher than that of Main. The shape of the survival probability for the two factors is similar up to the 38th year of their lifetime. After the 43rd year of their life, the survival probability of the Main factor becomes higher than that of Small Main. The probability for an element to reach this age is very low (3%).
- Fig. 2.24D corresponds to the survival plot for the "*Material*" variable of the WM class model (Distribution System). The survival plot from this analysis shows that, with the exception of the Galvanized material category, the number of events that are distributed to the different categories is very small, and the proportion of the completed incidents compared to the censored incidents is very low. Based on the presented results, the survival probability of the Galvanized factor is higher than that of A/C, as was also the case in Fig. 2.22E (C model).
- Fig. 2.24E corresponds to the survival plot for the "*Reason*" variable of the WM class model (Distribution System). Up to the age of 30 years, the order in percentage value of the survival probability for the three different factors is the same with that of Fig. 2.23E (HC model). The Damage factor has the highest survival probability while the Tree Roots factor has the lowest. From that point on, the order of severity changes continuously.

Survival Analysis for the Pipeline Network Model

The analysis series related to the Pipeline Network model uses, as a dataset, all incidents related with the underground buried elements of the WDN. This model of analysis is the one proposed by this work.

The survival probabilities and hazard rates of factors for the different variables are better (higher survival probability and hence lower hazard rate) than their counterparts of the Distribution System model due to the fact that Component and Misc. elements, which present a lot of successive complete incidents within a short time period, excluded from the analysis.

Fig. 2.25 presents the hazard rate plots of the model's classes for NOPB and DMAs.

- Both comments concerning Fig. 2.21 (Distribution System), on the successive complete incidents within a short time period as well on the

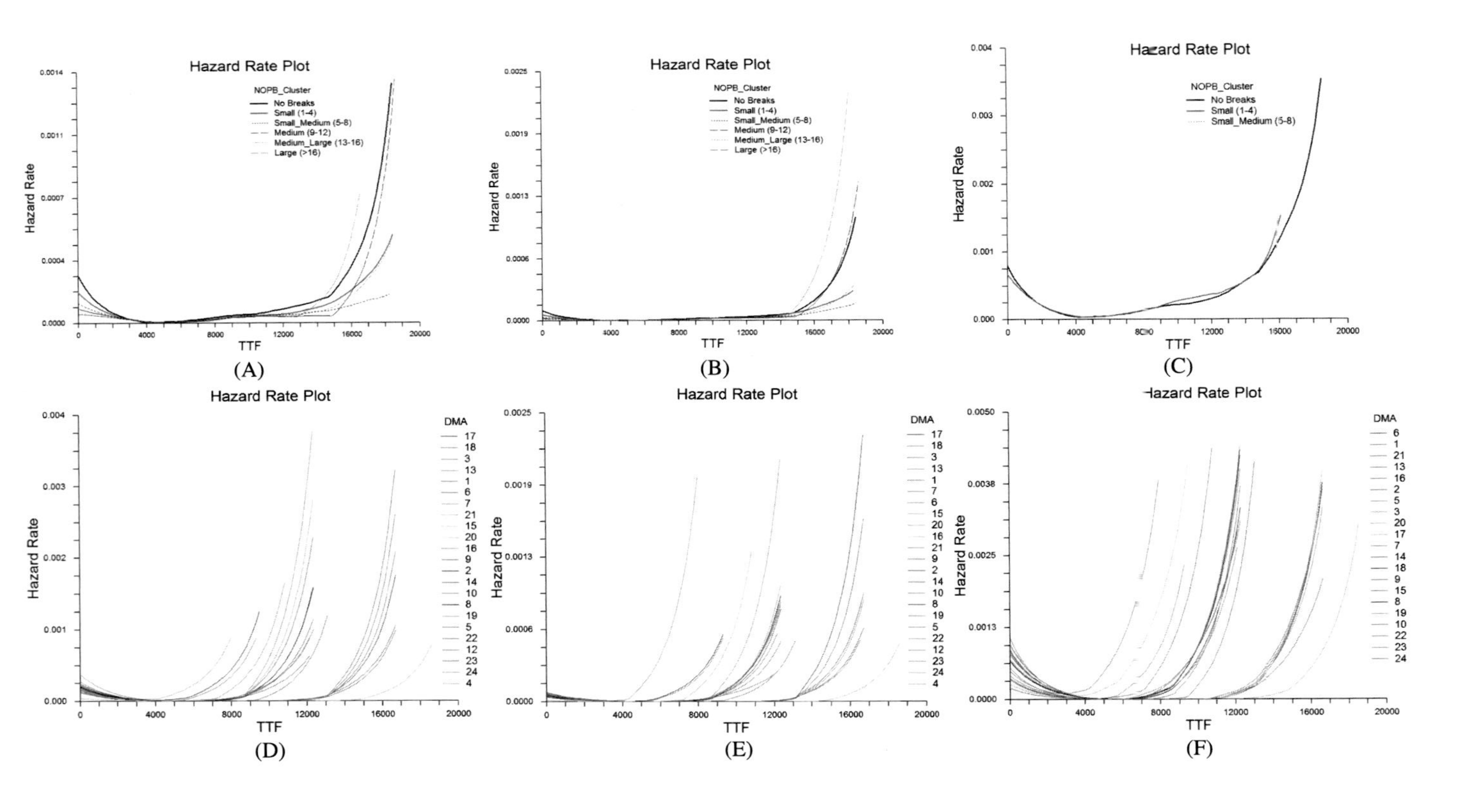

Figure 2.25 Hazard rate plots of the NOPB and DMA for the PN model (Street Level/All Incidents).

fact that the hazard rate of the No Breaks factor is not the lowest one, apply also to Fig. 2.25.

- The hazard rate plots of the NOPB Cluster variable show that the increase of NOPB is not highly correlated with an increasing mortality of the elements. For the C class model (Fig. 2.25A), the condition of the elements is considered very good during the first 29 years of their life. For the next 5 years, it is satisfactory and thereafter they enter into the time period of significant deterioration. The corresponding time intervals for the other two models (Figs. 2.25B–C) are 38 and 2 years for the HC class and 25 and 10 for the WM class, respectively. The classes of the Pipeline Network model present a slightly better hazard rate (lower) than those of the Distribution System model.
- The findings relating to the hazard rate of DMAs are exactly the same as those from Fig. 2.21.

(a) Couple Class

This analysis group is related to the C class of the Pipeline Network model. Fig. 2.26 presents the survival plots of the class for the different variables. The analysis used all incidents concerning the Pipeline Network model.

- Fig. 2.26A corresponds to the survival plot for the Part Type variable of the C class model (Pipeline Network). The decline rate of the survival probability of both factors is low up to the 38th year of their lifespan while after that point it suddenly increases. The pipes' lifetime has a high probability to exceed 45 years.
- Fig. 2.26B corresponds to the survival plot for the Pipeline class variable of the C class model (Pipeline Network). The factors of interest in this plot are those of HC and WM. The survival probability for the HC pipeline class is higher than that of WM. The decline rate of the survival probability of both parameters is low up to the 38th year of their lifespan while after that point it suddenly increases. The lifetime of HC elements has a high probability to exceed 45 years.
- Fig. 2.26C corresponds to the survival plot for the Pipeline Network model. The survival probability of the model's elements is reduced only by 17% (the corresponding value in Fig. 2.22C is 44%) during the first 5 years of their life. The shape of the model's survival line is similar to that of the HC (Fig. 2.26B) due to many incidents related to HC in the dataset. The decline rate of the survival probability is low up to the 38th year of the model's lifespan while after that point it suddenly increases.

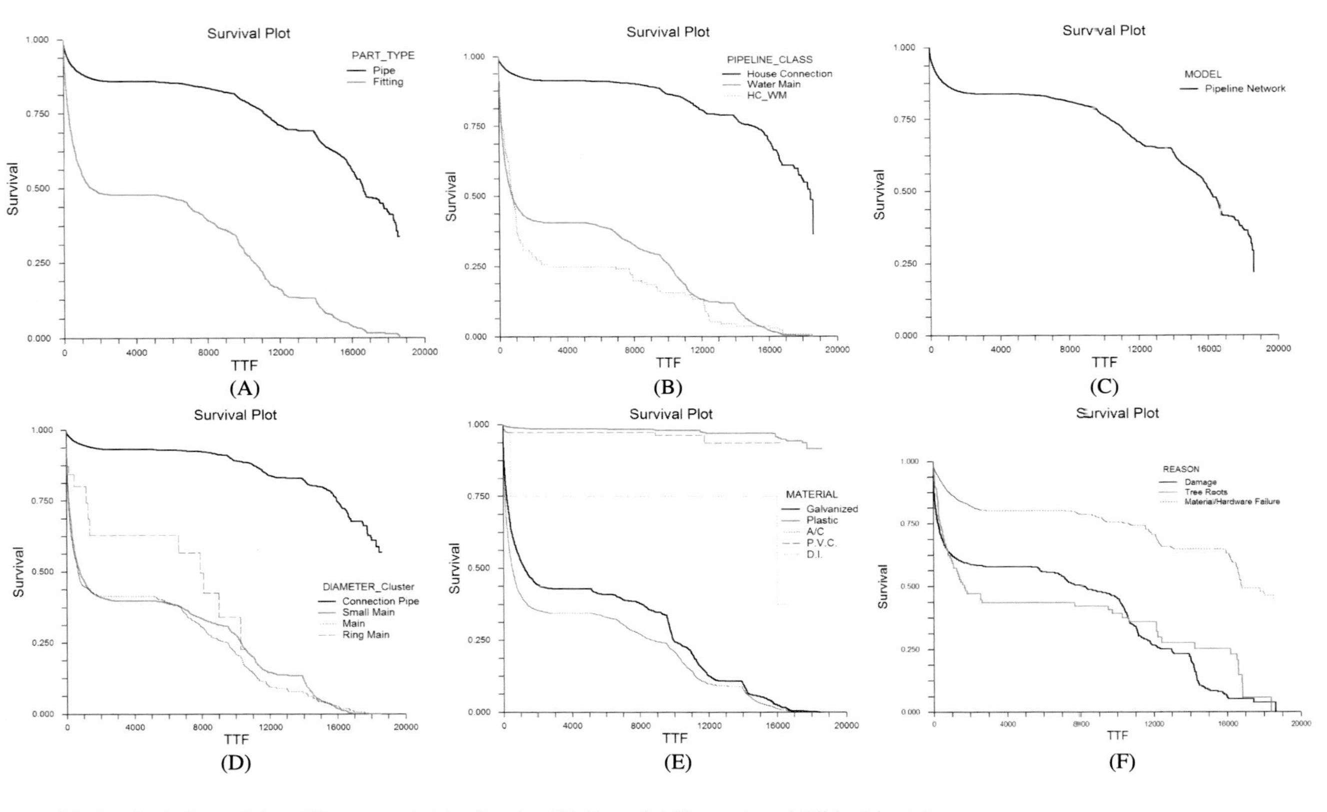

Figure 2.26 Survival plots of the different variables for the PN-C model (Street Level/All Incidents).

The elements lifetime has a probability of 65% (the corresponding value in Fig. 2.22C is 34%) to exceed this age.

- Fig. 2.26D corresponds to the survival plot for the Diameter Cluster variable of the C class model (Pipeline Network). The survival probability for the Connection Pipes is the highest. The decline rate of the survival probability for this factor is low up to the 48th year of its lifespan when the probability to survive is 67%. The incidents involving the Ring Main are very few in the total dataset hence conclusions export related with it will not be reliable. During the first 17.5 years of the element's life, those correlated with the factor Main have a greater chance to survive than those related to Small Main. From that point onwards, it happens the other way around.
- Fig. 2.26E corresponds to the survival plot for the Material variable of the C class model (Pipeline Network). The findings relating to this plot are exactly the same as those for Fig. 2.22E (Material variable of the C class – Distribution System model).
- Fig. 2.26F corresponds to the survival plot for the Reason variable of the C class model (Pipeline Network). Based on the existing incidents, the survival probability for the Material/Hardware Failure is the highest. Furthermore, during the first 30 years of the elements' life, those correlated with the Damage factor have a greater chance to survive than those related to Tree Roots. From that point onwards, it happens the other way around.

(b) House Connection Class

This analysis group is related to the HC class model (Pipeline Network). Fig. 2.27 presents the survival plots of the class for the different variables. The analysis used the incidents of Pipeline Network model that concern the HC class.

- Fig. 2.27A corresponds to the survival plot for the Part Type variable of the HC class model (Pipeline Network). The plot indicates that the HC pipes do not face a lot of failure incidents that have short time intervals between them. Survival probability of the Pipe factor is the highest. The decline rate of the survival probability for this factor is low up to the 48th year of its lifespan when the probability to survive is 67% (7% more than the corresponding probability from Fig. 2.23A). For the same lifespan, the survival probability of the Fitting factor is less than 5%.

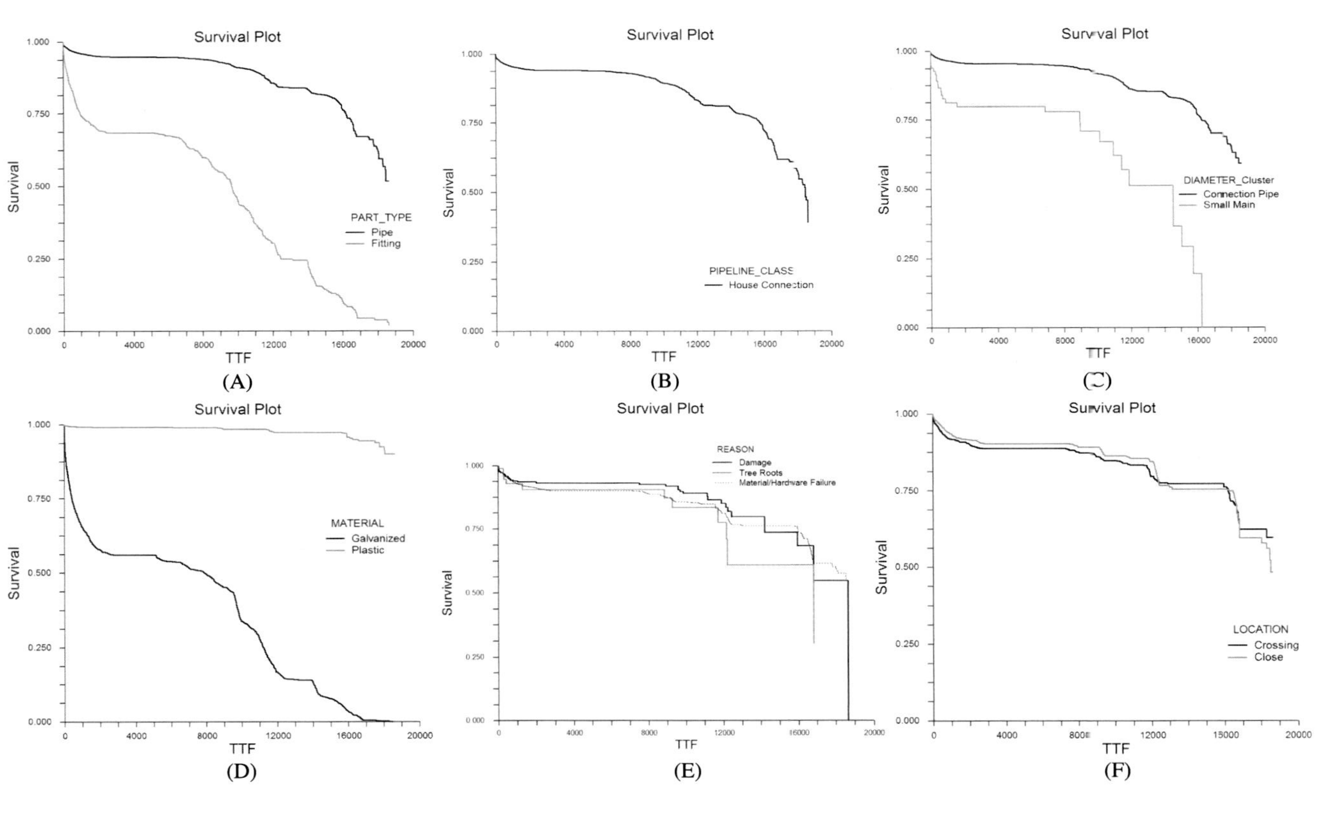

Figure 2.27 Survival plots of the different variables for the PN-HC model (Street Level/All Incidents).

- Fig. 2.27B corresponds to the survival plot for the Pipeline class variable of the HC class model (Pipeline Network). The survival probability of the model's elements is much higher compared to that of Fig. 2.23B. It is reduced only by 6% during the first 5 years of the elements' life (35% in Fig. 2.23B). The decline rate of the survival probability is low up to the 48th year of the model's lifespan; after that point it suddenly increases. The elements' lifetime has a probability of 60% to exceed this age (survival probability in Fig. 2.23B is 42.5% for 38 years of lifespan).
- Fig. 2.27C corresponds to the survival plot for the Diameter Cluster variable of the HC class model (Pipeline Network). The survival probability for the Connection Pipes is the highest. It has the same shape as that of the C model (it has been shifted a bit upwards). The decline rate of the survival probability for the Small Main factor is low up to the 40th year of its lifespan when the probability to survive is 52%. The line shape for the survival probability reveals that the facts related to this factor are few.
- Fig. 2.27D corresponds to the survival plot for the Material variable of the HC class model (Pipeline Network). The survival probability for the Plastic factor is much higher than that of Galvanized. The reason for this is the same with that of Fig. 2.22E (C class for Distribution System model).
- Fig. 2.27E corresponds to the survival plot for the Reason variable of the HC class model (Pipeline Network). For the first 39 years of the elements' lifetime, those associated with the damage factor have the highest survival probability. At that point it decreases dramatically and the elements associated with the Material/Hardware Failure factor have a higher probability to survive from that point onward. The Tree Roots factor has the lowest survival probability.
- Fig. 2.27F corresponds to the survival plot for the Location variable of the HC class model (Pipeline Network). This factor is only related to HC model. The survival probability of the factors is almost the same. The probability for an element's lifetime to surpass 46 years exceeds 58%.

(c) Water Main Class

This analysis group is related to the WM class model (Pipeline Network). Fig. 2.28 presents the survival plots of the class for the different variables. The analysis used the incidents of Pipeline Network model that concern the WM class.

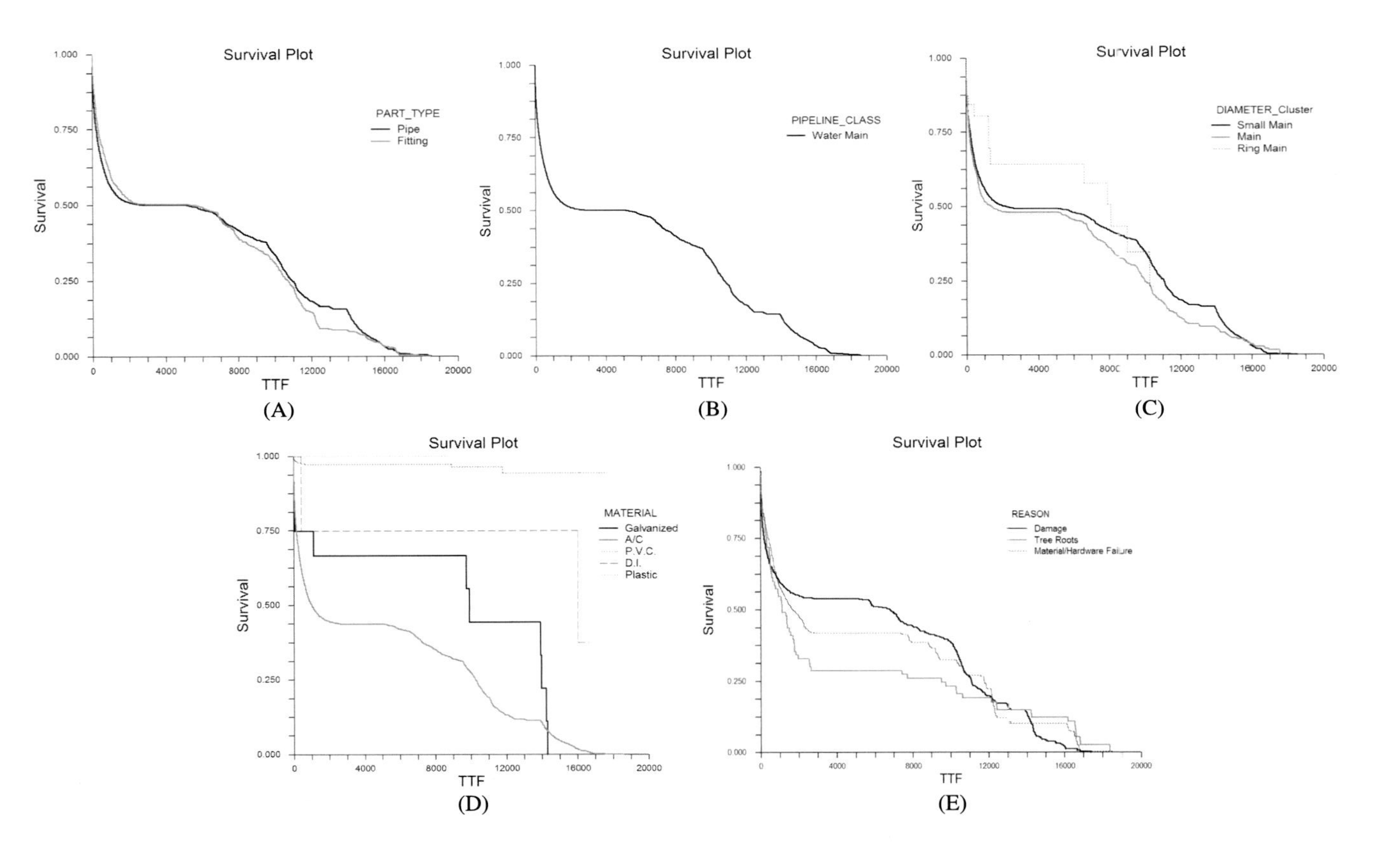

Figure 2.28 Survival plots of the different variables for the PN-WM model (Street Level/All Incidents).

- Fig. 2.28A corresponds to the survival plot for the Part Type variable of the WM class model (Pipeline Network). The survival probability of the factors Pipe and Fitting is almost the same. The probability for an element's lifetime to overcome the 30 years is only 23%. Hence the WM pipes are more vulnerable in time compared to HC pipes.
- Fig. 2.28B corresponds to the survival plot for the Pipeline class variable of the WM class model (Pipeline Network). The survival probability of the model's elements is reduced by 50% during the first 5 years of their life. The decline rate of the survival probability is low up to the 38th year of the model's lifespan while after that point it suddenly increases. The elements' lifetime has a probability of 15% to exceed this age. A comparison of the results with those for Fig. 2.24B shows that exclusion of the incidents associated with the Component factors has worsened the model's condition. Aggregating this with the previous findings, it is concluded that the WM pipes have a negative effect on the WDN condition compared to components and HC pipes, while the Component factor has a negative effect compared to HC Pipe factor.
- Fig. 2.28C corresponds to the survival plot for the Diameter Cluster variable of the WM class model (Pipeline Network). The plot is similar to that of Fig. 2.24C (WM class of the Distribution System model).
- Fig. 2.28D corresponds to the survival plot for the Material variable of the WM class model (Pipeline Network). The plot is similar to that of Fig. 2.24D (WM class of the Distribution System model).
- Fig. 2.28E corresponds to the survival plot for the Reason variable of the WM class model (Pipeline Network). The plot is similar to that of Fig. 2.24E (WM class of the Distribution System model).

Survival Analysis of the Pipeline Model

The analysis series related to the Pipeline model uses, as dataset, all incidents related only with the pipe elements of the WDN. This model of analysis is the one that is used today in cases when survival analysis is preferred.

The survival probabilities and hazard rate of factors for the different variables are better (higher survival probability and hence lower hazard rate) than their counterparts of the Distribution System and Pipeline network model due to the fact that Fitting, Component, and Misc. elements, which present a lot of successive complete incidents within a short time period, are excluded from the analysis.

The survival probabilities and hazard rates of factors for the different variables are very similar as their counterparts of the Pipeline Network models due to the fact that the proportion of the incident number related with the Fitting factor compared to that of the Pipe is very low.

Fig. 2.29 presents the hazard rate plots of the model's classes for NOPB and DMAs.

- Both comments concerning Fig. 2.13 (Distribution System), on the successive complete incidents within a short time period as well on the fact that the hazard rate of the No Breaks factor is not the lowest one, apply also to Fig. 2.29.
- The findings relating to the hazard rate of NOPB are similar to those for Fig. 2.25.
- The findings relating to the hazard rate of DMAs are exactly the same as those for Fig. 2.21.

(a) Couple Class

This analysis group is related to the C class of the Pipeline model. Fig. 2.30 presents the survival plots of the class for the different variables. The analysis used all incidents concerning the Pipeline Network model.

- Fig. 2.30A corresponds to the survival plot for the Pipeline class variable of the C class model (Pipeline). The survival probability for the HC pipeline class is higher than that for WM. The decline rate of the survival probability for the two factors, up to the 38th year of their lifespan, is 80% for the HC and 15% for the WM, respectively. The lifetime of HC elements has a high probability to exceed 50 years.
- Fig. 2.30B corresponds to the survival plot for the Pipeline Network model. The survival probability of the model's elements is reduced only by 13% during the first 5 years of their life. The decline rate of the survival probability is low up to the 38th year of the model's lifespan while after that point it suddenly increases. The elements' lifetime has a probability of 71% at that point.
- Fig. 2.30C corresponds to the survival plot for the Diameter Cluster variable of the C class model (Pipeline). The survival probability for the Connection Pipes is the highest. The decline rate of the survival probability for this factor is low up to the 48th year of its lifespan when the probability to survive is 71%. The incidents involving the Ring Main are very few in the total dataset hence drawn conclusions about it will not be reliable. During the first 18 years of the elements' life, those correlated with the factor Main have a greater chance to survive than those related to Small Main. From that point onwards, it happens the other way around.

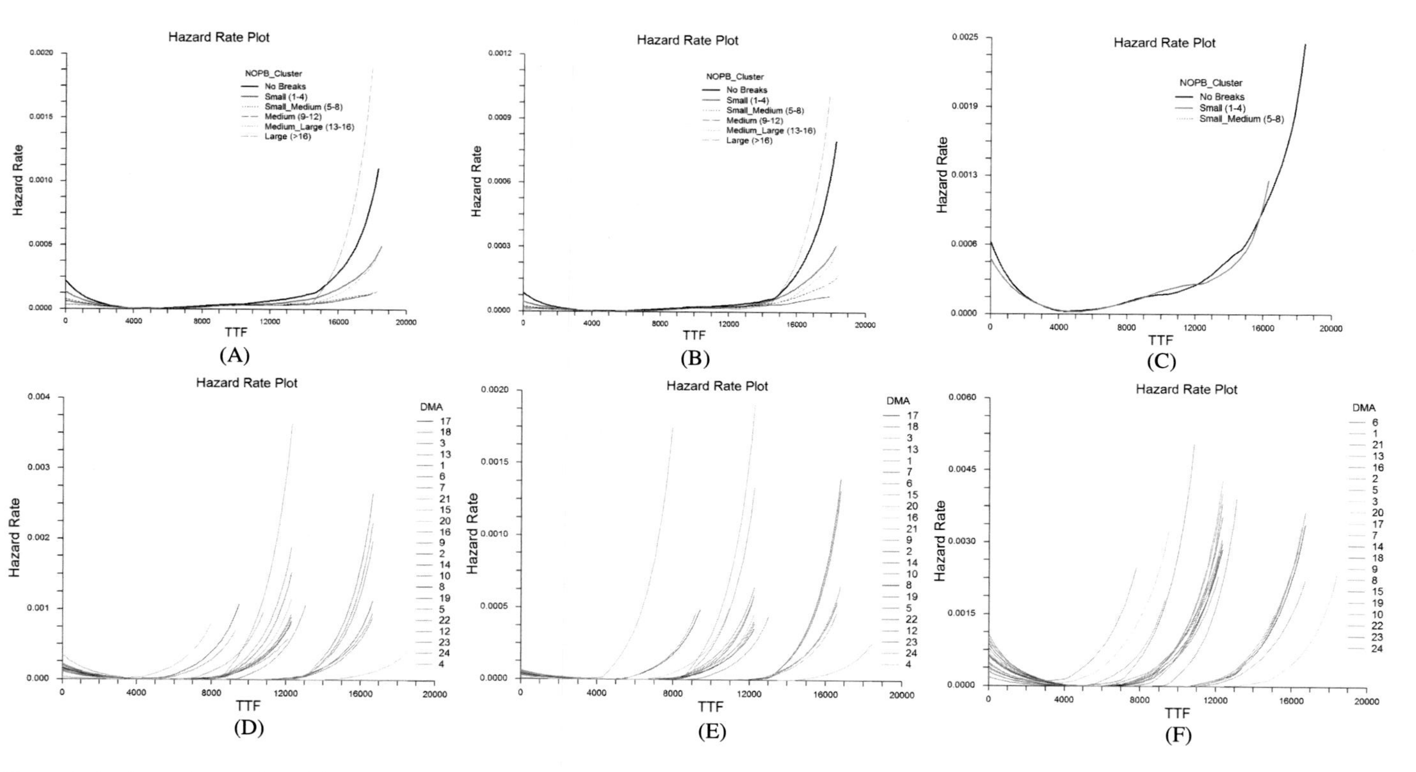

Figure 2.29 Hazard rate plot of the NOPB and DMA for the P model (Street Level/All Incidents).

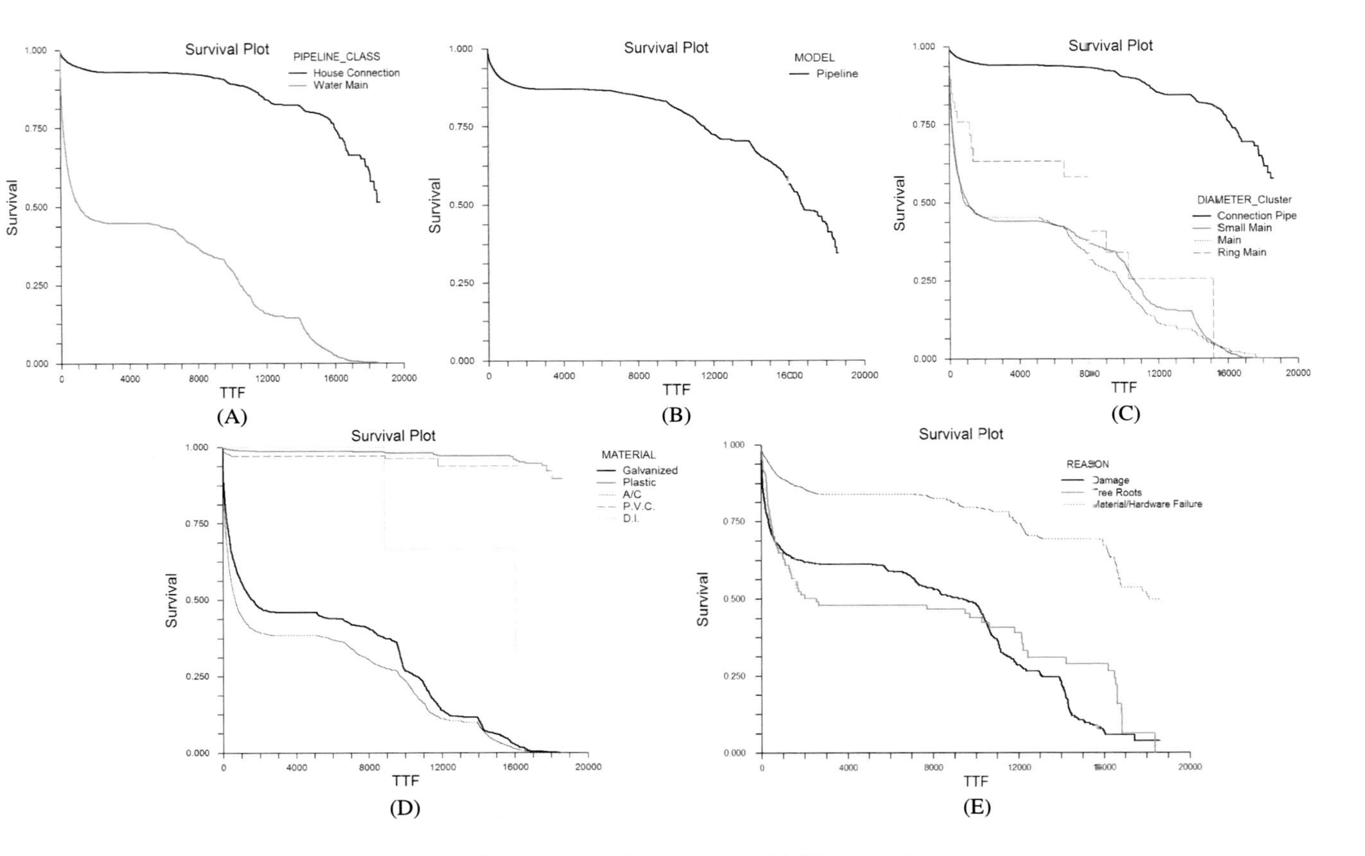

Figure 2.30 Survival plots of the different variables for the P-C model (Street Level/All Incidents).

- Fig. 2.30D corresponds to the survival plot for the Material variable of the C class model (Pipeline). The findings relating to this plot are exactly the same as those for Fig. 2.22E (Material variable of the C class – Distribution System model) and Fig. 2.26E (Material variable of the C class – Pipeline Network model).
- Fig. 2.30E corresponds to the survival plot for the Reason variable of the C class model (Pipeline). Based on the existing incidents, the survival probability for the Material/Hardware Failure is the highest. Furthermore, during the first 28.5 years of the elements' life, those correlated with the factor Damage have a greater chance to survive than those related to Tree Roots. From that point onwards, it happens the other way around.

(b) House Connection Class

This analysis group is related to the HC class model (Pipeline). Fig. 2.31 presents the survival plots of the class for the different variables. The analysis used the incidents of Pipeline Network model that concern the HC class.

- Fig. 2.31A corresponds to the survival plot for the Pipeline class variable of the HC class model (Pipeline). The survival probability of the model's elements is slightly higher compared to that of Fig. 2.27B. It is reduced only by 5% during the first 5 years of the elements' life. The decline rate of the survival probability is low up to the 48th year of the model's lifespan while after that point it suddenly increases. The elements' lifetime has a probability of 66% to exceed this age.
- Fig. 2.31B corresponds to the survival plot for the Diameter Cluster variable of the HC class model (Pipeline). The survival probability for the Connection Pipes is the highest. It has the same shape with the corresponding of the C model (it has been shifted a bit upwards). The decline rate of the survival probability for the Small Main factor is low up to the 40th year of its lifespan when the probability to survive is 56%.
- Fig. 2.31C corresponds to the survival plot for the Material variable of the HC class model (Pipeline). The findings relating to the variable are exactly the same as those for Fig. 2.22E (C class for Distribution System model).
- Fig. 2.31D corresponds to the survival plot for the Reason variable of the HC class model (Pipeline). The elements associated with the Damage factor have the highest survival probability. The elements associated with the Tree Roots have a higher probability than those of

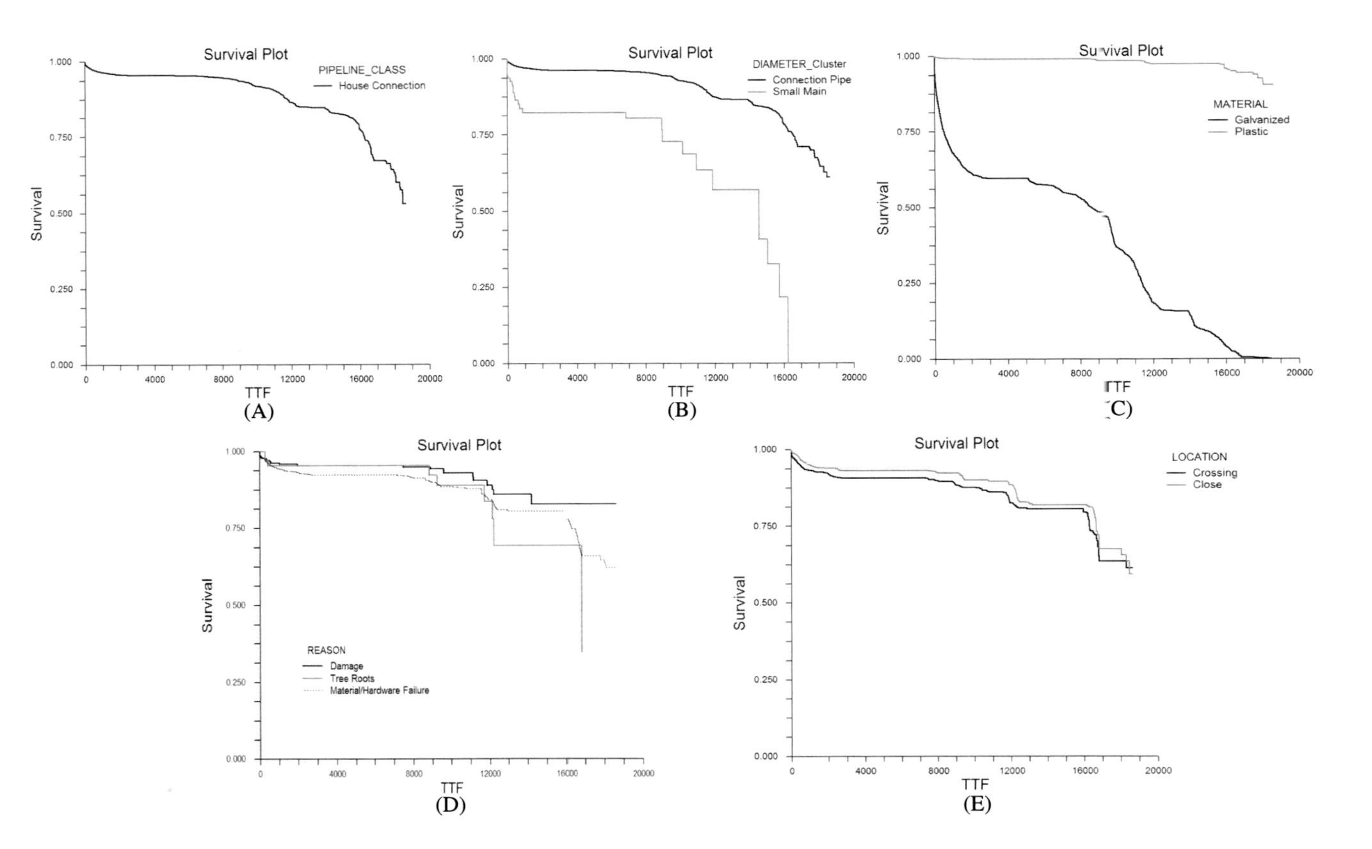

Figure 2.31 Survival plots of the different variables for the P-HC model (Street Level/All Incidents).

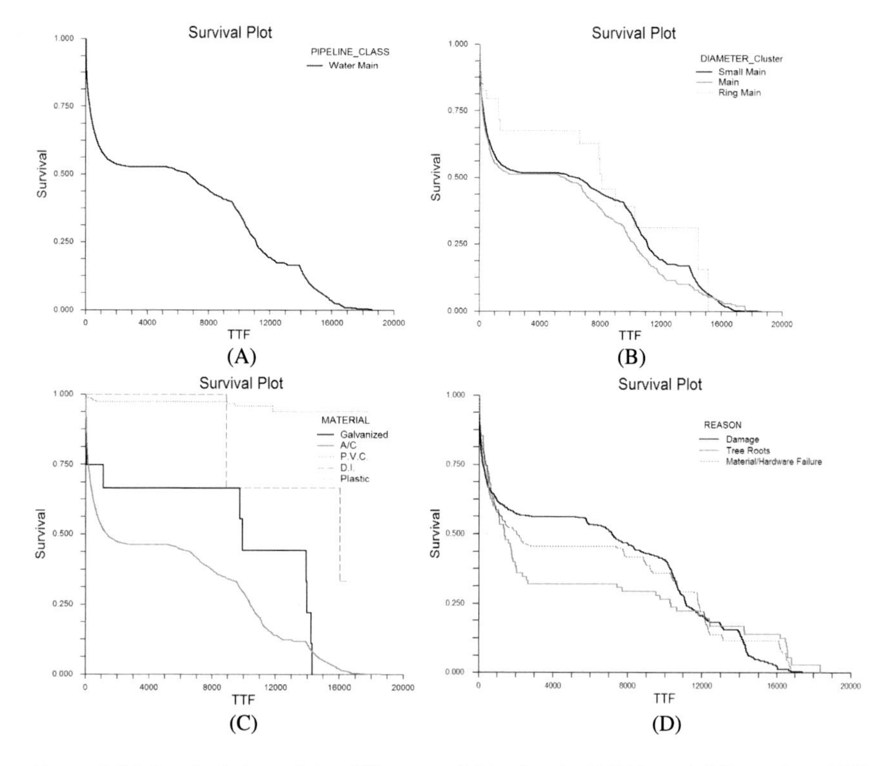

Figure 2.32 Survival plots of the different variables for the P-WM model (Street Level/All Incidents).

Material/Hardware Failure up to the 32nd year of their lifetime. From that point onward, the Tree Roots factor has the lowest survival probability.

- Fig. 2.31E corresponds to the survival plot for the Location variable of the HC class model (Pipeline). This factor is only related to HC model. The survival probability of the factors is almost the same. The probability for an element's lifetime to surpass 46 years exceeds 63%.

(c) Water Main Class

This analysis group is related to the WM Class model (Pipeline). Fig. 2.32 presents the survival plots of the class for the different variables. The analysis used the incidents of Pipeline Network model that concern the WM class.

- Fig. 2.32A corresponds to the survival plot for the Pipeline class variable of the WM class model (Pipeline). The line of the survival probability is similar to that of Fig. 2.28B. It is slightly shifted upwards, meaning that exclusion of the Fitting factor has improved the model's condition. Therefore the Fitting factor has a negative effect compared to WM Pipe factor.

- Fig. 2.32B corresponds to the survival plot for the Diameter Cluster variable of the WM class model (Pipeline). The plot is similar to that of Fig. 2.24C (WM class of the Distribution System model).
- Fig. 2.32C corresponds to the survival plot for the Material variable of the WM class model (Pipeline). The plot is similar to that of Fig. 2.24D (WM class of the Distribution System model).
- Fig. 2.32D corresponds to the survival plot for the Reason variable of the WM class model (Pipeline). The plot is similar to that of Fig. 2.24E (WM class of the Distribution System model).

Regression Analysis of the Pipeline Model

One conclusion of the results that were presented shows that the order in which the parameters of the variable Part Type affect the survival probability is something which is not enough for the analysis. A relation showing the correlation of the different variables and factors with respect to the total hazard rate as well as an indication for the effect on the hazard rate by each variable's factor, compared to the others, is very important for the analysis.

For this reason the analysis has been extended using the Cox regression, which provides added information on the proportional hazards of the variables providing a sophisticated analysis tool. The important information contains the Regression Coefficient (indicating the amount that the logarithm of the hazard rate changes when x_i is increased by one unit) and the Risk Ratio metric (indicating the ratio of two hazards whose only difference is that x_i is increased by one unit).

The results of Cox regression analysis for the three classes of the Pipeline model (street level) are presented in Table 2.14. The column *p-value* refers to the significance of the variable's factor in relation to the model. A factor of a variable is significant for the Cox regression model when this value is less than 0.05.

(a) Couple Class

The Regression Coefficient and Risk Ratio metric as well as the Cox regression model of the Couple class model are presented in the first part of Table 2.14. Based on this table, there are 15 variables that affect the

Table 2.14 Risk Ratio and Regression Models for the P models (Street Level/All Incidents).

Model	Variable	Regression coefficient (β_i)	Standard error of (β_i)	Risk ratio (exp[β_i])
C	(DIAMETER_ Cluster="MAIN")	1.058488	0.508644	2.8820
	(DIAMETER_Cluster="SMALL MAIN")	0.707625	0.501909	2.0292
	(DIAMETER_Cluster="RING MAIN")	0.601453	0.871928	1.8248
	(NOPB_Cluster="NO BREAKS")	0.560300	0.171097	1.7512
	(REASON="MATERIAL/HARDWARE FAILURE DETERIORATION")	0.216174	0.071059	1.2413
	(NOPB_Cluster="SMALL 1–4")	0.198172	0.167592	1.2192
	(NOPB_Cluster="SMALL_MEDIUM 5–8")	−0.015276	0.179011	0.9848
	(NOPB_Cluster="MEDIUM 9–12")	−0.054433	0.199077	0.9470
	(REASON="TREE ROOTS")	−0.142654	0.120775	0.8671
	(NOPB_Cluster="MEDIUM_LARGE 13–16")	−0.192932	0.252081	0.8245
	(MATERIAL="P.V.C.")	−4.305273	0.579006	0.0135
	(MATERIAL="D.I.")	−23.590139	0.000000	0.0000
	(MATERIAL="GALVANIZED")	−19.063558	0.503249	0.0000
	(MATERIAL="PLASTIC")	−22.629009	0.510350	0.0000
	(PIPELINE_CLASS="WATER MAIN")	−19.603693	0.000000	0.0000
	Estimated Cox Regression Model: Exp [+ 1.058488*(DIAMETER_Cluster="MAIN") + 0.601453*(DIAMETER_ Cluster="RING MAIN") + 0.707625*(DIAMETER_Cluster="SMALL MAIN") – 23.590139*(MATERIAL="D.I.") – 19.063558*(MATERIAL="GALVANIZED") – 4.305273*(MATERIAL="P.V.C.") – 22.629009*(MATERIAL="PLASTIC") – 0.054433*(NOPB_Cluster="MEDIUM 9-12") – 0.192933*(NOPB_Cluster="MEDIUM_LARGE 13-16") + 0.560300*(NOPB_ Cluster="NO BREAKS") + 0.198172*(NOPB_Cluster="SMALL 1–4") – 0.015276*(NOPB_ Cluster="SMALL_MEDIUM 5–8") – 19.603693*(PIPELINE_CLASS="WATER MAIN") + 0.216173*(REASON="MATERIAL/HARDWARE FAILURE DETERIORATION") – 0.142654*(REASON="TREE ROOTS")]			

continued on next page

Table 2.14 *(continued)*

Model	Variable	Regression coefficient (β_i)	Standard error of (β_i)	Risk ratio (exp[β_i])
HC	(NOPB_Cluster="NO BREAKS")	1.481292	0.273739	4.3986
	(NOPB_Cluster="SMALL (1–4)")	0.907599	0.260362	2.4784
	(REASON="TREE ROOTS")	0.720925	0.537587	2.0563
	(NOPB_Cluster="SMALL_MEDIUM (5–8)")	0.676417	0.278563	1.9668
	(NOPB_Cluster="MEDIUM (9–12)")	0.446661	0.318005	1.5631
	(REASON="MATERIAL/HARDWARE FAILURE (DETERIORATION)")	0.039994	0.384351	1.0408
	(LOCATION="CROSSING")	−0.181657	0.120054	0.8339
	(NOPB_Cluster="MEDIUM_LARGE (13–16)")	−0.190059	0.429179	0.8269
	(DIAMETER_Cluster="SMALL MAIN")	−3517.151075	0.000000	0.0000
	(MATERIAL="PLASTIC")	3502.382396	0.000000	
	Estimated Cox Regression Model: Exp [− 3517.1510748924*(DIAMETER_Cluster="SMALL MAIN") − 0.181657*(LOCATION="CROSSING") + 3502.382396*(MATERIAL="PLASTIC") + 0.446661*(NOPB_Cluster="MEDIUM 9–12") − 0.190059*(NOPB_Cluster="MEDIUM_LARGE 13–16") + 1.481292*(NOPB_Cluster="NO BREAKS") + 0.907599*(NOPB_Cluster="SMALL 1–4") + 0.676417*(NOPB_ Cluster="SMALL_MEDIUM 5–8") + 0.039994*(REASON="MATERIAL/HARDWARE FAILURE DETERIORATION") + 0.720925*(REASON="TREE ROOTS")]			
WM	B6 (NOPB_Cluster="SMALL (1–4)")	0.377984	0.131386	1.4593
	B1 (DIAMETER_Cluster="RING MAIN")	0.280649	0.713105	1.3240
	B8 (REASON="MATERIAL/HARDWARE FAILURE (DETERIORATION)")	−0.055906	0.074850	0.9456
	B9 (REASON="TREE ROOTS")	−0.284019	0.128217	0.7528
	B2 (DIAMETER_Cluster="SMALL MAIN")	−0.397022	0.080517	0.6723
	B4 (MATERIAL="P.V.C.")	−4.359902	0.579659	0.0128
	B3 (MATERIAL="D.I.")	−22.611136	0.000000	0.0000
	B5 (MATERIAL="PLASTIC")	−22.257124	0.000000	0.0000
	B7 (NOPB_Cluster="SMALL_MEDIUM (5–8)")	−20.895757	0.000000	0.0000
	Estimated Cox Regression Model: Exp [+ 0 .280649*(DIAMETER_Cluster="RING MAIN") − 0.397022*(DIAMETER_Cluster="SMALL MAIN") − 22.611136*(MATERIAL="D.I.") − 4.359902*(MATERIAL="P.V.C.") − 22.257124*(MATERIAL="PLASTIC") + 0.377984*(NOPB_Cluster="SMALL 1–4") − 20.895757*(NOPB_Cluster="SMALL_MEDIUM 5–8") − 0.055906*(REASON="MATERIAL/HARDWARE FAILURE DETERIORATION") − 0.284019*(REASON="TREE ROOTS")]			

Cox regression model, hence the HR. The variables Small Main (Diameter Cluster), Ring Main (Diameter Cluster), Small (NOPB Cluster), Small Medium (NOPB Cluster), Medium (NOPB Cluster), Tree Roots (Reason), Medium Large (NOPB Cluster), WM (Pipeline Class), and D.I. (Material) have a p-value higher than 0.05, therefore they can be excluded from the model since they don't affect its accuracy.

Therefore, without affecting the result accuracy, the Cox regression model can be reformulated as:

Exp [+ 1.0585 x (Main {Diameter Cluster})
– 19.0636 x (Galvanized {Material})
– 4.3053 x (P.V.C. {Material})
– 22.6290 x (Plastic {Material})
+ 0.5603 x (No Breaks {NOPB Cluster})
+ 0.2162 x (Material/Hardware Failure_ Deterioration {Reason})]

If the "No Breaks" factor of the "*NOPB Cluster*" variable increases by one, then the hazard rate is increased by 75.12% while if the P.V.C. factor of the "*Material*" variable is increased by one, then the hazard rate is decreased by 98.65%.

Overpopulation of the plastic pipes, the high proportion of HC incidents compared with those of WM, and the absence of information on some of the database fields, such as the Reason, are affecting the final outcome of the Cox regression model.

(b) House Connection Class

The Regression Coefficient and Risk Ratio metric as well as the Cox regression model of the HC class model are presented in the middle part of Table 2.14. Based on this table, there are 10 variables that affect the Cox regression model. Seven variables have a p-value higher than 0.05, therefore they can be excluded from the model.

Therefore, without affecting the accuracy of the results, the Cox regression model can be reformulated as:

Exp [+ 1.4813 x (No Breaks {NOPB Cluster})
+ 0.9076 x (Small {NOPB Cluster})
+ 0.6764 x (Small Medium {NOPB Cluster})]

If the "No Breaks" factor of the "*NOPB Cluster*" variable increases by one, then the hazard rate is increased by 439.86% while if the Small factor of the same variable is increased by one, then the hazard rate is increased by 247.84%.

Overpopulation of the Plastic (Material Cluster) and Connection Pipe (Diameter Cluster) as well as the absence of information on some of the database fields, such as the Reason, are affecting the final outcome of the Cox regression model.

(c) Water Main Class

The Regression Coefficient and Risk Ratio metric as well as the Cox regression model of the WM class model are presented in the last part of Table 2.14. Based on this table, there are 9 variables that affect the Cox regression model. Five variables have a p-value higher than 0.05, therefore they can be excluded from the model.

Therefore, without affecting the result accuracy, the Cox regression model can be reformulated as:

Exp [– 0.3970 x (Main Main {Diameter Cluster})
– 4.3599 x (P.V.C. {Material})
0.3780 x (Small {NOPB Cluster})
– 0.2840 x (Tree Roots {Reason Cluster})]

If the Small factor of the NOPB Cluster variable increases by one, then the hazard rate is increased by 45.93% while if the Tree Roots factor of the Reason variable increased by one, the hazard rate deceased by 24.72%.

Absence of information on some of the database fields, such as "*Reason*" and "*Material*", has an effect on the final outcome of the Cox regression model. Compared to the other two models, the data of the WM class are more qualitative since it consists of a variety of variables.

Segment Level Analysis

The second part of the survival analysis relates to the analysis at the segment level and includes only Pipeline Network models (the proposed one). For this type of analysis, only the incidents of the dataset that include the street number, which represents the point where an event has occurred, can be used. First, a data analysis at the street level is performed to develop a comparable model and then an analysis at the segment level is carried out.

These series of analysis do not include the C class model because the numbering of the various elements differs. The incidents correlated with the WM elements are associated only with odd street numbering while the incidents of HC elements do not have such a restriction.

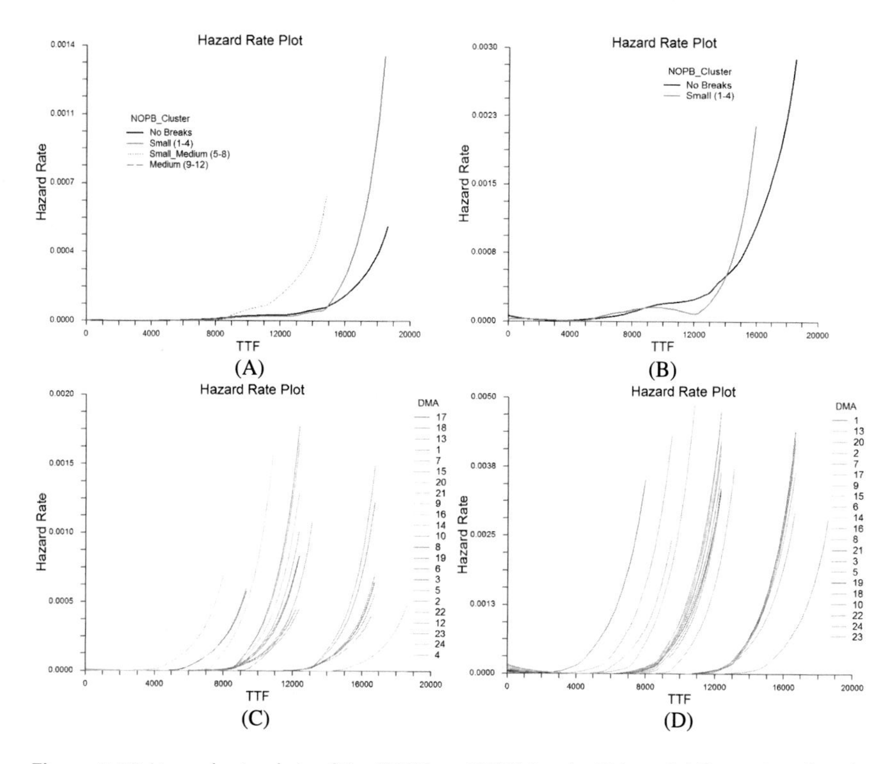

Figure 2.33 Hazard rate plots of the NOPB and DMA for the PN model (Street Level/Only incidents with street numbers).

Survival Analysis of the Pipeline Network at the Street Level

The first analysis series, which is related to the Pipeline Network model, uses all dataset incidents related to the underground buried elements of the WDN, and it is performed at the street level.

The survival probabilities and hazard rates of the different variables will not be similar to those given by the Pipeline Network model, which uses the complete dataset. The incident removal from the dataset will have a positive impact on the model. However, the distribution of data removed (not containing numbered streets) is not uniform (with respect to the various variables), which may positively or negatively influence the outcome.

Fig. 2.33 presents the hazard rate plots of the model's classes for NOPB and DMAs.

- The observations related with the successive complete incidents within a short time period and the fact that the hazard rate of the No Breaks factor is not the lowest one are much milder.

- The hazard rate plots of the NOPB Cluster variable show that the increase of NOPB is correlated with the increasing mortality of the elements. For the HC class model (Fig. 2.33A), the condition of the elements is considered very good during the first 23 years of their life. After that point the hazard rate of the elements associated with Small Medium factor dramatically increases. The hazard rate of the Small and No Breaks factors continues to be good for another 11 years. From this point onwards, the hazard rate is increasing quickly, while after the 40th year, the hazard rate of the Small factor increases too fast.
- For the WM class model (Fig. 2.33B), the condition of the elements is considered very good during the first 25 years of their life. For the next 10 years it is satisfactory, and thereafter they enter into the time period of significant deterioration.
- The findings relating to the hazard rate of DMAs (Figs. 2.33C–D) are exactly the same as those of Fig. 2.21.

(a) House Connection Class

This analysis group is related to the HC class model (street level). Fig. 2.34 presents the survival plots of the class for the different variables. The analysis used the incidents of Pipeline Network model that concern the HC class.

- Fig. 2.34A corresponds to the survival plot for the Part Type variable of the HC class model (street level). The plot indicates that the HC pipes face only few complete incidents that have short time intervals between them. The survival probability of the Pipe factor is the highest, especially after the 15th year. The decline rate of the survival probability for this factor is low up to the 48th year of its lifespan when the probability to survive is 67%. For the same lifespan, the survival probability of the Fitting factor is less than 5%. Initially, the model has a better probability of survival than that of Pipeline Class variable of HC class model (complete dataset), but as time passes the probabilities become equal.
- Fig. 2.34B corresponds to the survival plot for the Pipeline class variable of the HC class model (street level). The line of survival probability of the model is similar to that of Pipeline class variable of HC class model in the complete dataset. Initially, it is slightly shifted upwards, but as time passes it reaches about the same point as that of its counterpart.
- Fig. 2.34C corresponds to the survival plot for the Diameter Cluster variable of the HC class model (street level). Compared to the plot of Diameter Cluster variable of HC class model (complete dataset), the

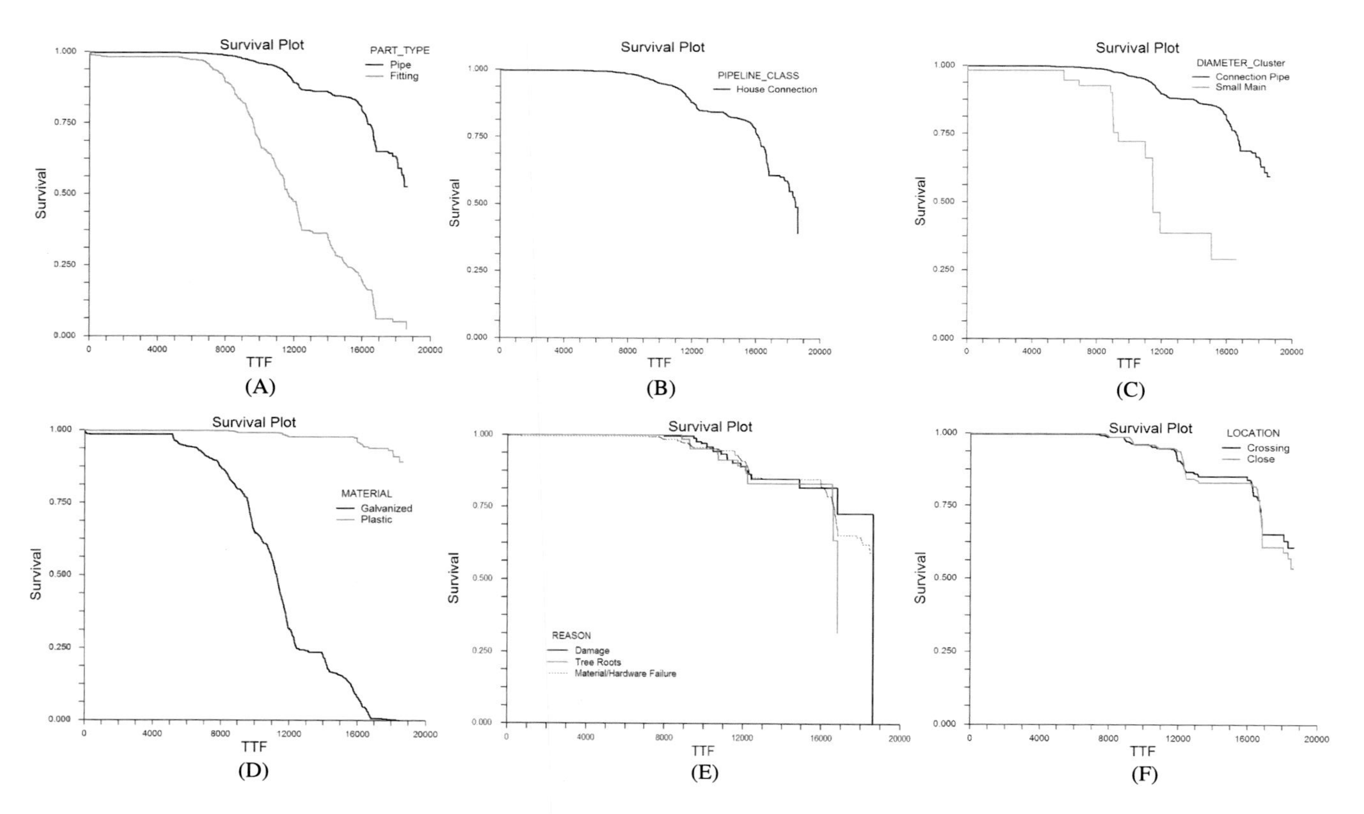

Figure 2.34 Survival plots of the different variables for the PN-HC model (Street Level/Only incidents with street numbers).

survival probability is slightly shifted upwards during the first 38 years while it is slightly shifted downwards after this point in time.

- Fig. 2.34D corresponds to the survival plot for the Material variable of the HC class model (street level). After the first 13 years, the survival probability for the Plastic factor is much higher than that of Galvanized. The reason for this is the same with that for Fig. 2.22E (C class for Distribution System model – complete dataset).
- Fig. 2.34E corresponds to the survival plot for the Reason variable of the HC class model (street level). For the first 45 years of the elements' lifetime, the survival probability of the three factors is similar and exceeds 75%.
- Fig. 2.34F corresponds to the survival plot for the Location variable of the HC class model (street level). This factor is only related to HC model. The survival probability of the factors is almost the same. The probability for an element's lifetime to overcome 47 years exceeds 60%.

(b) Water Main Class

This analysis group is related to the WM Class model (Street Level). Fig. 2.35 presents the survival plots of the class for the different variables. The analysis used the incidents of Pipeline Network model that concern the WM class.

- Fig. 2.35A corresponds to the survival plot for the Part Type variable of the WM class model (Street Level). The survival probability of the Pipe and Fitting factors is almost the same during the first years. The probability for an element's lifetime to overcome 30 years is 45%. Hence the WM elements are more vulnerable in time compared to those of HC.
- Fig. 2.35B corresponds to the survival plot for the Pipeline Class variable of the WM class model (street level). During the first 32 years of the model, the survival probability of the plot is much higher than that in Fig. 2.28B (Pipeline class variable of the WM class model – complete dataset). The decline rate of the survival probability is low up to the 38th year of the model's lifespan while after that point it suddenly increases. The elements lifetime has a probability of 27% to exceed this age.
- Fig. 2.35C corresponds to the survival plot for the Diameter Cluster variable of the WM class model (street level). The incidents involving the Ring Main are very few in the total dataset hence conclusions drawn about it will not be reliable. The survival probability for the Small Main factor is higher than that of Main. The survival probability

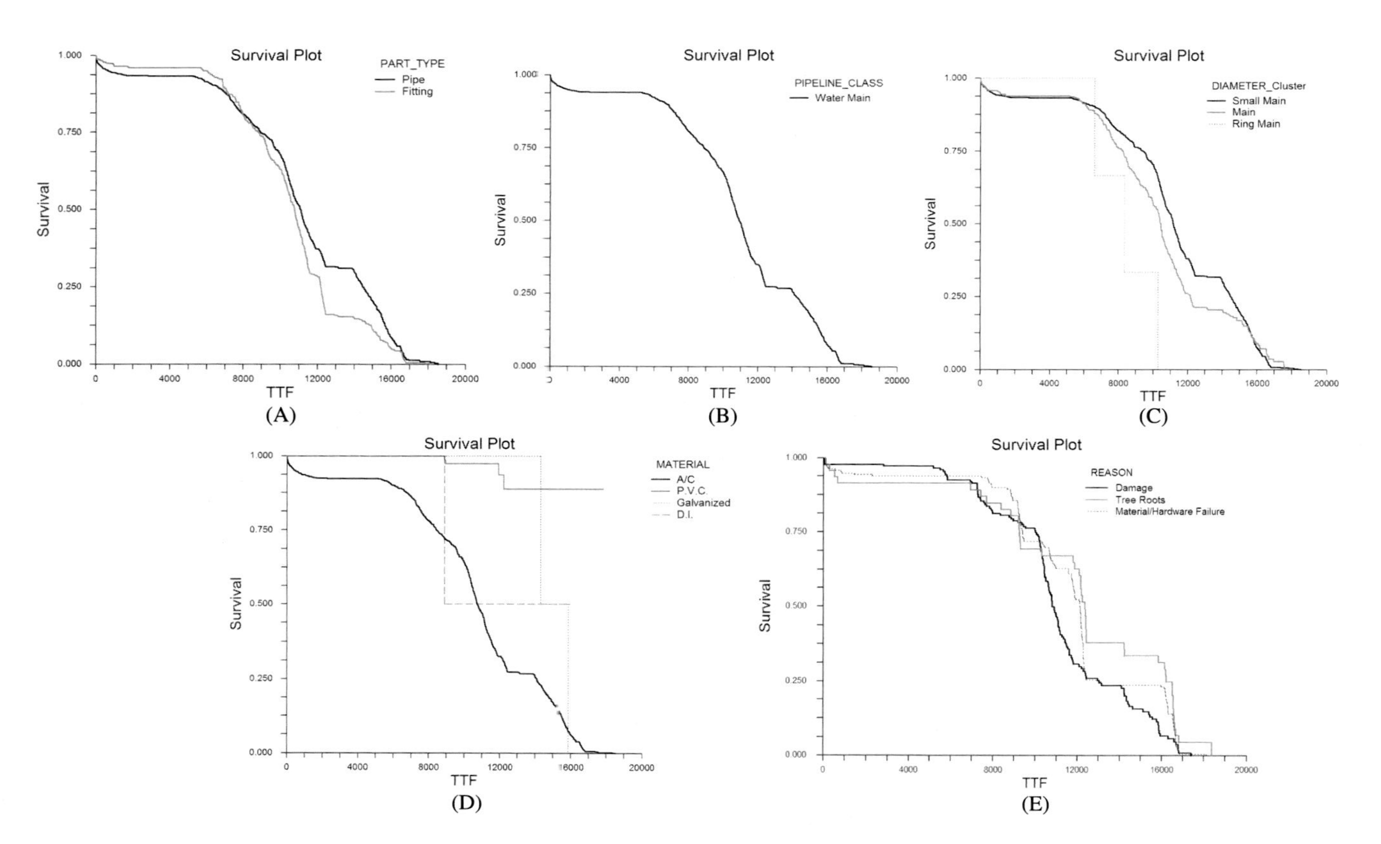

Figure 2.35 Survival plots of the different variables for the PN-WM model (Street Level/Only incidents with street numbers).

for the Small Main factor to overcome the 38th year of its lifetime is more that 30% while for the Main factor it is 20%.

- Fig. 2.35D corresponds to the survival plot for the Material variable of the WM class model (street level). The plot is similar to that of Fig. 2.24D (WM class of the Distribution System model – complete dataset).
- Fig. 2.35E corresponds to the survival plot for the Reason variable of the WM class model (street level). During the first 28 years of their life, the three factors have similar survival probabilities. After that point, the elements associated with the Damage factor have the highest probability to survive while those associated with the Damage factor have the lowest.

Survival Analysis of the Pipeline Network at the Segment Level

The second analysis series is also related to the Pipeline Network model and uses for data set all incidents related with the underground buried elements of the WDN. This analysis is performed at the segment level.

The survival probabilities and hazard rates of the different variables are expected to improve (increased survival probability and reduced hazard rate) since the volume of incidents is distributed to more elements.

Fig. 2.36 presents the hazard rate plots of the model's classes for NOPB and DMAs.

- Fig. 2.36A presents the hazard rate plot of the NOPB Cluster variable showing that the increase of NOPB is less correlated with the increasing mortality of the elements compared with the same plot of Fig. 2.33A (street level). The condition of the elements is considered very good during the first 33 years of their life. After that point the hazard rate of the elements associated with Small Medium factor dramatically increases. The hazard rate of the Small and No Breaks factors continues to be good for another 5 years. From this point onwards, the hazard rate is increasing fast.
- For the WM class model (Fig. 2.36B), the condition of the elements is considered very good during the first 25 years of their life. For the next 10 years, it is satisfactory, and thereafter they enter into the time period of significant deterioration. The picture is similar to that of Fig. 2.33B (street level).
- The findings relating to the hazard rate of DMAs (Figs. 2.33C–D) are exactly the same as those for Fig. 2.21.

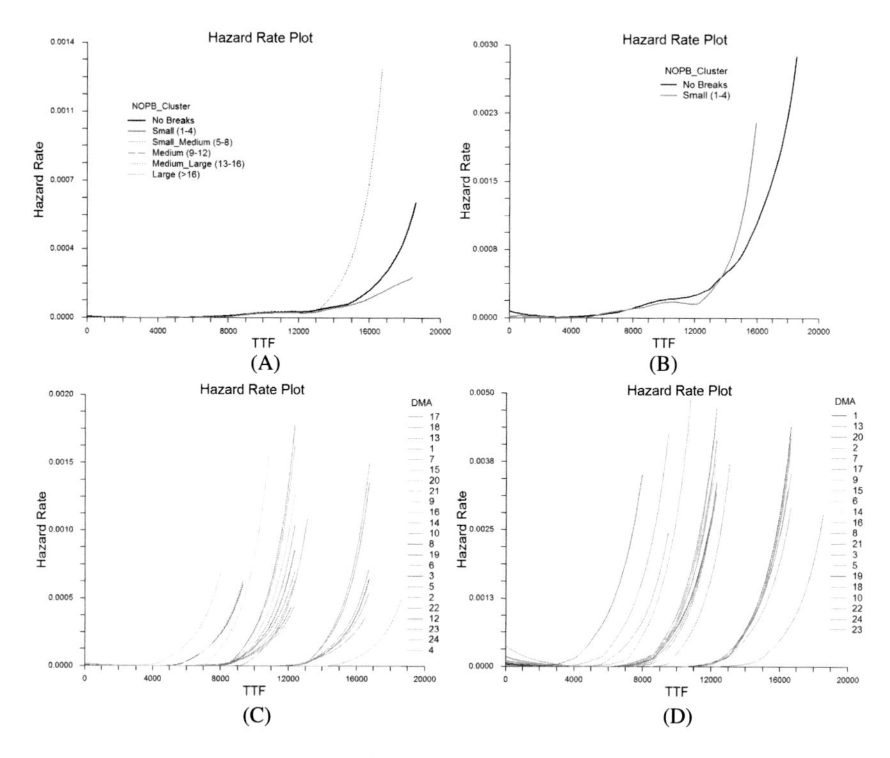

Figure 2.36 Hazard rate plot of the NOPB and DMA for the PN model (Segment Level/Only incidents with street numbers).

(a) House Connection Class

This analysis group is related to the HC class model (segment level). Fig. 2.37 presents the survival plots of the class for the different variables. The analysis used the incidents of Pipeline Network model that concern the HC class.

The plot concerning the survival probability of the HC elements at the segment level is the same as that at the street level (Fig. 2.34). This may be due to one or a combination of the reasons that follow.

- An explanation is that the number of cases of streets which have multiple incidents associated with the HC class is very small, even zero. If this is the case then the variation in the results within the segment analysis will not lead to major changes in the final result since the volume of affected incidents is be very small compared to the dataset.
- A second explanation is that this may be due to the fact that multiple incidents related with a street may be associated to the same or a very

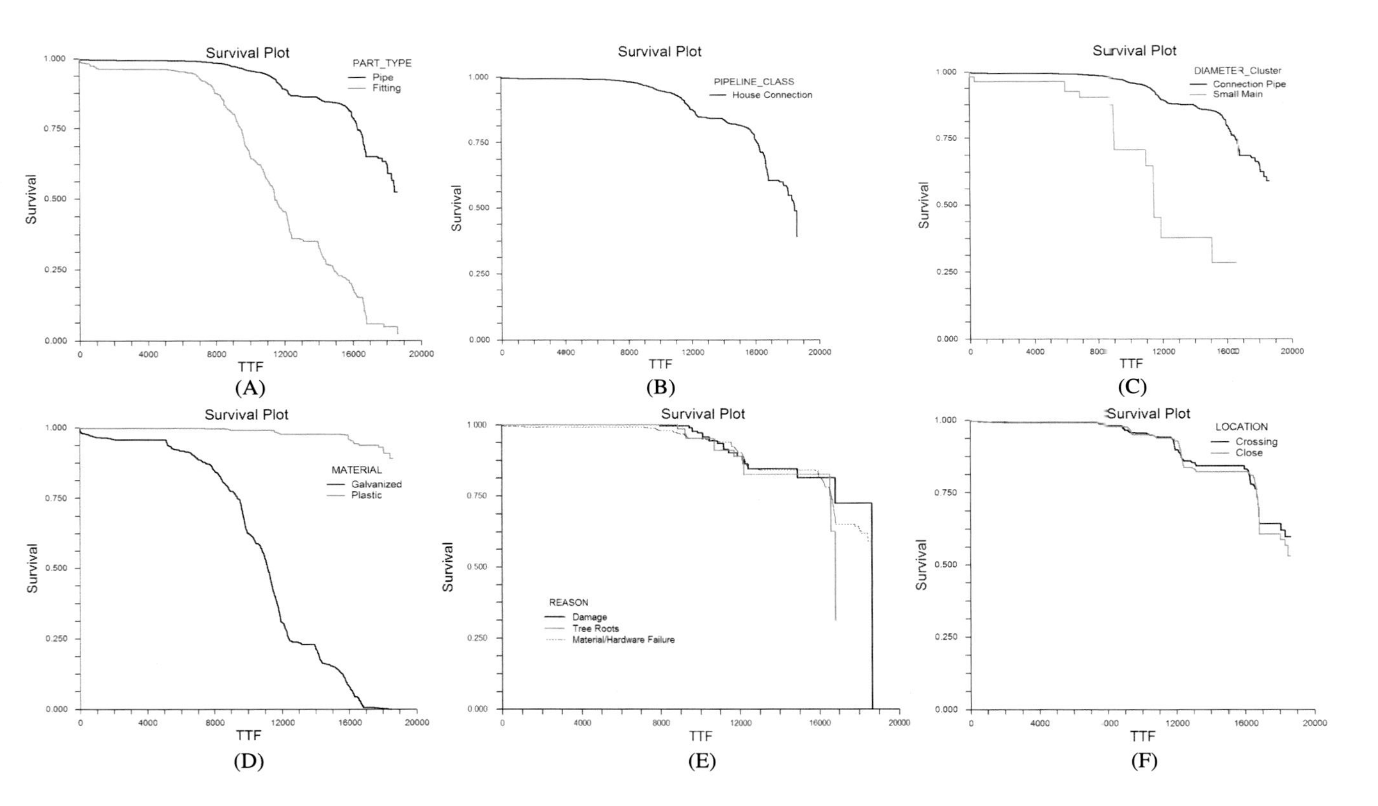

Figure 2.37 Survival plots of the different variables for the PN-HC model (Segment Level/Only incidents with street numbers).

small number of segments of the street. As a result, the analysis at the segment level will give the same result (if all incidents are associated to one specific segment) or it will not substantially alter the result (if the incidents are associated to a few segments).

- Yet another explanation is that this may be due to the scale of the left-censored data, i.e., the extension of the time period between the date of a pipe installation up to the date when its first incident occurred. Due to this long period, the estimation of the survival probability is overestimated (upward), and as a result the difference between analysis at the street and segment levels is irrelevant.

(b) Water Main Class

This analysis group is related to the WM class model (segment level). Fig. 2.38 presents the survival plots of the class for the different variables. The analysis used the incidents of Pipeline Network model that concern the WM class.

The plot concerning the survival probability of the WM elements at the segment level is the same as that at the street level (Fig. 2.35), therefore, what was reported for the HC class also applies here.

Given that HC elements behave differently than the WM elements, as revealed through the analysis presented above, it is most likely that the range of the left-censored data affects the level of the survival probability, so the difference between the street and segment levels is insignificant. This estimate is supported by the fact that analyses showed that the pipes have a high survival probability after the age of 35, which is deemed excessive.

2.4.3 Sensitivity Analysis

To test the reliability of the results, a series of sensitivity analyses were carried out. Note, at this point, that some data parameters may be likely sources of error in the analysis and thus a sensitivity analysis is warranted. Such parameters are the assumed (and/or specified) birth date for the pipelines, and the repair incidents prior to the start of the study period (i.e., before 01/01/2003) for which no information is provided.

The parameter related to the birth date of the pipelines, which affects the validity of the results, is the estimated age of the pipes in relation to the date that the first incident of the dataset was recorded (01/01/2003). Table 2.15 presents the estimated initiating period of water supply in the various DMAs, as they were used in the survival analysis. It is expected that

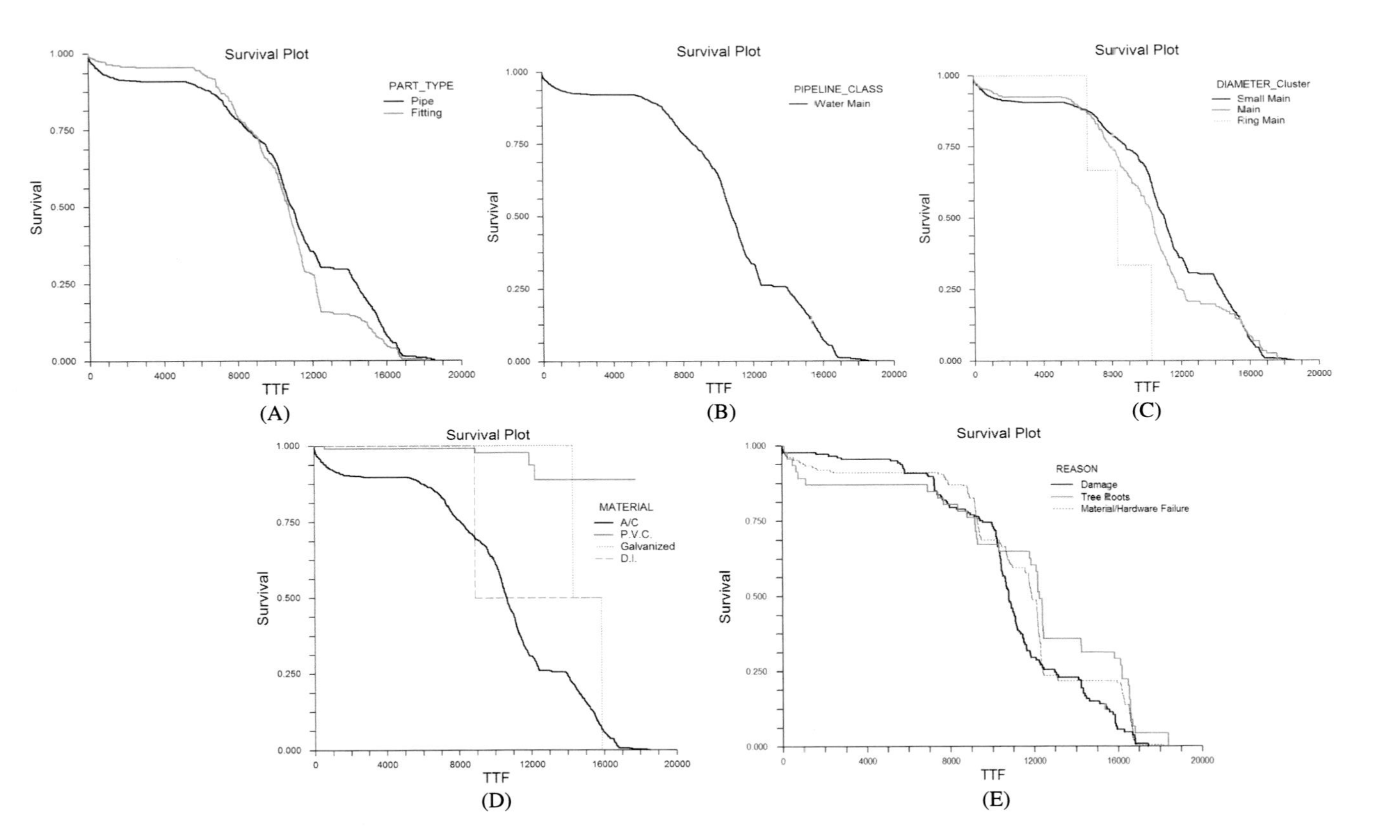

Figure 2.38 Survival plots of the different variables for the PN-WM model (Segment Level/Only incidents with street numbers).

Table 2.15 Estimated initiating period of water supply in the various DMAs.

DMA	Estimated period of birth	Year used in the analysis
20	1955–1960	1960
1	1960–1970	1965
2	1960–1970	1965
3	1960–1970	1965
4	1960–1970	1965
5	1960–1970	1965
6	1960–1970	1965
10	1960–1970	1965
23	1974–1975	1975
7	1974–1980	1977
8	1974–1980	1977
9	1974–1980	1977
11	1974–1980	1977
12	1974–1980	1977
13	1974–1980	1977
14	1974–1980	1977
15	1974–1980	1977
18	1974–1980	1977
19	1974–1980	1977
24	1976–1977	1977
16	1980–1981	1981
17	1984–1985	1985
22	1980–1990	1985
21	1987–1990	1989

a sensitivity analysis on this assumption will present variations on the birth date, compared to the mathematical analysis presented above. The scope of this sensitivity analysis is the investigation of the effect of the birth date of the various WDN components on the survival curves.

In designing the analysis, it was set that each WDN element first appearing in the dataset of incidents will be assigned a number of previous breaks equal to zero. This can affect the reliability of the results because incidents linking these pipelines with their past are eliminated. It is expected that the variations in the presumed/observed number of previous breaks will have a small impact on the shape of the resulting survival curves, compared to the baseline analysis.

Birth Date Assignment to Elements

The first performed sensitivity test comprised a variation in the presumed pipeline age. The change in the resulting shape of the survival curves was examined and compared to that obtained from the initial analysis, by shifting the presumed birth date of the elements by 5 and 10 years later, as well as 5 years earlier, for all nine models presented in the analysis.

As the sensitivity analysis shows, the survival plots for all the different parameters of the studied models exhibit the same behavior, with the survival curves retaining the original shape but having a shift in the time-to-failure, depending on whether the age of the element is shortened or lengthened.

When the birthday of an element is defined to be earlier than that used in the analyses, the time to failure lengthens. Conversely, if the birthday is defined later, the time to failure shortens. The change in the time to failure is logical and expected. The time-to-failure for the elements of the studied WDN is estimated based on the number and type (terminal or censor) of incidents as well as the range of the left-censored data period (i.e., the length of the period from the defined birthday of an element until the day when its first recorded incident occurred). Hence the time to failure of the element lengthens as the age of an element increases relative to the date when its first incident was recorded, keeping the same dataset. Furthermore, the extension of the time to failure indicates elongation of the element's lifetime as well as an improvement on the survival probability. An example of the sensitivity test that comprised variation of the pipeline age is presented in section on p. 125.

In summary, the sensitivity analysis suggests that the exactness or accuracy of the age designation for the elements composing a WDN does not affect the validity of the results since, apart from the expected shift in time to failure, there is no substantial variation in the profile of the survival probability curve.

Number of Previous Breaks

The last sensitivity analysis relates to the number of repair incidents that possibly have occurred before the start of the study period of 01/01/2003. The mathematical analysis presented above considers that the elements that were installed in the WDN before 1/1/2003 did not present any water loss incident prior that date. This affects the reliability of the results because incidents prior that date were not recorded, thus eliminating the link between related pipelines with their past performance. The sensitivity analysis comprises a variation of the initial NOPB for each element which was installed before 1/1/2003. The change in shape of the survival curves was examined, compared to the original analysis, by executing two additional analyses with NOPB = 2 and NOPB = 4 for all WDN pipes "born" prior to 2003, and for all the nine models presented in the analysis.

The extracted survival plots for the NOPBs, for all the studied models, do not show any significant change. The survival curves have the same shape and similar time-to-failure with the corresponding graphs of the original (baseline) analyses. Therefore, this sensitivity analysis series suggests that the exactness or accuracy of the NOPBs, of the elements composing a WDN, does not affect the validity of the results.

An example of the sensitivity test on the variation of the initial NOPB for each element installed prior to 1/1/2003 is presented in section below.

Water Main Class of the Pipeline Model

The following paragraphs present the main results of the performed sensitivity analysis. Since the sensitivity analyses of the different models are numerous and all lead to the same comments, the WM class of the Pipeline model was selected as a representative case for presenting the results of these sensitivity analyses.

Birth Date Shifting

The first sensitivity test comprised a variation of the elements' birth date, with the obtained results presented in Fig. 2.39. Fig. 2.39A depicts the survival plot for the material variable, as extracted by the mathematical analysis presented above (Fig. 2.32C, p. 107). The plots associated with the shift of the elements' birth date 5 and 10 years later, as well as 5 years earlier, in time are presented in Figs. 2.39B–D, respectively.

The first observation is that the time to failure shifts, depending on whether the age decreases or increases, while the curve profile of the survival probability remains the same. For drawing qualitative and reliable conclusions, the survival curves associated with the AC material of the four cases are plotted on the same graph (the curves of the other materials were not included due to the small dataset associated with them). The developed survival plot is presented in Fig. 2.40.

The survival plot in Fig. 2.40 illustrates the four survival curves associated with the AC material (WM pipes), as extracted from the mathematical model (curve "A") and the analysis of the three sensitivity studies (curves "B", "C", and "D"). The survival curves "B" and "C" correspond to a

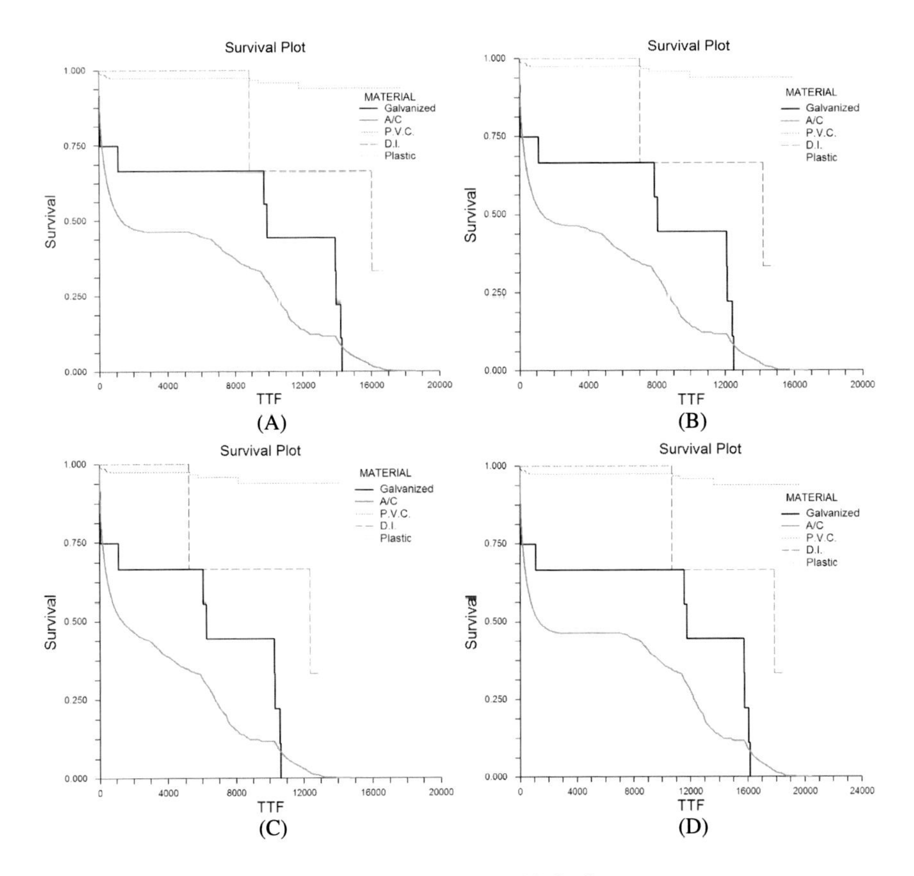

Figure 2.39 Survival plots for different DMA ages (shifted).

shift in the elements' age (which was applied in the mathematical analysis) of 5 and 10 years later, while curve "D" corresponds to a shift of 5 years earlier.

The survival plots suggest that elements' age does not affect the mathematical analysis. A shift in the age results in a corresponding horizontal shift of the survival curve without a change in the survival curve's shape, though.

Different Initial NOPB

The sensitivity test comprising the variation of the NOPB for the left-censored data elements is presented in Fig. 2.41. The plot of Fig. 2.41A is the survival plot for the NOPB extracted by the mathematical analysis presented above (Fig. 2.29C, p. 103). The plots associated with the increased

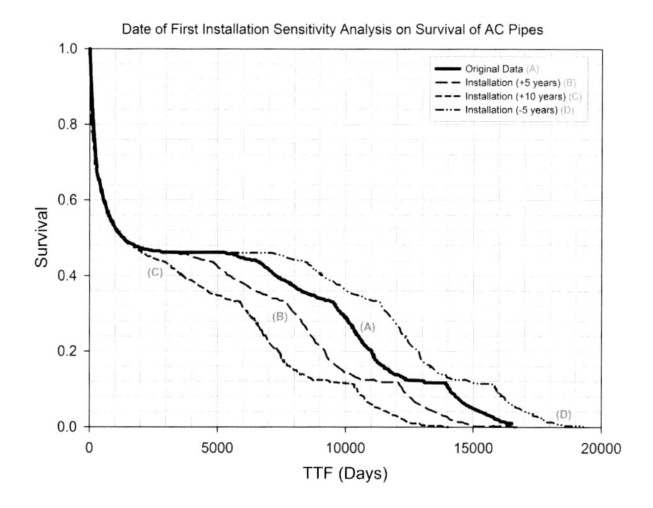

Figure 2.40 Survival plots for different time intervals between the DMAs.

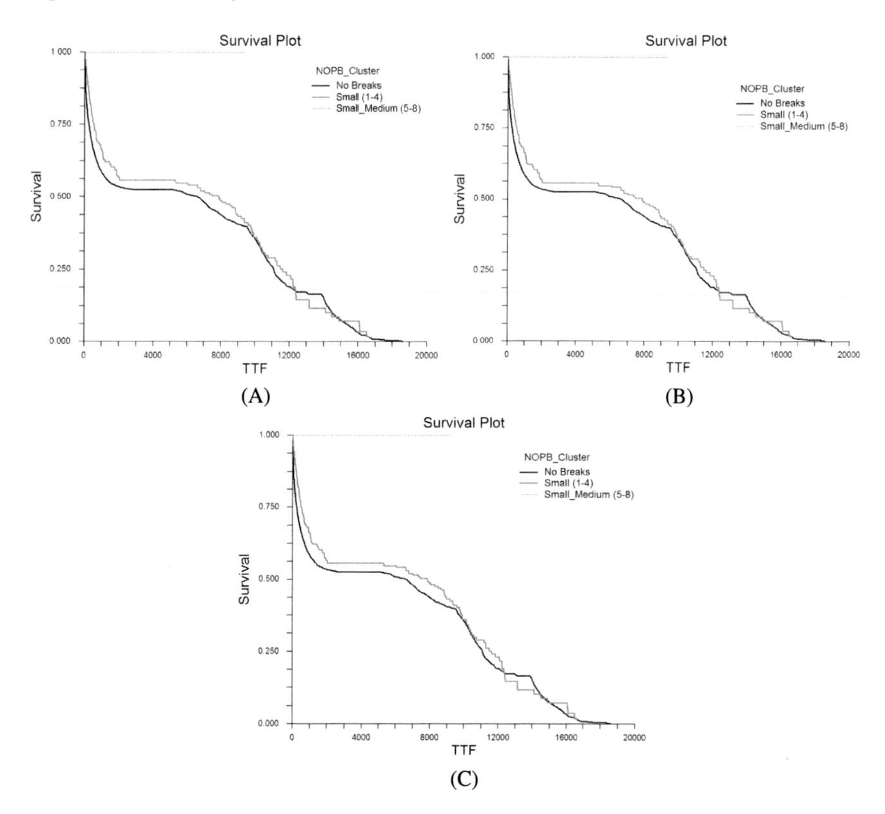

Figure 2.41 Survival plots for different initial NOPB used in the analysis.

NOPB of the left-censored incidents are presented in Fig. 2.41B (NOPB increased by 2) and Fig. 2.41C (NOPB increased by 4).

The survival plots generated by the sensitivity analysis are identical to the plot resulting from the survival analysis presented above. This suggests that the assumption for the values of NOPB for the left-censored data is not affecting the outcome of the analysis. The observation is in line with the results of the mathematical analysis related to the NOPB. The mathematical analysis shows that the survival probability of the elements is not affected by the NOPB, since the variation in time-to-failure for the different clusters is insignificant.

2.5 CONCLUDING REMARKS

Chapter 2 encompassed two thematic areas, which are interconnected. The first one is related to the development of a simplified database management system which can be used in assembling the knowledge related to the operational behavior of a WDN. The second one is related with the development of the mathematical models that process the information recorded in the database.

The developed mathematical model consists of two different approaches: (1) statistical analysis and (2) survival analysis. The statistical analysis uses common measures of central tendency for the quantitative analysis of the dataset in study. The survival analysis, which is employed for a more qualitative data analysis, utilizes the Kaplan–Meier estimator for the estimation of the probability of survival and of the hazard rate of various WDN components over time, while and the Cox semiparametric proportional hazards model is employed for the estimation of the hazard ratio between various WDN classes.

The novelty of the mathematical methodology, which is discussed in further detail in the work of Agathokleous [3], is the amalgamation of several tools and the enrichment with additional water loss incidents associated with the pipe fittings during data analysis. Furthermore, a specific methodology for carrying out survival analysis at the segment (and not at simply street) level is proposed, which enhances the quality of results and conclusions. A third contribution that is equally important is the study of the impact on the WDN condition of the implementation of IWS policies. It has been proved, through this research work, that the implementation of such policies, as a management measure of the national water resources, is negatively affecting the condition of the WDNs.

Mathematical modeling is a proactive management strategy that assists WDN owners in evaluating the condition of their network, assessing historical incident and risk of failure data and prioritizing the work, based on the knowledge of the inherent risk encompassed in the studied datasets. The ultimate goal is to device mathematical models for the prediction of water loss incidents and to extract useful information and draw conclusions about the studied WDN, so that the developed mathematical tool helps compose priority lists for the maintenance of pipelines (repairs and replacements) and eventually for the more economical and efficient management of WDNs. Furthermore, through the developed mathematical models a correlation of intermittent water supply periods with the frequency of future leak incidents can be made (discussed in detail in Chapter 3), so that the effects of the intermittent water supply (IWS) operations can be studied.

For the specific case of the WDN of the city of Nicosia, Cyprus, the results showed that the studied WDN experiences more leakage incidents on the house connections (HC) than on the water mains (WM). The sources of vulnerability and the reasons for WDN component failures vary from DMA to DMA. For example, comparing two areas which both faced leakage incidents related to the house connections, in the first area such incidents occurred because of the aging network and in the second area these were due to damages caused by infrastructure works in the area.

CHAPTER 3

Vulnerability Assessment of Water Distribution Networks Under Abnormal Operating Conditions and Nonseismic Loads – The Case of Intermittent Water Supply (IWS)

There is no water-supply in which some unnecessary waste does not exist and there are few supplies, if any, in which the saving of a substantial proportion of that waste would not bring pecuniary advantage to the water authority.

Hope, W. (1892),
The waste of water in public supplies and its prevention.
Minutes of the Proceedings 110 (1892): 260–275.

3.1 INTERMITTENT WATER SUPPLY PERIOD

Faced with extended periods of drought and short supply of water, arid-weather countries have turned to intermittent water supply (IWS) as a means to reduce water consumption and to prolong their national water reserves. Unfortunately, such drastic measures usually fail to consider the effects of intermittent supply on the condition of piping networks and the resulting water losses, inefficiencies, and overall maintenance cost on these networks.

The IWS period in study was applied on 31/3/2008 and was terminated on 31/1/2010. The IWS design foresaw separation of the DMAs into sub-DMAs and the sub-DMAs into 4 groups, to which water was supplied sequentially for a time period of 12 hours per sub-DMA. During the initial period, the program was fully implemented but thereafter the IWS measures with regard to the duration of the water supply were loosened, since the planned period of 12 hours of supply was not enough to fully serve the WDN areas at high altitude.

Urban Water Distribution Networks
DOI: http://dx.doi.org/10.1016/B978-0-12-813652-2.00003-7

Table 3.1 The water volume distribution that was channeled into Nicosia's WDN from 2005 until 2010.

Year	Unbilled water (%)	Water Channeled to the WBN's Network (m^3)	Water losses (m^3)	Water consumption (m^3)
2005	20.59	18,983,570	3,908,717	15,074,853
2006	19.5	19,181,450	3,740,383	15,441,067
2007	19.4	19,981,040	3,876,322	16,104,718
2008	13.7	17,221,910	2,359,402	14,862,508
2009	17.7	18,807,580	3,328,942	15,478,638
2010	26.11	23,840,640	6,224,791	17,615,849

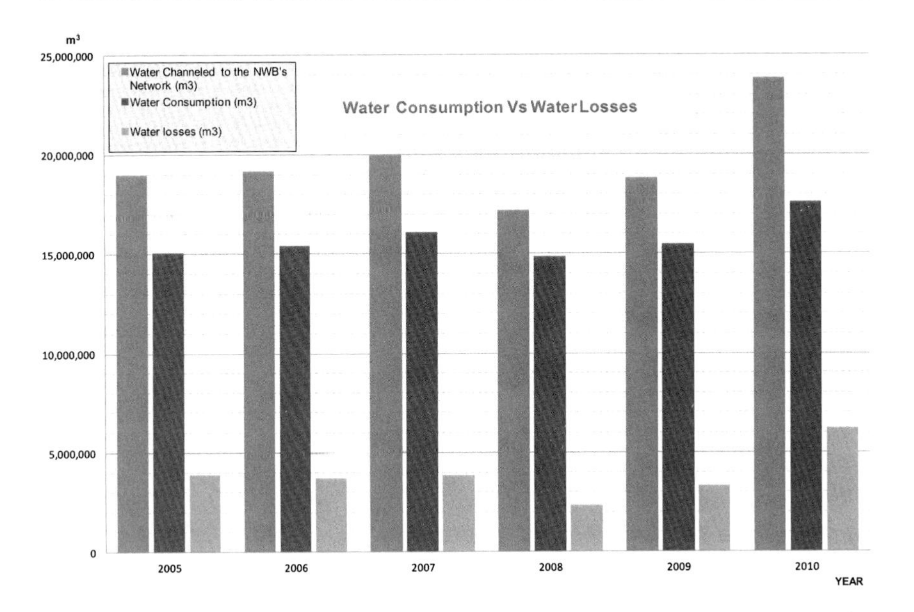

Figure 3.1 Water consumption and water losses for the time period between 2005 and 2010.

A quick statistical analysis of the results related to the effects of the IWS policy, as summarized in Table 3.1 and depicted in Fig. 3.1, captures the overall view on the percentage nonrevenue water and the volume of water channeled to the network in study.

Through the results, two important conclusions can be drawn in relation to the water consumption and to the water losses. The first is that water consumption reduced only during the first year of the IWS policy implementation, while the yearly consumption for the year that followed the termination of the IWS was significantly higher. This observed increase was partly due to an increase in justified water consumption, and partly due to an increase in unjustified ("wasted") water consumption attributed to consumer psychology and stemming from the desire of consumers to overutilize a social good which they lacked for two years. The second conclusion is that during the two years of IWS policy implementation, the level of the nonrevenue water volume was indeed reduced. The year after, though, the volume increase compared with the average of the previous years was much more than the savings that was achieved through the implementation of the IWS policy.

Thus, as an initial conclusion, an IWS policy should not be viewed as a policy of saving water, but rather it is a method of managing national water resources over time. Through such a measure, water volumes are saved during prolonged drought periods. However, the volume of nonrevenue water is expected to increase after the termination of the IWS policy.

3.1.1 Statistical Analysis

During the first year alone, the WBN reported a total of more than 12,000 service calls related to water loss or pipe inspection incidents. The number of daily incidents in the following two years rose to about 70–80 – a number disturbingly high. The majority of these incidents were related to house connections and small-diameter pipes, with an increase in incidents during the intermittent supply period of about 28% compared to the normal operating conditions period (uninterrupted supply). This finding is in agreement with the fourth finding of the analysis by Christodoulou et al. [38] on the increased risk of failure of house connection pipes.

The statistical analysis focuses on the comparison of the incidents during the IWS period in relation to the CWS dataset. Table 3.2 presents a comparison of the percentage increase of the leakage incidents per year per part type as to the year 2003 while Fig. 3.2 presents the number of incidents as well as number of leakages related to the part type during the first seven months of each year.

As per Table 3.2, the incidents related to WM pipes show a continuous increase from year to year, while all other WDN part types show a large increase in the incidents related to them following implementation of the IWS measures. Thus, as shown by the preliminary analysis, the IWS indeed affects the state of the network.

Fig. 3.2A, showing the number of incidents per year (Jan.–Jul.), confirms the observations from Table 3.2. WDN parts classified as "HC pipe",

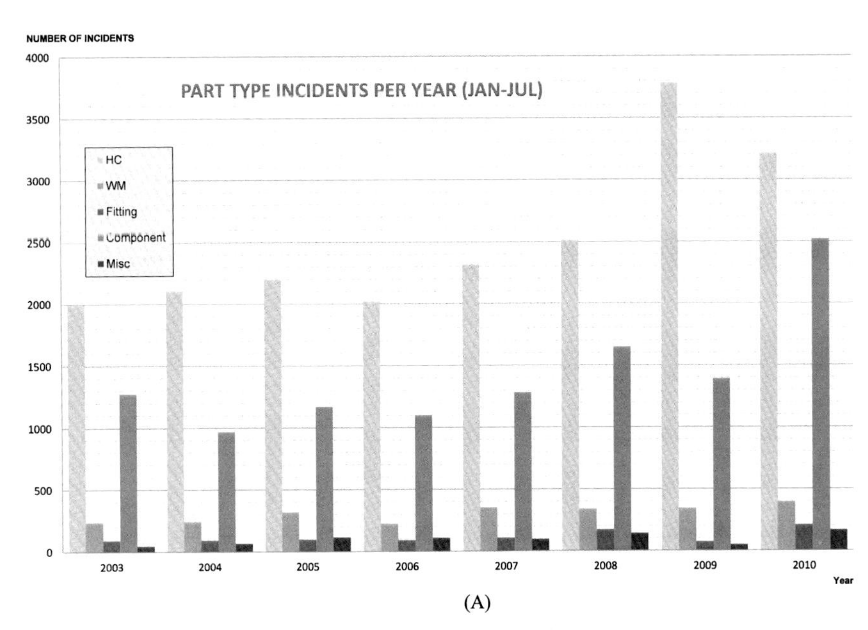

(A)

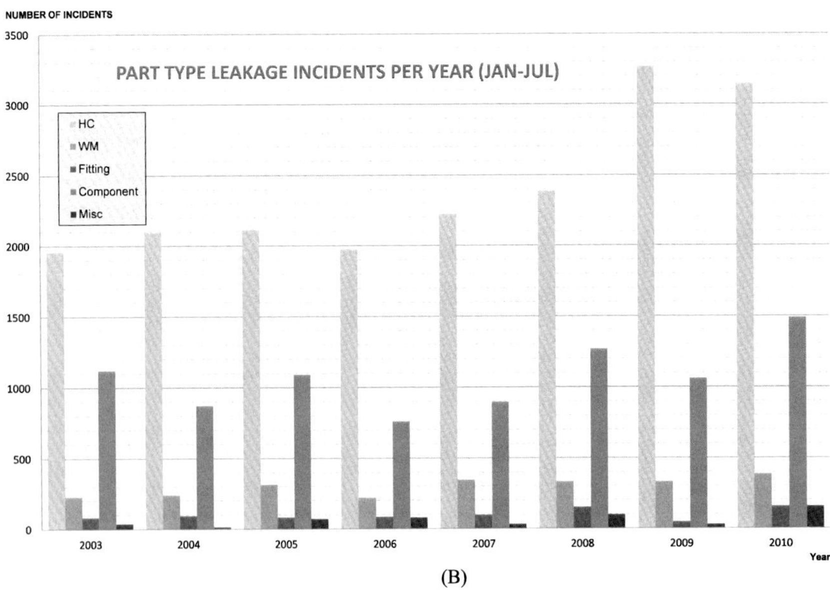

(B)

Figure 3.2 'Part Type' incidents and 'Part Type' leakage incidents in case-study WDN, for the 2003–2010 period.

Table 3.2 Change in the percentage of the number of incidents per year for the "*Part Type*" field (year 2003 is used as the baseline).

Year	Part type leakage incidents				
	Component	*Fitting*	*Misc.*	*Pipe*	
				HC	*WM*
2003	–	–	–	–	–
2004	−22.21	13.41	−63.16	7.05	5.26
2005	−2.59	0.00	84.21	8.02	38.16
2006	−32.20	3.66	113.16	0.87	−3.95
2007	−20.07	18.29	−18.42	13.69	51.32
2008	13.20	82.93	155.26	21.96	44.74
2009	−5.26	−43.90	−23.68	66.80	43.86
2010	32.74	87.80	305.26	60.52	67.54

"Fitting", "Component" or "Misc." are affected by the IWS measures, while the WM pipes exhibit a lower IWS impact on them.

The above observation stems from the analysis of all recorded incidents in the time period of interest. A further analysis, though, dealing with only incidents related to leakage was deemed necessary and for this reason Fig. 3.2B was developed.

Fig. 3.2B reinforces the findings on the vulnerability of the "HC pipe", "WM pipe", "Fitting", and "Misc." classes. The IWS measures seem, though, to affect the "Component" class in a different way than the others. The majority of the excess incidents observed during the IWS period are associated with "Inspection" and not with "Repair" or "Replacement" classifications (please refer to Fig. 2.5, p. 51), and as such they may be excluded from the analysis.

3.1.2 Survival Analysis

At first, a survival analysis specifically designed for investigating the impact of IWS on the condition of the WDN was run, without taking into account the elements' age. This analysis is based on a three-year period, and the leak data was left-censored at 01/01/2008 and right-censored at 31/01/2010. In addition, the "date of failure" and the "date of rebirth" of a WDN component were considered to be the date that a leak-related incident was reported, while the analysis performed was at street level (i.e., the pipe runs were considered to be of the same length as the street the pipes were laid in). Finally, only leak-observed incidents related to pipes (water mains and house connections) were used, while all other incidents (fittings,

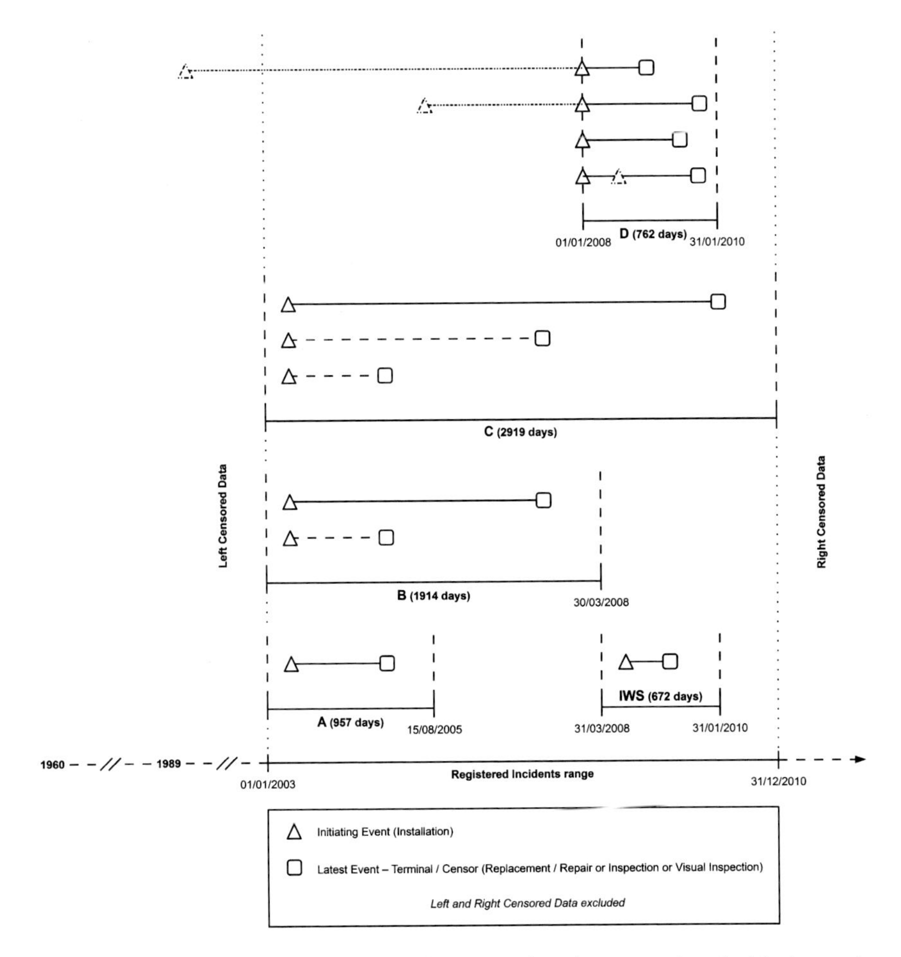

Figure 3.3 Diagram of the timelines for the survival analyses associated with the study of the impact of the IWS.

components and misc) were excluded from the analysis. The information related to the analysis period (duration and data included) are noted by the time period "D" in Fig. 3.3.

The nonparametric Kaplan–Meier survival analysis estimator was used for the analysis and the results are presented by the survival and hazard rate plots given below (Fig. 3.4 and Fig. 3.5).

Fig. 3.4 represents the survival plot for both the house connection pipes and the water mains. It can be seen from the survival curves that the behavior of the two different pipe classes is very similar, with a slightly better performance exhibited by the house connection pipes. Also, water mains

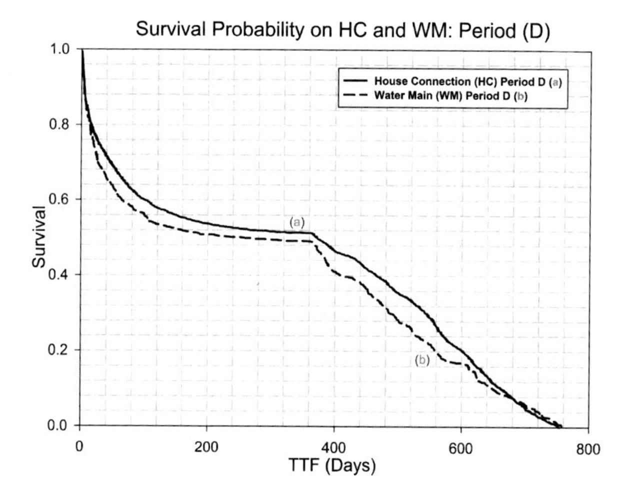

Time Period D

Time Range	Total No of Days	Parts Type	Class	Incidents	Terminal Ev.	Censored Ev.	Terminal to Censored Ratio
01/01/2008 - 31/01/2010	762	Pipes	HC	8725	1218	7507	16.2%
			WM	949	329	620	53.1%

Figure 3.4 Survival plot of the Pipeline class variable for the P-C model (intermittent water supply period incidents).

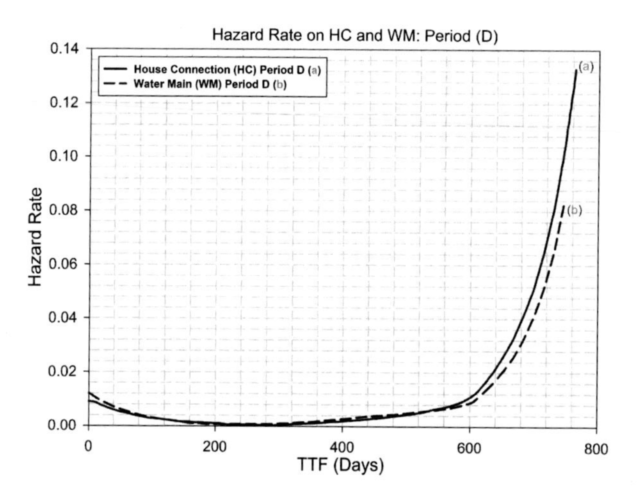

Figure 3.5 Hazard rate plot of the Pipeline class variable for the P-C model (intermittent water supply period incidents).

have a slightly faster drop in their survival rate at the beginning, while house connections have a faster drop at the end of the period examined. Notable is the change in the curves' slopes at about $t = 370$ (about a year after the implementation of the IWS policy) and the increased pace in the reduction of survivability.

Fig. 3.5 depicts the hazard rate plot for the period in study. As in the case of the survival plot, the curves show an almost identical behavior by the two pipe classes. The water mains curve has a slower hazard rate than the curve for the house connections, despite the fact that initially it has a higher hazard rate.

The results show that the HC and WM pipes have similar survival probability through time, which is in contrast to the results of the mathematical analysis of the corresponding model (Fig. 2.30A, p. 104). This finding suggests that the IWS indeed affects the condition of WDNs, and a comprehensive mathematical analysis is needed to investigate the magnitude of the effect.

The mathematical framework employed in the impact study of the IWS period on the condition of the WDN, consists of four survival analyses that use data from three different time periods. The fist analysis uses recorded incidents from 01/01/2003 until 15/08/2005 (the first half of the time period before the day when the IWS policy had come into effect). The second analysis uses recorded incidents from 16/08/2005 until 30/03/2008, which is the time period before the day when the IWS started. The third analysis uses incidents from the total time period for which there are recorded data. The incidents used in the last analysis are only associated with the IWS period.

Since the aforementioned analyses examine the IWS effect on the condition of the WDN in study for a specific time period, and the existence of left-censored data causes shifting of the underlying survival curves, it is prudent that elements whose birth date (i.e., installation date) is unknown be excluded from the analysis. Additionally, (1) the "date of failure" and the "date of rebirth" are considered to be the date that a leak-related incident was reported, (2) the analysis performed was at street (and not segment) level, and (3) only leak-observed incidents related to pipes and fittings were used, while all other incidents (components and misc) were excluded.

Fig. 3.3 illustrates the information related to the different time periods, their duration and the data included in the different survival analyses employed by the mathematical framework studying the impact of the IWS measures on the vulnerability of the WDN in study.

It is expected that the survival curves for the various time periods will have the same "shape", and that the time to failure and the probability

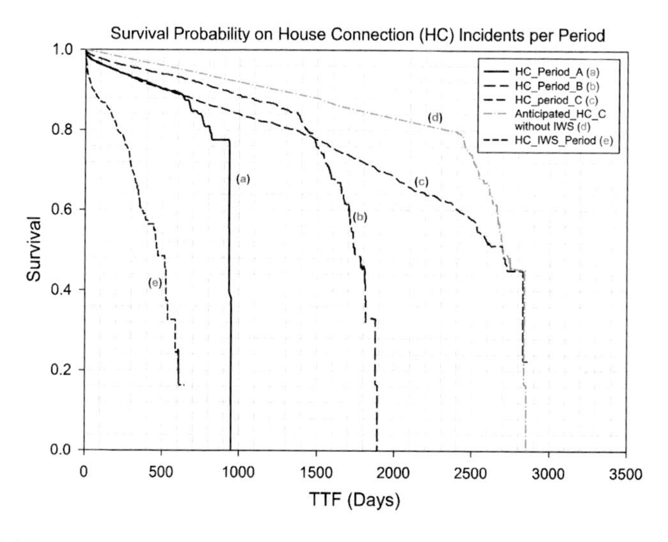

Time Period A

Time Range	Total No of Days	Parts Type	Class	Incidents	Terminal Ev.	Censored Ev.	Terminal to Censored Ratio
01/01/2003 - 15/08/2005	957	Pipes	HC	2520	188	2332	8.1%
		Fittings	WM	563	527	36	1463.9%

Time Period B

Time Range	Total No of Days	Parts Type	Class	Incidents	Terminal Ev.	Censored Ev.	Terminal to Censored Ratio
01/01/2003 - 30/03/2008	1914	Pipes	HC	6129	498	5631	8.8%
		Fittings	WM	1526	1426	100	1426.0%

Time Period C

Time Range	Total No of Days	Parts Type	Class	Incidents	Terminal Ev.	Censored Ev.	Terminal to Censored Ratio
01/01/2003 - 31/12/2010	2919	Pipes	HC	13831	1648	12183	13.5%
		Fittings	WM	2556	2193	363	604.1%

Time Period IWS

Time Range	Total No of Days	Parts Type	Class	Incidents	Terminal Ev.	Censored Ev.	Terminal to Censored Ratio
31/03/2008 - 31/01/2010	672	Pipes	HC	1922	356	1566	22.7%
		Fittings	WM	249	169	80	211.3%

Figure 3.6 Survival plots on HC incidents per period.

of survival should be improved, when the time period of analysis and the number of incidents associated with it increase.

House Connections

Fig. 3.6 illustrates the survival probability curves on the HC incidents for the defined time periods. The survival probability curve (a) represents the survival curve of the first time period (01/01/2003–15/08/2005), which lasts 957 days. The second survival curve (b), which lasts 1914 days, has the same shape but it has higher survival probability than curve (a). This

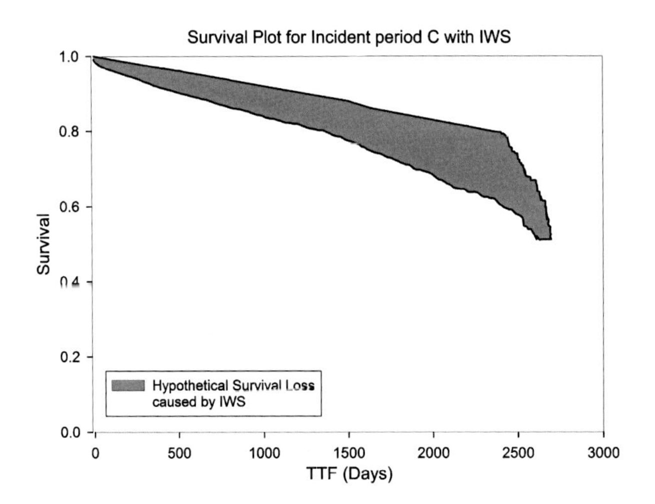

Figure 3.7 Survival plot for incident period C with IWS.

increase is expected due to the fact that older elements, compared to those of the first time period, are incorporated into the analysis dataset which affect positively the survival probability.

The second analysis uses recorded incidents up to the day when the IWS was applied, while the third time period includes all incidents that occurred during the IWS period. Should this period not cause any additional stress on the WDN, the relationship of the data analysis associated with it, compared to the corresponding of the second one, should be similar to that of the second as to the first. Thus, the analysis of the data related to the third period is expected to have resulted in a survival probability curve similar to curve (d) of Fig. 3.6. Instead of that, the survival analysis of the data associated with the third time period has as a result the curve (c). This result further reinforces the findings that IWS policies negatively affect the condition of WDNs and increase their vulnerability. Fig. 3.7 illustrates the hypothetical survival loss caused by IWS.

To further study the effect of the IWS period, a final survival analysis was performed using data that only related to this specific period. The resulting plot is represented by the curve (e) in Fig. 3.6. As shown, the shape of the curve is not the same with that of (a) and (b) as it should be, but it has a large drop in the early lifetime stage of the elements. Also, the survival probability of this curve will be even lower, especially towards the end of time to failure, if the data regarding the incidents that occurred from 08/01/2009–31/12/2009 were available to be included in the analysis.

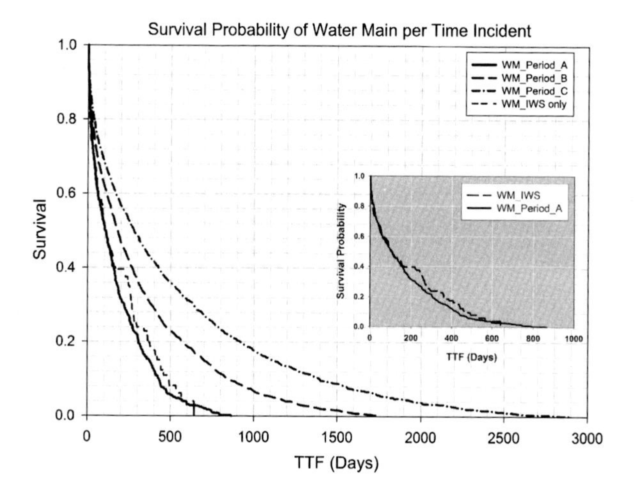

Time Period A

Time Range	Total No of Days	Parts Type	Class	Incidents	Terminal Ev.	Censored Ev.	Terminal to Censored Ratio
01/01/2003 - 15/08/2005	957	Pipes	HC	2520	188	2332	8.1%
		Fittings	WM	563	527	36	1463.9%

Time Period B

Time Range	Total No of Days	Parts Type	Class	Incidents	Terminal Ev.	Censored Ev.	Terminal to Censored Ratio
01/01/2003 - 30/03/2008	1914	Pipes	HC	6129	498	5631	8.8%
		Fittings	WM	1526	1426	100	1426.0%

Time Period C

Time Range	Total No of Days	Parts Type	Class	Incidents	Terminal Ev.	Censored Ev.	Terminal to Censored Ratio
01/01/2003 - 31/12/2010	2919	Pipes	HC	13831	1648	12183	13.5%
		Fittings	WM	2556	2193	363	604.1%

Time Period IWS

Time Range	Total No of Days	Parts Type	Class	Incidents	Terminal Ev.	Censored Ev.	Terminal to Censored Ratio
31/03/2008 - 31/01/2010	672	Pipes	HC	1922	356	1566	22.7%
		Fittings	WM	249	169	80	211.3%

Figure 3.8 Survival plot of WM incidents per time period.

The conclusions of this analysis reinforce the overall presumption about the effect of the IWS in the WDNs condition.

Water Mains

A similar series of analyses, on the impact of IWS, is also performed for the WM elements. Fig. 3.8 illustrates the survival probability curves on the WM incidents for the defined time periods. The survival probability curve representing the survival curve for the second time period has the same format with the curve of the first time period, having an increase in the time to failure and the survival probability as expected.

Analysis of the data associated with the third time period resulted in a survival probability curve that has the same shape as that of the other two time periods. In addition the correlation between this curve and the one associated with the second time period, as to the time to failure and the survival probability of the elements, follows the same pattern with the corresponding correlation between the curves that resulted from the data analysis associated with the first two time periods. These observations suggest that IWS do not affect the WM elements.

Finally, a survival analysis associated with the recorded data associated with the WM during the IWS period is performed, aiming the further study of the effect of the IWS period. The resulting plot is represented by the curve "WM-IWS only" in Fig. 3.8. As shown, the shape of the curve is similar with that of the other three survival probability curves, as it should be. The time to failure of the curve is shorter than that of period "A", as expected, but the probability of survival of the elements is at higher levels, which is not normal. This is due to the fact that the data regarding the incidents that occurred from 08/01/2009–31/12/2009 were not available to be included in the analysis. In that case observations extracted from this analysis reinforce the conclusion that IWS do not affect condition of the WM elements.

3.2 FURTHER DISCUSSION ON FINDINGS

3.2.1 Results

Several conclusions derived from the results of the mathematical analysis relate to the operational mode of a WDN, as well as to characteristics of a WDN's structural design and construction.

"Incidents No." per Group of Pipes (HC and WM)

The main observation/conclusion is that the HC incidents are much more than those of WM (Fig. 2.7, p. 55). The increased vulnerability is related to the hydraulic forces developed within the WDN during IWS. The simultaneous fluctuation of water demand across the network develops pressure and suction forces at various points in the network. These forces cause fatigue of the pipelines, especially at points where there is a change in the diameter of the pipes. The larger the alternation is in the pipe diameter the more intense the caused phenomenon is. The pipes that are affected the most are those of smaller diameter.

The observed seasonality in the increasing leakage incidents during the warmer months is also associated with the above phenomenon. As summer approaches, water consumption increases and the fluctuations in water demand are more intense. This results in an increase in leakage incidents during the summer months. As the small-diameter pipes are those most affected, the seasonal increase in leakage incidents is stronger evident in HC elements while it is not as strongly evidenced in the case of the WM elements (Fig. 3.15B, p. 153).

Therefore, it is very important that during the design of a WDN particular attention is given to the dimensioning of the various WDN components (especially of pipes), as well as to the preservation of the system's operation pressure at low and uniform levels, so as to reduce the vulnerability of the WDN to the fluctuation in water pressure.

The stress in the network stemming from the application of IWS policies is similar in nature to that of simultaneous variations of water demand in it, albeit to a much greater degree (hydraulic transient/hammer effect). For this reason, as shown by the analysis results, the HC elements are negatively affected to a greater extent by the implementation of such measures, while the WM elements are not influenced as much (Fig. 3.6 and Fig. 3.8).

The Ratio of "Repair" Over "Replacement" Actions

Another conclusion relates to the ratio of repair/replacement for the HC and the WM elements. Most data involving WM elements mainly relates to "replacement" incidents rather than to "repair" ones. On the other hand, in the data related to the HC elements the repair incidents dominate.

The majority of the HC elements group consists of one-piece plastic flexible pipelines, whose age is small compared to the main pipeline system. The early-age stage implies that the impact of physical deterioration is still very small. Further, the material type of the HC piping allows for the easy treatment of any problem in the WDN with a repair action, by replacing the defective section using fittings. In fact, the reported cases associated with HC incidents that needed replacement of the whole pipe are very rare.

The majority of the elements composing the main pipeline system of the WDN are of greater age and hence the impact of physical deterioration is high. For this reason most of the failures lead to leakage problems that require replacement of the element. Furthermore, the material type of the older WM elements is outdated (asbestos cement, cast iron, ductile iron, galvanized) and for that they are replaced by plastic even if the failure is repairable.

3.2.2 Database and Mathematical Modeling

The last part of the chapter lists observations and conclusions derived from the previously presented analysis results.

Data Quantity

The full complement of the information relating to an incident is very important to the proper completion of the analysis. Frequently, though, some information associated with the database fields of "*Reason*" and "*Location*" is missing and, as a result, some of the analysis results may be incomplete or biased. In such cases, survival analysis may provide a means to account for this ellipsis and to arrive at meaningful results with censored data sources.

Data Quality

The major disadvantage of the studied database is the quality level of its data, primarily due to the absence of additional datafields, the time horizon of the included data, as well as the lack of detailed classification of the included analysis factors.

A major omission of the database is a field which indicates the type of failure of an element. For example, if a pipe failure is longitudinal or transverse. This extra information would enable a more qualitative analysis and a better correlation of the events with the source of their vulnerability.

Another omission is the fact that some areas are not classified correctly. For example, the category "Material/Hardware Failure" ("*Reason*" datafield) involves three options in a single choice. This category should be split up into three options related with failure, which are the failure due to aging, failure due to material failure, and failure due to exogenous factors (i.e., IWS). A further important example is how the "*Material*" field is categorized. For instance, for the HC pipes made of plastic there is only one available material type. This does not help the scientific analysis of the data because the correlation of the material with the survival probability cannot actually be studied in depth. There should be a separation of the plastic material in the various categories that exist and the different batches. This will allow an in-depth study on whether there is a correlation of the incidents occurrence with the type and the batch number of the plastic pipe.

Level of Analysis

As revealed by the results, if the time period of left-censored data is much more extended than those of recorded incidents, then the segment-level analysis does not contribute to the final result of the analysis.

Model Classes

In the case when the dataset is biased with the one of the two pipeline classes (i.e., the majority of the incidents associated with HC or WM) then the analysis of the Coupled model is not necessary, since the result will be similar to those of the class that has the majority of the incidents it is associated with.

3.2.3 Water Board of Limassol

The succeeding sections present, for comparison purposes with the results on the WDN of Nicosia, part of the results from an analysis of the WDN of Limassol that were carried out during previous research works.

The WDN of the Water Board of Limassol (WBL) is the second largest water network in Cyprus. It is over 50 years of age and serves approximately 170,000 residents through approximately 64,000 consumer meters in an area of 70 square km. The annual volume of potable water distributed through the network of pipes, of approximate length 795 km, is about 13.7×10^6 m^3 and of value €7.0 million. The network is divided into pressure zones, each subdivided into DMAs having a single metered source with physical discontinuity of pipe network between DMA boundaries. Meter readings at water sources (boreholes and treatment plant) are connected via an SCADA telemetry system to the control room. This enables continuous monitoring of the water source outputs and accurate recording of flows. The continuous monitoring of the DMA meters combines information technology and telecommunication networks to transfer the data via the World Wide Web.

The recorded pipe incidents of the WBL for the time period between 01/01/2007 until 31/01/2011 are presented in Table 3.3. For the WBL network, the total number of incidents per annum increased from 2092 in 2007 (the presumed steady state condition) to 2565 in 2008, 2853 in 2009, and 3566 in 2010. This, compared to the steady state condition (year 2007), translates to an increase of 23% for 2008, 36% for 2009, and 70% for 2010. Also there was an increase in the incidents on water mains of about 85% in year 2010 compared to the base year of 2007 (9700 vs. 5236). A similar

Table 3.3 Annual number of pipe-related incidents (WBL dataset).

Incident type	2007	2008	2009	2010	2011	Total
Leak – Water Main	1010	1497	2328	2603	201	7639
Leak – House Connection	350	334	270	598	44	1596
Leak – Apparent Loss	164	3	2	32	10	211
Faulty Water Meter	2	2	2	0	0	6
No Water	83	93	92	118	12	398
Water Waste	428	345	89	109	4	975
Low Water Pressure	48	37	55	103	5	248
Miscellaneous	7	254	15	3	2	281
Total	2092	2565	2853	3566	278	11,354

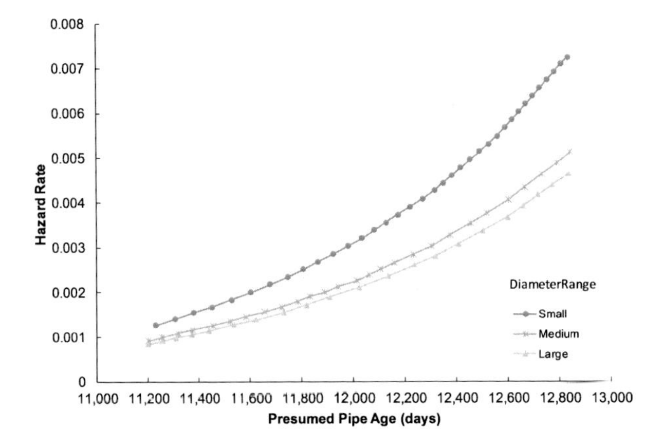

Figure 3.9 Hazard rate plot for different diameters of pipes (WBL dataset).

situation is observed on the incidents pertaining to house connections (71% increase, 598 vs. 350 incidents).

From Table 3.3 the difference in vulnerability and in behavior between the two WDNs is evident. In the WDN of Limassol the WM incidents dominate, while it was shown that in Nicosia the main volume of incidents relates to HC elements.

Fig. 3.9 illustrates the presumed pipe age for the different diameter clusters of the pipes. The presumed pipe age for the three different types of diameter range is almost the same but the hazard rate increases as the diameter of the pipe reduces. The corresponding plots involving the WDN of Nicosia show that the hazard rate is increasing as the diameter of the pipe increases.

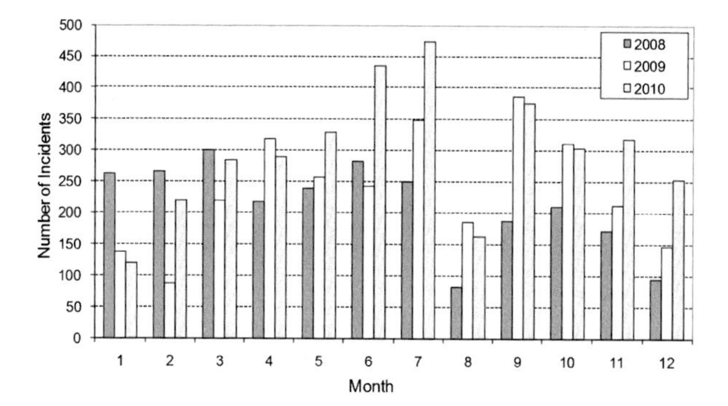

Figure 3.10 Incidents per month, for three-year period of intermittent supply (WBL dataset).

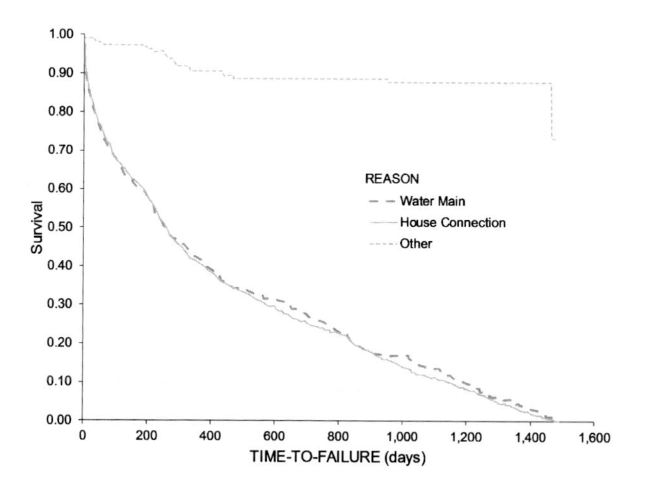

Figure 3.11 Survival plot for the time period of the intermittent water supply (WBL dataset).

The total sum of the incidents per month, for the three-year period of IWS is presented in Fig. 3.10. The seasonal distribution of the incidents is not associated with the corresponding graphs of Nicosia's WDN.

The survival analysis that examines the IWS period used the aforementioned 3-year dataset (2007–2010) and water loss incident information on the date/time of event, the location (street), the type of pipe (diameter, material), the type of pipe (house connection, water main), and the number of previously observed breaks. Furthermore, the method allowed for incomplete data through the incorporation of left-censored and right-censored flags. The results of the analysis are captured in Figs. 3.11 and 3.12.

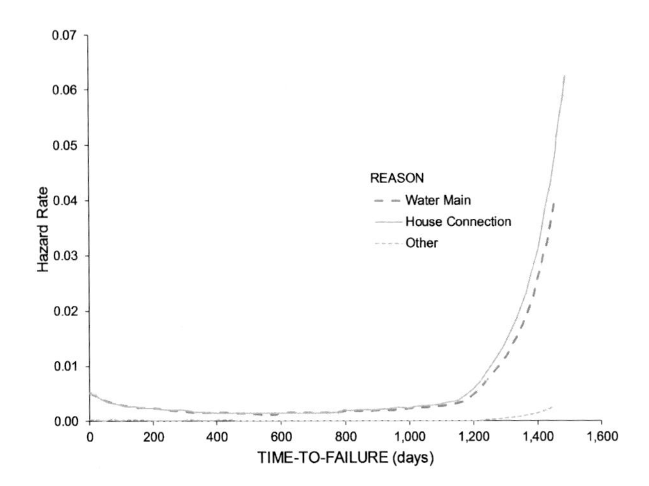

Figure 3.12 Hazard rate plot for the time period of the intermittent water supply (WBL dataset).

Fig. 3.11, showing the survival of water mains and of house connection pipes, indicates an almost identical behavior over time for these two piping classes. A closer look at Fig. 3.12, though, indicates a faster hazard rate (for $t > 1125$ days) for water mains compared to house connections. This, in effect, translates to a higher vulnerability for water mains.

The comparison of these two plots with the corresponding plots for Nicosia's WDN proves, as all previous analyses and findings, that IWS measures negatively affect the vulnerability of WDNs, albeit differently as each WDN has its own characteristics that make it different from any other WDN.

3.3 REGRESSION TREES

The same case-study WDN (of Nicosia, Cyprus) and underlying dataset, regarding the impacts of IWS operations on WDN vulnerability, were also examined by means of regression-tree analysis.

As a recap, the WDN in study is divided into 21 District Metered Areas (DMAs) and, according to data of year 2013, it requires for its operation about 19.3 million m^3 of water per year in continuous flow, while the maximum daily demand in the summer is around 63,700 m^3. The minimum consumption per day is 39,000 m^3, while the daily average water consumption is 52,700 m^3. The Water Board of Nicosia (WBN) has in operation about 112,000 water meters and a total length of water mains of more than 1400 km. Further to the network's spatial division into DMAs and sub-DMAs, the WBN's WDN is remotely monitored through a supervisory control and data acquisition (SCADA) system consisting of 32 electronic stations across the network, which continually collect and transmit operational information. Apart from the SCADA telemetry system database, the WBN has recently developed and put in operation a database where all information related with the network operations is stored. Finally, all leakage incident data reports since 2003 are maintained in a related DBMS.

The data utilized for the development of the mathematical model covers a time period of approximately eight years (01/01/2003 to 31/12/2010, with a small gap in the recorded data from 01/08/2009 to 31/12/2009), and includes 38,346 incidents. The mathematical framework utilized in studying the impact of the IWS policies on the condition of the WDN consists of four survival analyses that use data from four different time periods, as shown schematically in Fig. 3.3 (p. 136), and listed below:

- Time period "A" – The analysis associated with this period uses recorded incidents from 01/01/2003 until 15/08/2005 (the first half of the time period before the day when the IWS policy had come into effect).
- Time period "B" – The analysis associated with this period uses recorded incidents from 01/01/2003 until 30/03/2008 (the time period before the day when the IWS started).
- Time period "C" – The analysis associated with this period uses incidents from the total time period for which there are recorded data (the time period before, during and after IWS application).
- Time period "IWS" – The analysis associated with this period uses incidents that happened during the IWS period.

The analyses of the first three time periods ("A", "B", and "C") are the essential analyses of the case-study WDN, while the analysis that is associated with the IWS period is exploratory, and is used to confirm the findings. The case-study dataset was examined by use of decision trees and survival analysis tools, and by use of data stratifications of various levels.

3.3.1 Regression Trees – Methodology

Decision-tree analysis is a branch of statistics that is used for classification and regression, with *Classification and Regression Trees* (CART) being used for the development of a decision support tool that classifies a dataset or predicts an outcome by analyzing historical data. Whilst, though, in the case of classification trees the response variable could be either binary ("true" or "false" values) or could have more than two categories, in the case of regression trees the response variable is numeric or continuous. Another important difference between the two models is that classification trees are used when the objective of the analysis is to split a dataset (based on homogeneity of data) into smaller classes, looking for categorical solutions that explain behavior or describe the characteristics of a specific data set, whereas regression trees result in numerical answers that predict a future event or action using historical data.

This study aims to investigate whether WDNs are negatively affected by the implementation of IWS policies, and the approach taken in investigating this issue is the study of the occurrence rate of WDN failures during several periods of significance. The hypothesis examined is that if the implementation of IWS negatively affects the condition of a WDN, then the occurrence rate of WDN failures will be increased, i.e., the time period between two successive failure incidents will shorten during and following IWS operations. The analysis utilizes the *Regression Tree* algorithm by Breiman et al. [19] which has the ability to analyze historical data and classified it into a tree diagram that predicts the remaining time for a possible future incident to take place.

The historical database used in the analysis and classification consisted of leakage incidents related to two pipe classes (water mains, WM; and house connections, HC), and each pipe class consisted of data of two part types ("*Pipe*" and "*Fitting*" elements). Since the aforementioned analyses examine the effect of IWS on the condition of the WDN for a specific time period, and because the installation date of each element is important to be known, all elements whose "birth date" was unknown were excluded from the analysis. Additionally, the "date of failure" and the "date of rebirth" were considered to be the date that a leak-related incident was reported. The analysis performed was at the street (and not component or segment) level and the time from previous action (TFPA) was used as a response variable. Fig. 3.13 presents all variables and values used in the Regression Tree analysis.

The variables are the different characteristic properties of the WDN's elements associated with the reported leakage incidents. Only variables for which there was sufficient data volume have been used, so that the mathematical analysis of models lead to reliable results. The variable labeled

SYMBOL	VARIABLE	VALUE	DESCRIPTION	Incidents No											
				HC[2]-Period A		HC[2]-Period B		HC[2]-Period C		WM[3]-Period A		WM[3]-Period B		WM[3]-Period C	
x1	Action	0	Repair	2277	2466	5440	5938	11501	13149	27	554	77	1503	300	2493
		1	Replace	189		498		1648		527		1426		2193	
x2	NOPB[1]	0	No Breaks	595		1129		2886		534		1466		2321	
		1	(1-4) Breaks	1076		2328		5544		20		37		165	
		2	(5-8) Breaks	376	2466	1002	5938	2097	13149	0	554	0	1503	7	2493
		3	(9-12) Breaks	151		518		1034		0		0		0	
		4	(13-16) Breaks	86		300		555		0		0		0	
		5	(16>) Breaks	182		661		1033		0		0		0	
x3	Part Type	0	Empty Cell	0		0		0		0		0		0	
		1	Pipe	2426	2466	5837	5938	12883	13149	494	554	1285	1503	2122	2493
		2	Fitting	40		101		266		60		218		371	
x4	Diameter	0	Empty Cell	120		374		1162		43		188		316	
		1	Connection Pipe	2335		5544		11937		0		0		0	
		2	Main	0	2466	0	5938	0	13149	140	554	355	1503	623	2493
		3	Ring Main	0		0		0		10		24		28	
		4	Small Main	11		20		50		361		936		1526	
x5	Material	0	Empty Cell	46		204		1043		66		251		436	
		1	Galvanized	147		365		814		3		3		5	
		2	Plastic	2273	2466	5369	5938	11292	13149	0	554	0	1503	2	2493
		3	AC	0		0		0		465		1207		1814	
		4	PVC	0		0		0		20		42		234	
		5	DI	0		0		0		0		0		2	

[1] Number of Previous Breaks
[2] House Connection Elements
[3] Water Main Elements

Period A: 01/01/2003 - 15/08/2005
Period B: 01/01/2003 - 30/03/2008
Period C: 01/01/2003 - 31/12/2010

Figure 3.13 Dataset for the Regression Tree analysis.

"*Action*" states the decision, which is taken by the repair group, associated with the rehabilitation action of a damaged element. The possible options are "replacement" or a "repair" of the element. The "*NOPB*" variable denotes the number of observed previous breaks of an element, i.e., the number of repair actions up to that time (NOPB). The "*Part Type*" variable denotes the type of the element, i.e., if the leakage incident is associated with a pipe or a fitting. Additionally, the "*Material*" and "*Diameter*" variables are associated with the physical properties of the element. Since the "*NOPB*" and "*Diameter*" variables may take one of many possible values, which makes the analysis highly complex and possibly unreliable (due to the small number of incidents per variable value), the selections were grouped into clusters in order to facilitate the analysis and the extraction of useful and generalized results. Finally, the classification "empty cell" for a specific variable denotes the lack of information under this variable for the specific leakage incident.

It should be noted that if the IWS operations are of no impact to the condition of the WDN then the time to failure for a specific failure incident type (element of certain material, diameter, part type, etc.) will significantly

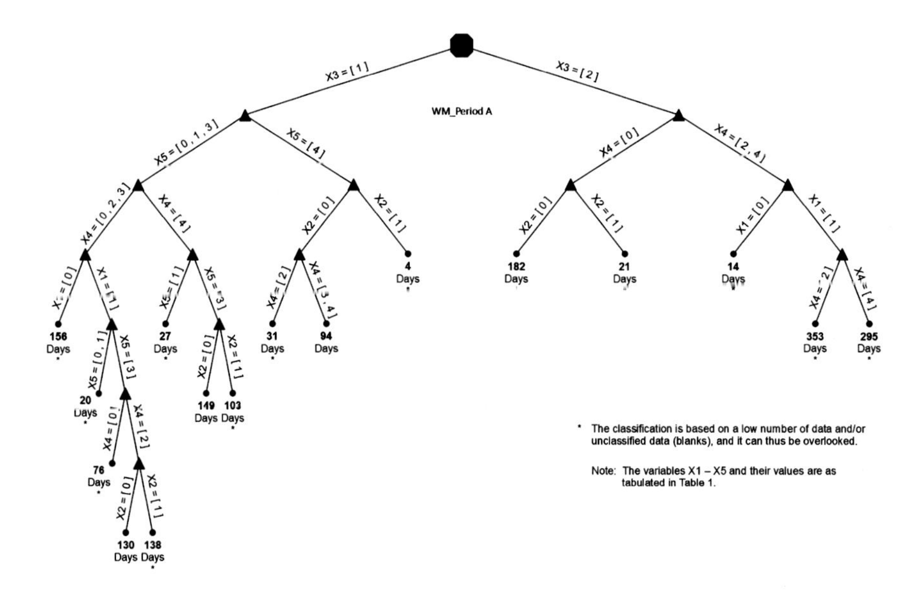

Figure 3.14 Regression tree of the "WM" Pipe class incidents for time period "A."

increase as the length of the time period is increased (i.e., when expanding period A to period B, and then to period C).

3.3.2 Regression Trees – Analysis

WM Class

Fig. 3.14 illustrates the resulting regression tree diagram for the "WM" Pipe class incidents in time period "A" (01/01/2003–15/08/2005), which lasted 957 days. Due to the small volume of data, this regression tree is not sufficiently comprehensive on information. The most important conclusions that can be extracted from the regression tree are:

(i) Next replacement action for an asbestos-cement (AC) water-main pipe element which didn't experience any previous breaks is expected in 130 days.

(ii) Next action for a small-diameter asbestos-cement (AC) water-main pipe element which didn't experience any previous breaks is expected in 149 days. This result is in line with the previous one since pipes of the same material and of smaller diameter live longer.

Fig. 3.15 illustrates the resulting regression tree diagram for the "WM" Pipe class incidents in time period "B" (01/01/2003–30/03/2008), which

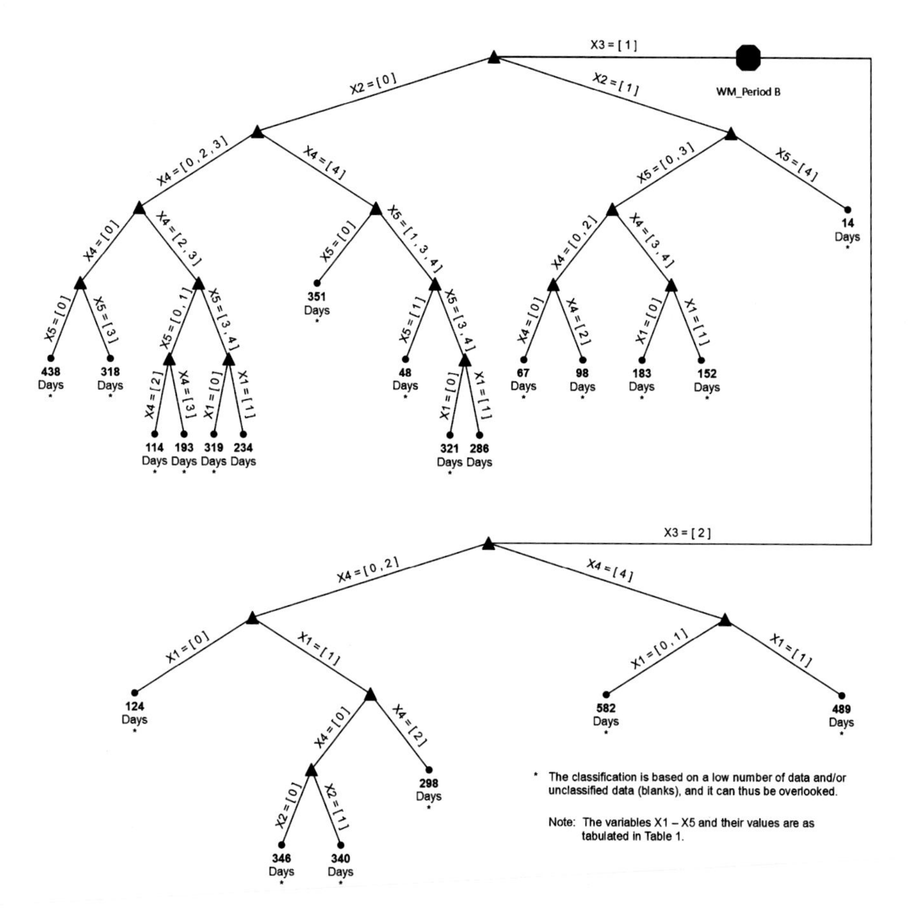

Figure 3.15 Regression tree of "WM" Pipe class incidents for time period "B."

lasted 1914 days. It is expected that as the expanded dataset has WDN elements of longer lifespans, the observed time-to-failure (TTF) in period "B" will be on the average higher than that observed in time period "A." The most important information extracted from the regression tree analysis for this period is as shown below.

(i) Next replacement action for an AC (or PVC) main (or Ring-Main) pipe element which didn't experience any previous breaks is expected in 234 days.

(ii) Next replacement action for an AC (or PVC) Small-main pipe element which didn't experience any previous breaks is expected in 286 days. This result is in line with the previous one since pipes of the same material and of smaller diameter live longer.

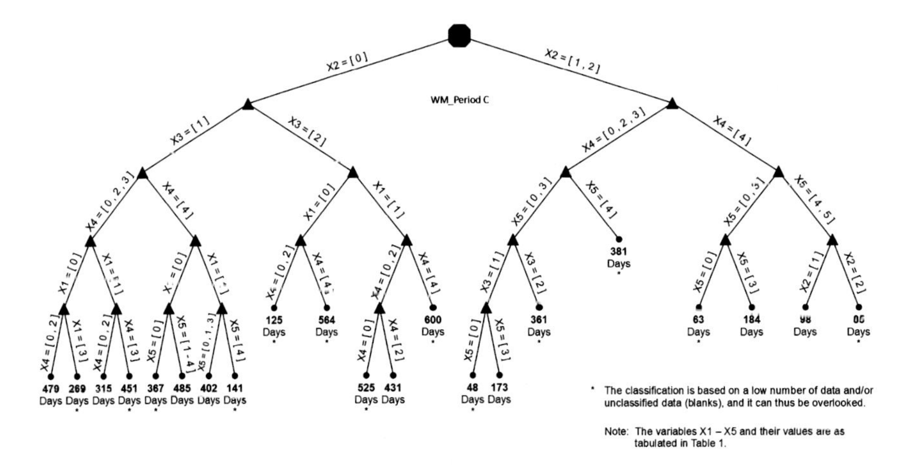

Figure 3.16 Regression tree of "WM" Pipe class incidents for time period "C."

The above results reinforce the hypothesis that the introduction of data associated with the time period from 16/08/2005 to 30/03/2008 has significantly increased the observed TTF. So during that time period, the vulnerability of the network condition has not worsened. Hence, this time period is not negatively affecting the condition of WDNs.

Fig. 3.16 illustrates the resulting regression tree diagram for the "WM" Pipe class incidents in time period "C" (01/01/2003–31/12/2010), which lasted 2919 days. As with the previous case, it is expected that as the expanded dataset has WDN elements of longer lifespans the observed time-to-failure (TTF) in period "C" will be on the average higher than that observed in time periods "A" and "B." The most important information deduced from the analysis is the following:

(i) Next repair action for a Main pipe element which didn't experience any previous breaks is expected in 479 days.

(ii) Next replacement action for a main pipe element which didn't experience any previous breaks is expected in 315 days.

(iii) Next repair action for a Small-main pipe element which didn't experience any previous breaks is expected in 485 days. This result is in line with the other two, since pipes of smaller diameter live longer.

(iv) Next replacement action for an AC Small-main pipe element which didn't experience any previous breaks is expected in 402 days. This result is in line with the other two since pipes of smaller diameter live longer.

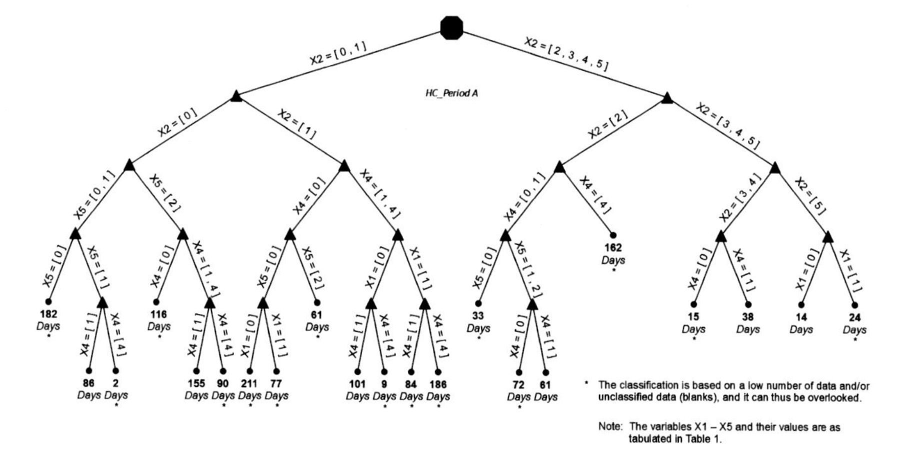

Figure 3.17 Regression tree of "HC" Pipe class incidents for time period "A."

(v) Next replacement action for a Small-main fitting element which didn't experience any previous breaks is expected in 431 days.

(vi) Next action for an AC Main pipe element that has up to 4 previous breaks is expected in 173 days. This result is in line with the other conclusions because as the number of previous breaks (NOPB) increases, the TTF decreases dramatically.

(vii) Next action for an AC Small-main pipe element that has up to 4 previous breaks is expected in 184 days. This result is in line with the finding (vi), since pipes of smaller diameter live longer.

(viii) Next action for a Ductile-Iron (DI) Small-main pipe element that has 1–4 previous breaks is expected in 98 days.

From the above results it can be deduced that the introduction of data associated with the time period from 31/03/2008 to 31/12/2010 (IWS period) has, as expected, significantly increased the observed TTF for these elements. Hence, "WM" Pipe class elements are not affected by the IWS policies.

HC Class

Fig. 3.17 illustrates the resulting regression tree diagram for the "HC" Pipe class incidents in time period "A" (01/01/2003–15/08/2005), which lasted 957 days. The most important conclusions drawn from the analysis are:

(i) Next action for a Galvanized Connection-pipe element which didn't experience any previous breaks is expected in 86 days.

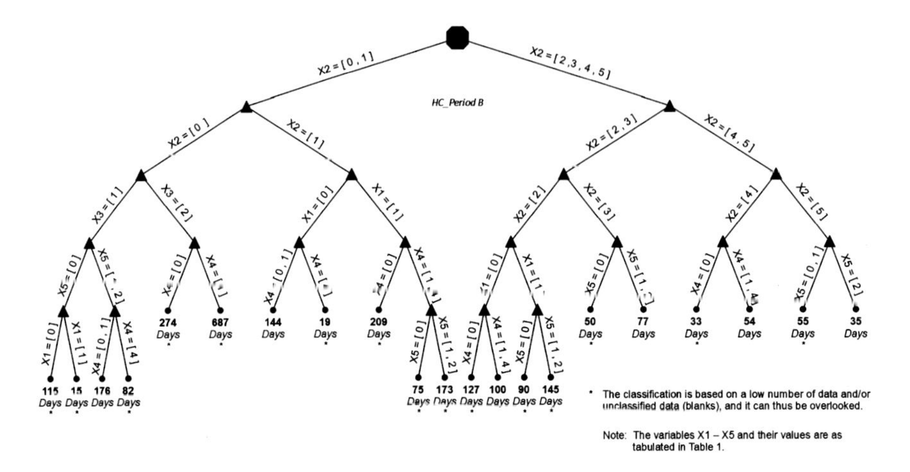

Figure 3.18 Regression tree of "HC" Pipe class incidents for time period "B."

(ii) Next action for a Plastic Connection-pipe element which didn't experience any previous breaks is expected in 155 days. This result is in line with the previous one since plastic pipes live longer.

(iii) Next repair action for a Connection-pipe element that has 1–4 previous breaks is expected in 101 days.

(iv) Next action for a HC element that has 5–8 previous breaks is expected in 61 days. This result is in line with the previous conclusions because as the NOPB increases, the TTF decreases dramatically.

(v) Next action for a HC element that has 9–16 previous breaks is expected in 38 days. This result is in line with the previous statements because as the NOPB increases, the TTF decreases dramatically.

(vi) Next action for a HC element that has 16+ previous breaks is expected in 24 days. This result is in line with the previous statements because as the NOPB increases, the TTF decreases dramatically.

A regression tree diagram for the "HC" Pipe class incidents of the time period "B" (01/01/2003–30/03/2008; 1914 days) is depicted in Fig. 3.18. The most important information stemming from the analysis is as follows:

(i) Next action for a Connection-pipe which didn't experience any previous breaks is expected in 176 days.

(ii) Next repair action for a Connection-pipe element that has 1–4 previous breaks is expected in 144 days. This result is in line with the previous statement because as the NOPB increases, the TTF decreases dramatically.

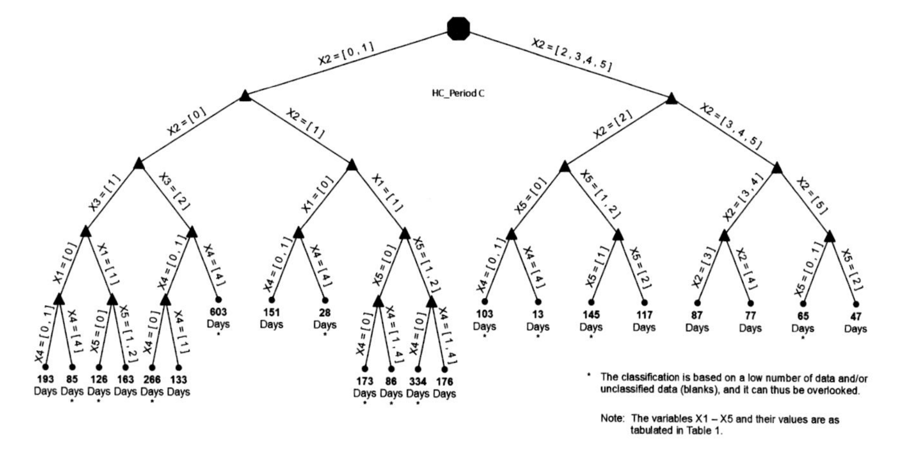

Figure 3.19 Regression tree of "HC" Pipe class incidents for time period "C."

(iii) Next repair action for an HC element that has 5–8 previous breaks is expected in 90 days. This result is in line with the previous statements because as the NOPB increases, the TTF decreases dramatically.

(iv) Next action for an HC element that has 9–12 previous breaks is expected in 77 days. This result is in line with the previous statements because as the NOPB increases, the TTF decreases dramatically.

(v) Next repair action for a HC element that has 13–16 previous breaks is expected in 54 days. This result is in line with the previous statements because as the NOPB increases, the TTF decreases dramatically.

(vi) Next repair action for a plastic element that has more than 16 previous breaks is expected in 35 days. This result is in line with the previous statements because as the NOPB increases, the TTF decreases dramatically.

From the aforementioned results it can be deduced that the inclusion in the analysis of data associated with the time period from 16/08/2005 to 30/03/2008 has increased the TTF. Thus, the aforementioned time period did not negatively affect the condition of house connections in the case-study WDN.

A regression tree diagram for the "HC" Pipe class incidents of time period "C" (01/01/2003–31/12/2010; 2919 days) is depicted in Fig. 3.19. Since time period "C" includes additional data and covers a longer period, it is expected that the TTF will significantly increase compared to the other two periods (periods "A" and "B"). The most important information stemming from the analysis is as follows:

(i) Next repair action for a Connection-pipe which didn't experience any previous breaks is expected in 192 days.

(ii) Next replacement action for a pipe element which didn't experience any previous breaks is expected in 163 days.

(iii) Next action for a Connection-pipe fitting which didn't experience any previous breaks is expected in 133 days. This result is in line with the other two since pipes live longer than fittings.

(iv) Next repair action for a Connection pipe element that has 1–4 previous breaks is expected in 151 days. This result is in line with finding (i), because as the NOPB increases the TTF decreases.

(v) Next replacement action for a HC element that has 1–4 previous breaks is expected in 176 days.

(vi) Next action for a plastic HC element that has 5–8 previous breaks is expected in 117 days.

(vii) Next action for a HC element that has 9–12 previous breaks is expected in 87 days.

(viii) Next action for a HC element that has 13–16 previous breaks is expected in 77 days. This result is in line with the previous one because as the NOPB increases, the TTF decreases. Next action for a plastic HC element that has more than 16 previous breaks is expected in 47 days. This result is in line with the conclusion (vi) because as the NOPB increases, the TTF decreases

From the above results it can deduced that the introduction of data associated with the time period from 31/03/2008 to 31/12/2010 (IWS period) has not increased significantly the TTF, and thus the IWS period negatively affects the condition of the HC elements within a WDN.

3.3.3 Regression Trees – Discussion

The work presented in the preceding sections examines, by use of regression trees, the behavior of WDNs during a period of IWS. It investigates the effects of this policy on the behavior, condition, and failure rate of the pipeline system by studying the change in the rate of occurrence of failures, using both regression trees and survival analysis. The analyses show that during the period of implementation of IWS and right after that, there is a significant increase in the number of water-leak incidents and a deterioration of the network condition, indicating that IWS operations negatively impact the vulnerability of WDNs. This is particularly evidenced in the case of house-connection pipes ("HC" Pipe class).

Regression-tree analysis showed that the "WM" Pipe class elements are not significantly affected by the IWS policies. The time to failure for this element class is significantly increased as the length of the time period is increased, which means that vulnerability rate of this class is not increasing during the IWS period, but remains at expected levels. From the other hand, the regression tree diagrams associated with the "HC" Pipe class elements show that this class of elements is negatively affected by the IWS policies. The time to failure for this element class is not significantly increased as the length of the time period is increased (importing data of the IWS period), which means that the vulnerability rate of the "HC" Pipe class elements increases during the IWS period.

3.4 CONCLUDING REMARKS

Chapter 3 is a natural progression of the analysis presented in Chapter 2, discussing how the WDN in study performs under intermittent water supply (IWS) operations. As aforementioned in Chapter 2, through the developed mathematical models a correlation of intermittent water supply periods with the frequency of future leak incidents can be made (Chapter 3), and the effects of the IWS operations on the sustainability of networks and on the increase of the number of leakage incidents can be examined.

An evaluation and comparison of data over the IWS period has shown that at the same time that the intermittent water supply policy was enacted the number of leakage incidents was increased dramatically. Specifically, an increase in the occurrence of leakages (per day) was observed when water was circulated in the pipelines. At the end, a comparison between the results for the Water Board of Nicosia (WBN) with those of the Water Board of Limassol (WBL) proved that each water distribution network, because of their particular characteristics and properties, exhibited different behavior over time and under varying operating conditions. Nonetheless, not only can the detrimental impacts of IWS on the condition of WDNs be presumed, but such behavior can also be modeled and the reduced survival levels be forecasted.

CHAPTER 4

Topological Vulnerability

The reliability of a water distribution network (WDN) is a function of several time-invariant and time-dependent factors affecting its components and connectivity... In terms of network topology, this attribute even though generally thought as time-invariant it actually is time-dependent, as the paths in a water distribution network change over time based on the hydraulics in the network (water demand and water pressure/flow alter the way water flows in the piping network).

Agathokleous, A., et al. (2016),
Journal of Water Resources Management.

4.1 INTRODUCTION

One of the biggest challenges faced by water authorities is the management of their WDN's topology, especially in light of the increasing network complexity and of the increasing rate of pipe deterioration. The term "topology" refers to the arrangement of the various WDN elements (pipes, junctions, valves, etc.), their connectivity (and sizing), and the routing of the networks.

At the outset of this increasing WDN complexity, increasingly important has become the ability to assess, by use of effective and easy-to-use management tools, a network's vulnerability and its resilience to catastrophic events. Such management tools include and integrate a number of technologies: decision support systems (DSS), monitoring devices and sensors, hydraulic simulation models, extended database management systems, geographical information systems (GIS), and mathematical models that simulate the behavior of a WDN through time (Fig. 4.1).

The development and deployment, though, of an integrated tool for the condition assessment of WDNs is an expensive, time-consuming, and multifaceted procedure which, despite its complexity, most of the time fails to reflect in great detail the actual real-time behavior of network operation. Further, the development, the operations and the maintenance of the

Urban Water Distribution Networks
DOI: http://dx.doi.org/10.1016/B978-0-12-813652-2.00004-9

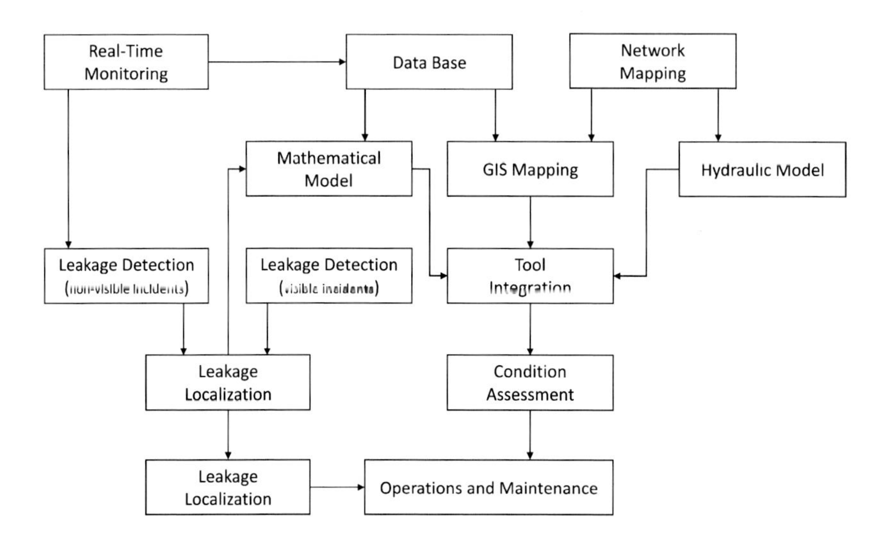

Figure 4.1 Schematic framework of a WDN management system.

underlying hydraulic and mathematical tools require specialized knowledge not typically found within the water agencies, as well as the analysis of real-time WDN operational data. These difficulties hinder the applicability of the models and give rise to the need for automated early-stage appraisal systems of a WDN's vulnerability based on the WDN's topological characteristics, which could then be easily amended with additional information for a more holistic vulnerability model. Such additional information are the hydraulic characteristics and the past network performance under normal and abnormal operating conditions.

The goal is thus the investigation of alternative approaches for early-stage topology-based vulnerability assessment, and the development of a methodology that: (a) it is cost efficient; (b) its use does not require hydraulic and mathematical knowledge and advanced software skills; (c) it does not require operational data beyond the topography of the WDN; (d) it can lead to reliable conclusions within a short time; (e) it can also be used as a management tool for designing the operation mode of a WDN.

The succeeding sections of Chapter 4 propose such a methodology, based on the principle that the sections of a WDN which are hubs and support large segments of the pipeline system are subject to stresses whose size is of greater intensity and frequency relative to other network sections, thus being more prone to failure (i.e., of higher vulnerability). The importance

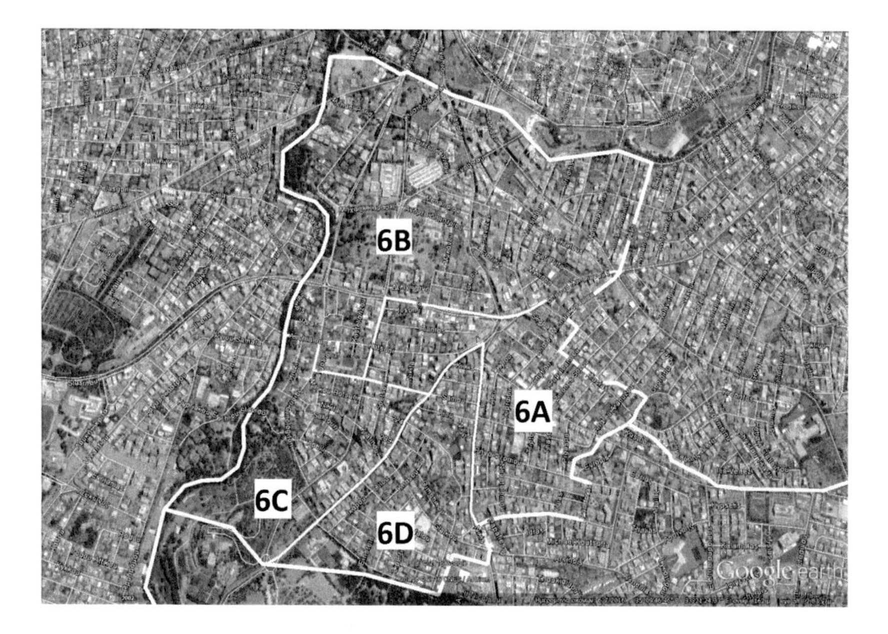

Figure 4.2 Case-study network.

of each node relative to the other nodes, with regards to its contribution to the provided network service, is computed by factoring-in the topology of the WDN as described by a betweenness centrality (BC) metric. Finally, for validating the applicability and accuracy of the proposed model, the deduced by the model spatial risk levels (a BC-weighted network graph) are compared to corresponding maps created by the use of actual WDN failure data (the observed pipe bursts in the network).

4.1.1 Case-Study Network

The case-study network utilized in the analysis is a District Metered Area (DMA) from the WDN of the city of Nicosia (Cyprus). The studied DMA (DMA 6, Fig. 4.2) is divided into 4 subdistrict metered areas (sub-DMAs) and it is remotely monitored through a supervisory control and data acquisition (SCADA) system across the DMA, which continually collects and transmits operational information.

All leakage incident data reports since 2003 had been maintained in a specially designed database. The data utilized for verification and for validation purposes of the proposed model covers the time period from

01/01/2003 to 31/12/2010, and includes 548 incidents associated with water mains (WM).

4.2 STATE OF KNOWLEDGE

4.2.1 Condition Assessment Models for WDNs

To date, several mathematical models simulating the operations and condition of WDNs have been developed. Initial research efforts were based on the development of a single-objective mathematical model expressing the failure pattern or system reliability, while subsequent studies expanded the work by presenting multiobjective failure models. More recent efforts focused on the development of models simulating WDN behavior. Finally some recent research activities focused on abnormal operating conditions due to exogenous factors that affect the condition of the network.

With respect to topological vulnerability (Fig. 4.3), relevant are the works of, among others, Yang et al. [175], Haimes et al. [68], Xu and Goulter [171], Tanyimboh et al. [157], Michaud and Apostolakis [114], Pinto et al. [137], Martínez-Rodríguez et al. [110], Aydin et al. [14], Shuang et al. [153], Christodoulou and Fragiadakis [35].

In summary,

- Yang et al. [175] presented a methodology to analyze the reliability of a water distribution network, based on the impact of link failures on source–demand connectivity, which is used as a measure of the mechanical reliability of the network. The mechanical reliability index is computed using the minimum cut-set method and following a four-stage process (for source–demand pairs, for individual demand nodes, for a group of demand nodes, and for all demand nodes in the system). The results from the reliability analysis are then used to locate crucial pipelines in the network whose failure will severely impair the source–demand connectivity.
- Haimes et al. [68] reviewed needs and opportunities to reduce the vulnerability of water distribution networks systems to potential physical, chemical–biological, and cyberattacks to these network. A hierarchical holographic model was introduced, including security, robustness, resilience, and redundancy.
- Xu and Goulter [171] presented an approach for reliability-based optimization of water distribution networks utilizing a probabilistic hydraulic model to account for uncertainty in nodal demands and pipe

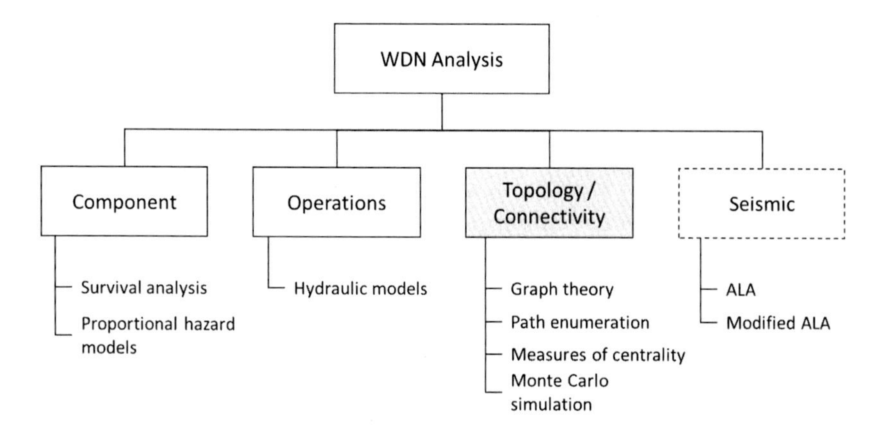

Figure 4.3 Water distribution network analysis (schematic).

capacity. Their model uses a first-order reliability-method-based algorithm to compute approximate values of the capacity reliability of water distribution networks, defined as the probability that the nodal demand is met at the prescribed minimum pressure for a fixed network configuration under random nodal demands and random pipe roughnesses. The model also incorporates a strategy for identifying the critical nodes on which the reliability constraints are imposed.

- Tanyimboh et al. [157] proposed a method for interpreting the results of demand-driven network analysis as an approximation to pressure-driven simulation. The researchers' formulation can be used to calculate the reliability of either an entire water distribution network or that of the individual demand nodes.
- Michaud and Apostolakis [114] presented a scenario-based methodology for the ranking of the elements of a water-supply network according to their value to the network's owner, based on multi-attribute utility theory and a graph theory-based network analysis algorithm. The researchers modeled the system as a network and, by use of simulated scenarios, evaluated the consequences of the failure of each of the network's elements to the supply level, considering the capacity of the elements' connection to the available resources.
- Pinto et al. [137] proposed a theory of vulnerability of water pipe networks (TVWPN), to help design water pipe networks which are more robust against damage to the pipelines by use of an analysis of the network's topology.

- Martínez-Rodríguez et al. [110] noted that the concept of reliability had been introduced in an attempt to quantitatively measure the possibility of maintaining an adequate service for a given period, and that numerous researchers had considered reliability as a measure of redundancy. The researchers then discussed why reliability cannot be considered a measure of redundancy and proposed two quantitative indices (reliability and tolerance) for measuring expected network behavior.
- Aydin et al. [14] used performance criteria (resiliency, reliability, and vulnerability) for evaluating sustainability indices for each node, in terms of nodal pressure and water age. Their calculation approach identified problematic nodes individually, and the sustainability indices for each zone were calculated by weighting each node by its demand.
- Shuang et al. [153] evaluated the nodal vulnerability of WDNs under cascading failures, by use of monitored pressures in different nodes and flows in different pipes which are then used to estimate the network topological structure and the consequences of nodal failure. The researchers note that the WDN's nodal vulnerability is evaluated based on the connectivity loss of the topological structure, and that their results show that the network vulnerability should be evaluated with the consideration of hydraulic analysis and network topology.
- Christodoulou and Fragiadakis [35] presented a methodology for the seismic assessment of the reliability of water distribution networks based on general seismic assessment standards, localized historical records of critical risk-of-failure metrics, and the topology of a WDN. The network reliability was assessed using Graph Theory and Monte Carlo simulation (MCS).

4.2.2 Betweenness Centrality

The centrality metric is the most widely used indicator for identifying important nodes in a network, based on topological characteristics. Centrality is given in terms of a real-valued function for each network node (vertices of a graph), and the resulting values provide a ranking which identifies the most important nodes [17,18]. The ease in calculation of the centrality metric depends on the network size (number of vertices and edges), the connectivity of vertices (number of edges linked on each vertex, directed/undirected edges and connected/disconnected vertices) and the weights assigned to the edges [123].

The betweenness centrality index is an indicator of network vertices that have high contribution to the transfer of items within the network, and it is equal to the degree of which a vertex falls on the shortest path between the other vertices [63]. In essence, betweenness centrality is related to a network's connectivity, in so much as high betweenness vertices have the potential to disconnect graphs if removed. Mathematically, the betweenness centrality (C_B) of a vertex, u, is defined by [10] as:

$$C_B(u) = \sum_{s \neq u \neq t} \frac{\sigma_{st}(u)}{\sigma_{st}} \tag{4.1}$$

where σ_{st} is the total number of shortest paths from vertex s to vertex t and $\sigma_{st}(u)$ is the number of those paths passing through vertex u.

It should be noted that, as implied by Eq. (4.1) and the indices of the summation in it, the betweenness centrality of a node scales with the number of pairs of nodes. Therefore, the calculation may be rescaled and normalized without a loss of precision in the range of $C_B \in [0, 1]$ by dividing through by the number of pairs of nodes not including u. The division is done by $(N-1)(N-2)$ for directed graphs and by $(N-1)(N-2)/2$ for undirected graphs, where N is the number of nodes.

Calculation of the betweenness centrality for all the vertices in a graph involves calculating the shortest paths between all pairs of graph vertices, typically by use of either the Floyd–Warshall algorithm or, in the case of unweighted graphs, Brandes algorithm.

4.3 ANALYSIS, RESULTS, AND DISCUSSION

4.3.1 Pipeline System of DMA 6 and Vulnerability During Continuous Water Supply

The WDN in study is first analyzed topologically, with BC metrics computed for every node in the network and geographically mapped. The result of the analysis is then spatially mapped using a weighted graph, showing the levels of the BC index across the case-study DMA (Fig. 4.4A).

The areas of high BC values (shown in red (dark grey in print version)) indicate the nodes through which a high number of origin-to-destination paths pass, when considering only the topology of the network. As expected, BC numbers are lower on the periphery of the network, and increase towards its center.

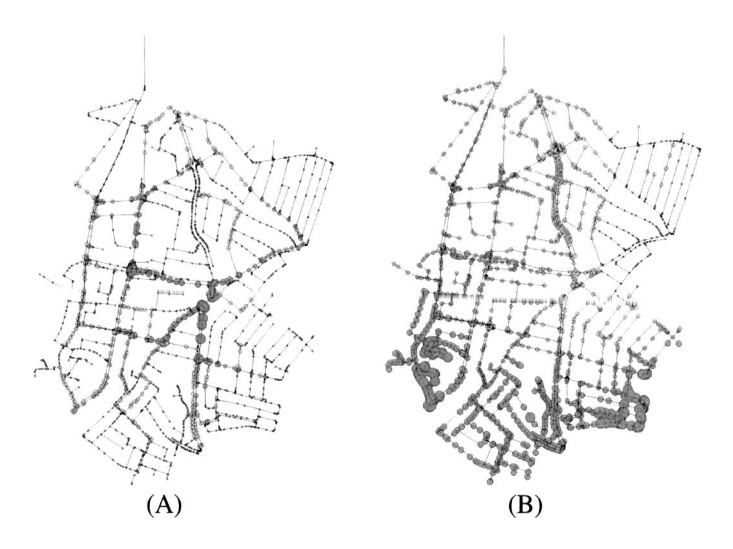

Figure 4.4 Betweenness centrality weighted graph for: (A) network topology, (B) continuous water supply (CWS) operations.

The WDN in study is then analyzed under normal (continuous water flow) operating conditions (data period before March 2008). For this, the continuous water supply (CWS) mode, the BC metrics are computed in conjunction with the underlying hydraulic model for the DMA. That is, the hydraulic model and the deduced nodal pressures and pipe water flows in essence dictate the origin-to-destination paths for every calculation and thus affect the BC computation. As with the previous case, the result of the analysis is then spatially mapped by use of a heatmap, showing the levels of the BC index across the case-study DMA (Fig. 4.4B).

A comparison between Fig. 4.4A and Fig. 4.4B shows the difference in the spatial allocation of the nodal BC indices and of the resulting network vulnerability, when the hydraulic behavior of the DMA under continuous water supply operations is considered. Now the BC values are higher along major water mains, as the deduced water flows at the nodes along these pipes create more water pathways passing through these nodes. Noteworthy is also the fact that the change in network vulnerability (Fig. 4.4B) is actually obtained by opening/closing 9 valves (nodes) in the original network (Fig. 4.4A). This change in vulnerability indicates the effects of hydraulics on the network performance.

4.3.2 Vulnerability Heatmap of Intermittent Water Supply (IWS)

When in abnormal operating conditions (such as IWS operations), the network in study exhibits a different behavior, with the BC indices varying compared to the ones deduced for the CWS operations (Fig. 4.4B). Under IWS operations (data period after March 2008), the DMA in study is subdivided into four sub-DMAs, each receiving water for 12 hours every 48 hours. During this operation mode the BC indices change, depending on which sub-DMA is active (Fig. 4.5) and how the varying water flow affects the water pathways and the origin–destination pairs in the network.

As with the CWS case (Fig. 4.4B), the BC indices are computed in conjunction with the underlying hydraulic model for the sub-DMA, and the deduced water flows in the piping network dictate the origin-to-destination paths in the network, thus affecting the computation of the nodal BC indices.

4.3.3 Spatial Distribution of the Failure Incidents

The results of the shift in the BC indices as a result of the change in the water supply operations (from CWS to IWS) can be seen in Fig. 4.6, depicting the failure incidents in the case-study WDN during the time periods of CWS and IWS.

As shown in the resulting heatmaps of Fig. 4.7, the spatial distribution of the failure incidents changes as the mode of operations changes, and the center of gravity shifts as the failure incidents cluster around regions of high BC indices.

This spatial shift and the observed clustering can be attributed to the "reorganization" of the WDN, as evidenced by the BC weighted graphs, stemming from the changes in water flow/pressure across the network.

A spatial analysis of the distribution of failure incidents during CWS and IWS operations by use of the Nearest Neighbor and the Spatial Autocorrelation methods attests to the clustering of incidents. In the case of the Nearest Neighbor Analysis, the observed spatial distribution is compared against a synthetic distribution characterized by complete spatial randomness (CSR), with the null hypothesis being that the observed distribution is a CSR pattern. Negative z-scores indicate clustering where positive scores indicate dispersion or evenness. The nearest neighbor index, defined as the ratio of the observed mean distance to the expected mean distance (for the CSR pattern) between neighboring incidents, points to clustered patterns when its value is below unity, and to dispersed data otherwise.

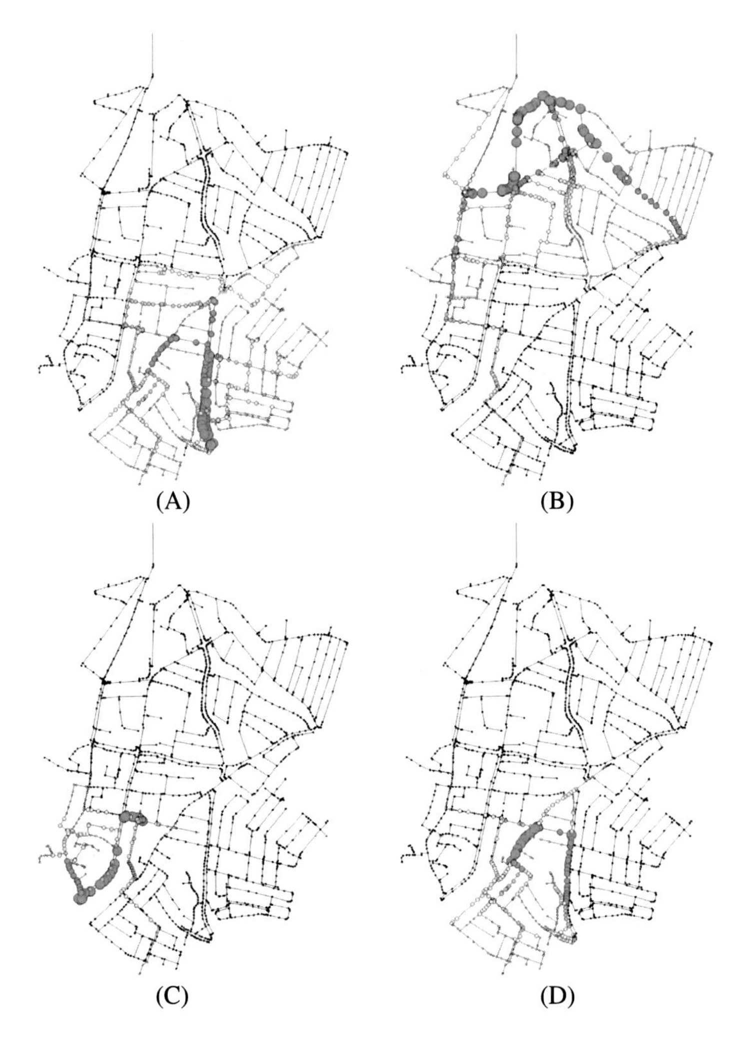

Figure 4.5 Betweenness centrality weighted graph for the intermittent water supply operation: (A) sub-DMA 6A, (B) sub-DMA 6B, (C) sub-DMA 6C, (D) sub-DMA 6D.

Similarly, spatial autocorrelation (typically measured by use of Moran's Index) measures the correlation of a variable with itself through space and the degree to which one point is similar to other nearby points. Positive spatial autocorrelation ($\leq +1$) is when similar values cluster together in a map, whilst negative spatial autocorrelation (≥ -1) is when dissimilar values cluster. In this method, the z-score indicates the statistical significance given the number of features in the dataset, and the p-value indicates the likelihood that the observed pattern is the result of random choice.

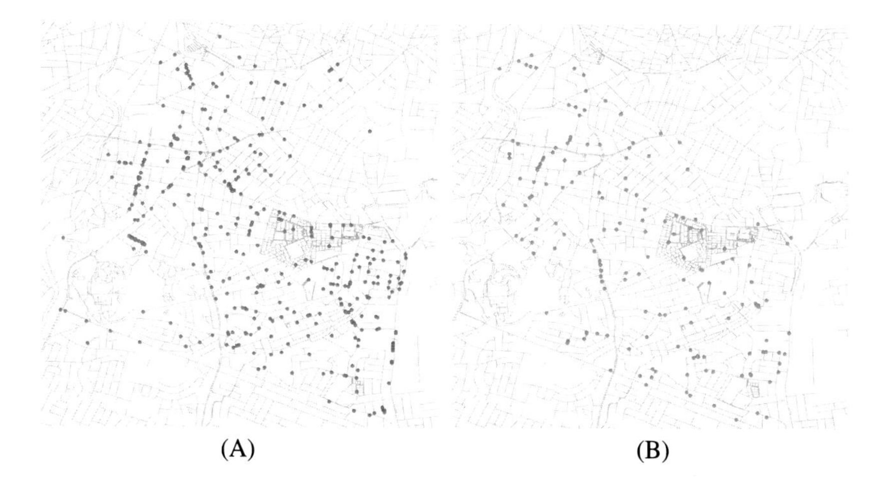

Figure 4.6 Failure incidents (pipe breaks) in the case-study WDN: (A) during CWS operations (before March 2008); and (B) during IWS operations (after March 2008).

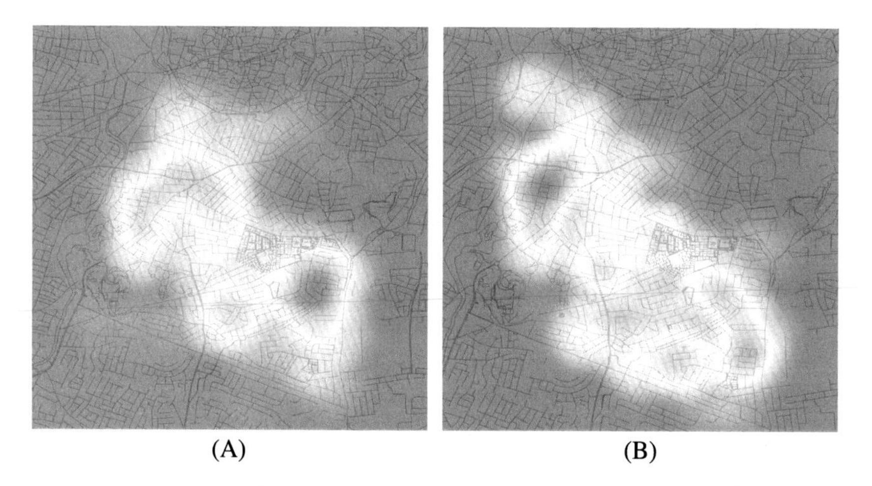

Figure 4.7 Heatmaps of failure incidents (pipe breaks) in the case-study WDN: (A) during CWS operations (before March 2008); and (B) during IWS operations (after March 2008).

The results from the aforementioned analyses attest to both the CWS and IWS datasets exhibiting clustering, having nearest neighbor indices smaller than one (0.444 and 0.745, respectively), with corresponding negative z-scores (−20.500 and −6.130, respectively). Further, the spatial autocorrelation analysis shows that there is less than 3% (p-value = 0.027) and less than 1% likelihood (p-value = 0.000) that the CWS and IWS clustered patterns, respectively, could be the result of random chance.

4.4 CONCLUDING REMARKS

The work presented in the preceding sections provides evidence of the links between the behavior of a WDN under varying operating conditions, the betweenness centrality indices of the network's nodes, and the network's vulnerability. Given these links, WDN operators can forecast a WDN's behavior under several scenarios and plan for them, optimizing the behavior of the WDN and minimizing its vulnerability against endogenous and exogenous threats, without the need for dynamic hydraulic models.

Further research work on statistically documenting the aforementioned links between the observed WDN fragility and its betweenness centrality characteristics, by use of a larger dataset (higher number of incidents, and longer time periods), various network topologies and spatial autocorrelation analysis, should further attest to these findings.

CHAPTER 5

Vulnerability Assessment of Water Distribution Networks Under Seismic Loads

The most commonly available fragility formulations for pipelines are based on damage and repair data collected after past earthquakes on segmented pipelines, and are primarily empirical.

Rajah, S., et al. (2012),
Pipelines Conference 2012.

5.1 INTRODUCTION

Past earthquakes (e.g., the New Zealand sequence) have shown that, following a strong earthquake, damage of the lifelines may cause a series of problems immediately after the event, while the residential, commercial, and industrial activities may be disrupted for a long time, causing severe economic losses. Such losses may be either direct or indirect. Direct losses are related to the cost of repair, while indirect losses usually depend on the way the economy is affected by the disruption of the lifeline and are usually larger in value for more developed communities.

The reliability of WDNs, though, is an amalgam of risk-of-failure factors. As discussed in previous chapters, when referring to vulnerability assessment and analysis of water distribution networks one should note that the topic is multifaceted and encompasses as a minimum the following three aspects: (1) component analysis (i.e., pipes, valves, reservoirs, etc.); (2) operations (i.e., the operating parameters of a network, such as water pressure and flow); and (3) topology and connectivity (i.e., the number of arcs/nodes in a network, elevations, arc lengths, etc.). For every aspect, a number of possible analysis methods exist. The problem is compounded, though, when the WDN in study is located in earthquake-prone regions and thus seismic vulnerability needs to be added to the mix of the other three sources of vulnerability (Fig. 5.1).

The chapter discusses such factors and their interactions, and proposes a methodology which acknowledges that the "Repair Rate" and the re-

Urban Water Distribution Networks
DOI: http://dx.doi.org/10.1016/B978-0-12-813652-2.00005-0

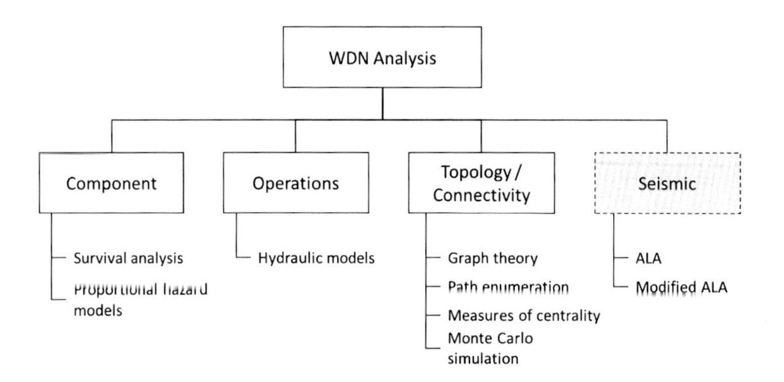

Figure 5.1 Water distribution network analysis (schematic).

sulting probability-of-failure metric proposed by the ALA guidelines are only a starting point, as these metrics are based on statistical observations from the performance of a generic UWDN whose historical nonseismic performance is not known. The proposed methodology then improves on these metrics by introducing the utilization of localized nonseismic performance data through survival analysis. A spatial analysis is then employed to demonstrate the effects of such nonseismic performance data on the seismic reliability of the network.

Presented herein is a methodology for the seismic risk assessment of water distribution (pipe) networks based on general seismic assessment standards, as per the American Lifelines Alliance (ALA) 2001 guidelines [5], and localized historical records of critical risk-of-failure metrics pertaining to the specific WDN under assessment. The assessment of reliability incorporates data of past nonseismic damage, the vulnerabilities of the network components against seismic loading, and the topology of an WDN. Historical data obtained using records of pipe burst incidents are processed to produce clustered "survival curves", depicting the pipes' estimated survival rate over time. The survival curves are then used to localize the generalized fragility values of the network components (primarily pipes), as assessed using the approach suggested by the ALA guidelines. The network reliability is subsequently assessed using Graph Theory, while the system reliability is calculated using Monte Carlo simulation. The methodology proposed is demonstrated on a simple small-scale network and on a real-scale district metered area (DMA). The proposed approach allows the estimation of the probability that the network fails to provide the desired level of service and allows for the prioritization of retrofit interventions and of capacity-upgrade actions pertaining to existing water pipe networks.

5.2 NETWORK VULNERABILITY

The following sections provide, as a prelude to the sections on seismic vulnerability of WDNs, a brief overview of related sources of vulnerability (as depicted in Fig. 5.1) and of the methods used for evaluating them.

5.2.1 Component-Based Analysis

As aforementioned in Section 1.6.1 (p. 13), most of the reported research work on the vulnerability of WDNs focuses on component analysis, i.e., the estimation of either the probability of failure or of the time-to-failure for the various pipe segments in the network. The most commonly used methods for such an analysis are the proportional hazards model, the survival analysis (see section below), neurofuzzy systems, and data mining techniques leading to intelligent decision-support systems. Even though what is reported upon are general rules and knowledge related to vulnerability (such as the prominent risk-of-failure factors) the estimation of vulnerability is then localized based on the historical performance, operations and maintenance of each network. To this effect, the "*number of observed previous breaks (NOPB)*" risk factor is of great importance to the appraisal of the vulnerability of a system component (e.g., a pipe).

Aparently, the component-based approach fails to provide vulnerability metrics on a node-to-node basis since it does not enumerate the various paths in a network, or does not incorporate the hydraulic model of the network.

Survival Analysis

Survival analysis is a branch of statistics dealing with deterioration and failure over time and involves the modeling of the elapsed time between an initiating and a terminal event [70]. In the case of piping networks, an initiating event can be the installation of a pipe, a water-leak observation, or the start of a pipe treatment. Cases of terminal events can be a relapse of a previous leak, a fix, or a failure. The method is based on estimating the reliability of a component and its lifetime, subject to multiple risk factors. The aim is to provide answers on the population fraction (e.g., pipes) that survives past an expected lifetime, on the effect of the various risk factors on the system's lifetime, and on the probability of survival and the expected mean time to failures [70,74]. The data values used in the analysis combine both complete and censored observations. In the former case, a terminal event is thought to have occurred, while in the latter case a terminal event has not yet occurred. A terminal event is assumed to occur just once for every subject.

In mathematical terms, a pipe's survival function S for elapsed time T until the occurrence of a pipe failure, is given by the expression.

$$S(t) = \int_{T}^{\infty} p(t)\,dt = 1 - P(t). \tag{5.1}$$

Thus, the survival function is the probability that the time to failure is longer than some specified time t. Furthermore, $P(t)$ is the cumulative distribution function that denotes the probability that a pipe survives until time t and $p(t)$ is the corresponding probability density function. The survival function S is numerically calculated using kernels (such as the Epanechnikov kernel and the Kaplan–Meier estimator (Kaplan and Meier 1958). The Kaplan–Meier estimator is of particular importance because it is nonparametric, thus relying on data rather than on analytical equations and probability density functions in order to produce the survival curves. Another important advantage of the Kaplan–Meier curve is that the method can take into account both left- and right-censored data.

To illustrate the method, a hypothetical set of 15 events is shown in Table 5.1, for pipes of two different material types and of varying age. The events' outcome ("failure", "survival", "unknown"), and the computed survival times are also tabulated. The survival times, including the censored values (indicated with a "+" sign in Table 5.2), are then computed and ordered in increasing duration. The dataset is then clustered by material type and the data for pipes of material type "2" are used to compute the survival values and the cumulative proportion surviving, $S(t)$. In the case of a censored time, the proportion surviving is taken as 1.

In terms of piping networks, the survival function has been shown to be dependent on several factors, the most important of which are the "number of observed previous breaks (NOPB)", the age and the material of the pipes [33]. These risk factors have been studied extensively [27,33] both when acting separately or when acting in tandem. For example, a survival analysis reported by Christodoulou and Ellinas [27] of an urban water distribution network under abnormal operating conditions revealed almost

Table 5.1 Sample pipe-breakage data used in demonstrating the survival analysis method.

Pipe ID	Survival time (months)	Outcome	NOPB	Material type	Pipe age (years)
1	4	Failure	0	2	35
2	4	Failure	0	2	39
3	7	Failure	0	2	45
4	8	Failure	0	2	36
5	9	Unknown	0	2	26
6	11	Failure	0	1	35
7	12	Survived	0	2	32
8	12	Failure	0	2	30
9	15	Failure	0	1	31
10	18	Unknown	0	1	33
11	25	Failure	0	2	26
12	28	Survived	0	1	33
13	40	Failure	0	1	28
14	58	Failure	0	1	19
15	75	Survived	0	1	21

identical survival curves for the network mains and its house connections, but when clustered by the "number of observed previous breaks (NOPB)" the survival curves varied substantially.

A typical set of survival curves is shown in Figs. 5.2A and 5.2B. The curves have been derived from a real network's performance data reported upon in the literature [27]. The former figure refers to pipes of different material, while the latter figure specifically refers to asbestos cement (AC) pipes. Additionally, the data in Fig. 5.2B have been clustered according to the "number of observed previous breaks (NOPB)", with four survival curves derived based on NOPB clusters of size "zero", "small" ($1 \leq NOPB \leq 4$), "medium" ($5 \leq NOPB \leq 8$), and "large" ($9 \leq NOPB$). As Fig. 5.2B depicts, even in the "NOPB = 0" case, a pipe will have to be eventually replaced after approximately 20,000 days ($\approx$ 55 years), while a pipe that has already broken more than 8 times[1] is not expected to survive more than 18 years. Moreover, a pipe that has broken at least once is very vulnerable, since its survival curve reduces rapidly compared to a pipe that has never been damaged.

[1] The NOPB metric in this case refers to street-length pipe runs; hence the high values.

Table 5.2 Sample survival analysis calculations.

Pipe ID (material type "2")	Survival time (months)	Number known to have survived (r_i)	Failures (f_i)	Proportion surviving (p_i)	Cumul. proportion surviving ($S(t)$)
	0				1.000
1	4	8			
2	4	8	2	$(8-2)/8=0.750$	$1.000 \times 0.750=0.750$
3	7	6	1	$(6-1)/6=0.833$	$0.750 \times 0.833=0.625$
4	8	5	1	$(5-1)/5=0.800$	$0.625 \times 0.800=0.500$
5	9+				
7	12	3	1	$(3-1)/3=0.667$	$0.500 \times 0.667=0.333$
8	12+				
11	25	1	1	$(1-1)/1=0.000$	$0.333 \times 0.000=0.000$

The "+" sign in the "Survival time" column indicates a censored time.

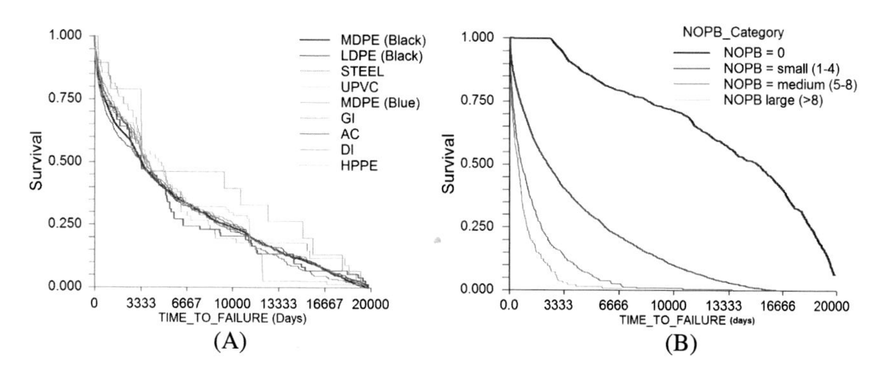

Figure 5.2 Survival curves: (A) for pipes, as a function of material type, (B) for asbestos cement (AC) pipes, as a function of the number of previous breaks (NOPB).

5.2.2 Network Analysis

A network's reliability is a function of not only the reliability of the network's components but also of its topology (connectivity) and operation (hydraulic parameters). Past research has put forward several network-based approaches, such as vulnerability theory [137], graph theory [177], ant colony optimization [34], and Monte Carlo simulation [61]. In general, the reliability of a water pipe network can be calculated if the vulnerability (also termed fragility) of every element of the water network is known. Although UWDNs consist of several elements (pipes, house connections, tanks, pumps, etc.), focus has traditionally been placed on pipes, since they are not only the most important component in a piping network but they are also the most difficult component to inspect and replace, and for which many possible risk-of-failure parameters come into play [27,33,146].

Path Enumeration Method

A direct calculation of the network failure probability, $\widehat{P}_{f,ij}$, can be obtained using the path enumeration [69] or the minimum cut-set [4,81] method. The path enumeration method involves identifying all minimal paths of the network. A minimal path (F_p) is defined as a set of links, which when are all reliable the network functions. A path is minimal if it has no subpaths and therefore each vertex appears in the path once. We use the subscript p for the paths of the path enumeration method. If the mth path between i and j has np pipes then its failure probability is that of a series system and is

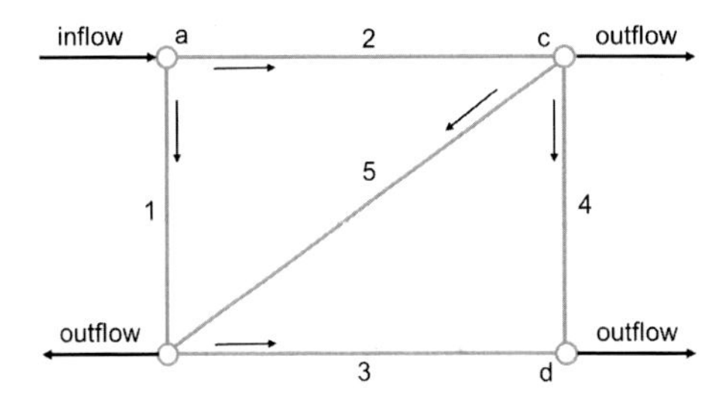

Figure 5.3 Network reliability by use of the Monte Carlo method (sample network 1).

calculated as [113]:

$$P\left(F_{p,m}\right) = 1 - \bar{P}_{f,1} \cap \bar{P}_{f,2} \cap \cdots \cap \bar{P}_{f,np} = 1 - \prod_{n=1}^{np}\left(1 - \bar{P}_{f,n}\right) \qquad (5.2)$$

where $\bar{P}_f$ is the failure probability of a single pipe and $\cap$ denotes intersection.

The direct calculation of the network failure probability, $\widehat{P}_{f,ij}$, involves identifying all minimal paths of a network [69]. If the probability of failure of nodes and links is known, the reliability can be calculated using Eq. (5.2) or other techniques (e.g., [4]). Once the minimal paths between every vertex i and j have been identified, the network reliability between i and j, $\widehat{R}_{f,ij}$, is equal to the probability that at least one path between i and j is successful. Thus, if M is the number of minimal paths between i and j, and $F_{p,m}$ is the mth minimal path, the exact calculation of the failure probability $\widehat{P}_{f,ij}$ is ([113]):

$$\begin{aligned}
\widehat{P}_{f,ij} = {} & P\left(F_{p,1} \cup F_{p,2} \cup \cdots \cup F_{p,m}\right) = \\
& P\left(F_{p,1}\right) + P\left(F_{p,2}\right) - P\left(F_{p,1} \cap F_{p,2}\right) \\
& + P\left(F_{p,3}\right) - P\left(F_{p,1} \cap F_{p,3}\right) - P\left(F_{p,2} \cap F_{p,3}\right) + P\left(F_{p,1} \cap F_{p,2} \cap F_{p,3}\right) \\
& + P\left(F_{p,4}\right) - P\left(F_{p,1} \cap F_{p,4}\right) - P\left(F_{p,2} \cap F_{p,4}\right) - P\left(F_{p,3} \cap F_{p,4}\right) \\
& + P\left(F_{p,1} \cap F_{p,2} \cap F_{p,4}\right) + P\left(F_{p,1} \cap F_{p,3} \cap F_{p,4}\right) + P\left(F_{p,2} \cap F_{p,3} \cap F_{p,4}\right) \\
& + P\left(F_{p,1} \cap F_{p,2} \cap F_{p,3} \cap F_{p,4}\right) + P\left(F_{p,5}\right) - \cdots \\
& = \textstyle\sum_{k=1}^{M} P\left(F_{p,k}\right) - \sum\sum_{k<l}^{M} P\left(F_{p,k} \cap F_{p,l}\right) \\
& + \textstyle\sum\sum\sum_{k<l<m}^{M} P\left(F_{p,k} \cap F_{p,l} \cap F_{p,m}\right) - \cdots
\end{aligned} \qquad (5.3)$$

where $\cap$ denotes intersection and $\cup$ is the symbol for union.

In the case of the network shown in Fig. 5.3 (adopted from [98]) we assume that the reliability ($R_i = 1 - P_{f,i}$) of all pipe segments (i) is the same (i.e., $R_1 = R_2 = R_3 = R_4 = R_5 \equiv R$). The network has four paths: $F_{p,1}, 1-3$; $F_{p,2}, 2-4$; $F_{p,3}, 1-5-4$; $F_{p,4}, 2-5-3$, and their failure probabilities are $P(F_{p,1}) = P(F_{p,2}) = 1 - R^2$, and $P(F_{p,3}) = P(F_{p,4}) = 1 - R^3$, respectively. The probabilities of the intersection terms are also easily calculated, e.g., $P(F_{p,1} \cap F_{p,2}) = 1 - R_1 R_3 R_4 R_5 = 1 - R^4$. The analytical, exact, reliability of the network for a water flow from node "a" to node "d", will be: $\widehat{R}_{s,ad} = 1 - \widehat{P}_{f,ad} = 2R^2 + 2R^3 - 5R^4 + 2R^5$ [98].

The size of Eq. (5.3) grows exponentially as M increases, since 2^{M-1} terms need to be calculated. For example, for a network with 10 paths the probability is the sum of $2^{10} - 1 = 1023$ terms, thus proving that the exact calculation becomes extremely time-consuming even for small networks, since finding all paths becomes too costly and inefficient. To overcome this problem, first- or second-order approximation or bounds of the failure probability can be calculated instead.

A first order approximation of Eq. (5.3) can be obtained as $\widehat{P}_{f,ij} \approx \sum_{k=1}^{m} P(F_k)$, assuming that $P(F_k) \ll 1$ and the $P(F_k \cap F_l)$ terms are negligible. Alternatively, more reliable, second-order bounds can be obtained as [49]:

$$P(F_1) + \sum_{k=2}^{M} \max \left[\left(P(F_k) - \sum_{l=1}^{k-1} P(F_k \cap F_l) \right); 0 \right]$$
$$\leq P_{f,ij} \leq \sum_{k=1}^{M} P(F_k) - \sum_{k=2}^{M} \max_{l<k} P(F_k \cap F_l). \tag{5.4}$$

The above equation provides estimates of $\widehat{P}_{f,ij}$ which are of sufficient accuracy for engineering purposes. Attention should be paid to the fact that the bounds depend on the ordering of the M paths. Approaches for optimal ordering and best results are discussed in [143] and [154]. Instead of second-order bounds, formulas that use third-order bounds also exist in the literature [180]. These also depend on the path numbering and although the bounds are much narrower (compared to Eq. (5.4)), the computing effort to calculate the three-term joint probabilities is significant.

Minimal Cut-Set Method

The minimal cut-set method is another alternative for the analytical calculation of the network's reliability. A minimal cut-set (F_c) of a graph is a cut whose cut-set has the smallest number of elements. In a water network a cut-set is a set of edges (pipes) that when removed the water flow between node i and j is not possible. The failure probability of a cut-set is equal to the probability that all pipes of the cut-set have failed, thus it is a parallel system and its probability is equal to:

$$P\left(F_{c,m}\right) = \bar{P}_{f,1} \cap \bar{P}_{f,2} \cap \cdots \cap \bar{P}_{f,nc} = \prod_{n=1}^{nc} \bar{P}_{f,n} \tag{5.5}$$

where nc is the number of pipes of the mth cut-set. There are various algorithms for finding the minimum cut-set of a graph [65]. According to Eq. (5.5), cut-sets with many pipes have very small probabilities $P(F_{c,m})$ and therefore can be omitted.

If the minimum cut-set method is followed, the network reliability is equal to the probability of occurrence of at least one cut. Moreover, the terms with many unions (e.g., greater or equal to three) of cut-sets can be neglected, since their probabilities are very small. Therefore, if M is the number of minimal cuts between i and j and if we use F to denote the minimum cuts, the exact calculation of the failure probability $\widehat{P}_{f,ij}$ can be obtained using Eq. (5.3), taking into account that

$$P_{f,ij}^{cut} = 1 - P_{f,ij}^{path} = 1 - P\left(F_{c,1} \cup F_{c,2} \cup \cdots \cup F_{c,m}\right). \tag{5.6}$$

Moving back to the network of Fig. 5.3, the network has four possible cuts: $F_{c,1}, 1-3$; $F_{c,2}, 2-4$; $F_{c,3}, 1-5-4$; and $F_{c,4}, 2-5-3$, while the failure probability of every pipe is $\bar{P}_f = p = 1 - R$. The cuts are the same with the paths of the path enumeration method, but they have been selected following a different rationale. The failure probability of each cut is $P(F_1) = P(F_2) = p^2$, $P(F_{c,3}) = P(F_{c,4}) = p^3$ and $P(F_{c,3} \cap F_{c,4}) = p^4$. Cuts $F_{c,3}$ and $F_{c,4}$ can be neglected since they have very small probabilities and thus the failure probability will be $\widehat{P}_{f,ij} = 1 - P\left(F_{c,1}\right) - P\left(F_{c,2}\right) + P\left(F_{c,1} \cap F_{c,2}\right) = 1 - 2p^2 + p^4$. Assuming $\bar{P}_f = p = 0.1$ or $R = 0.9$, the path enumeration method produces an exact value equal to $P_f^{path} = 0.978$, while the minimal cut method yields $P_f^{cut} = 0.980$.

Monte Carlo Simulation

The Monte Carlo simulation (MCS) method is often employed when the analytical solution is difficult or computationally costly to obtain. This is mainly the case of problems of complex nature, or of problems with a large number of variables. Successful applications of MCS can be found in [1,81, 121,139,144,169]. For the problem at hand, if the number of simulation N_∞ is large enough, then an unbiased estimator of the probability of failure will be

$$\widehat{P}_{f,ij} = \frac{1}{N_\infty} \sum_{j=1}^{N_\infty} I(\mathbf{x}_j) \approx \frac{N_H}{N_{\text{MCS}}} \tag{5.7}$$

where $I(\mathbf{x}_j)$ is a Boolean vector indicating successful or unsuccessful simulations; N_H is the number of failed simulations; N_{MCS} is the total number of simulations necessary to obtain an accurate estimation of the probability $\widehat{P}_{f,ij}$. Thus, for the calculation of $\widehat{P}_{f,ij}$ a sufficient number of N_{MCS} independent random samples is produced using a specific probability density function for each component of the array $\mathbf{x}$. If a given accuracy δ_0 is required, the sample size can be approximated as [170]:

$$N_{\text{MCS}} = \frac{1}{P_{f,ij}\delta_0^2}. \tag{5.8}$$

Thus, if the desired accuracy is $\delta_0 = 10\%$ and the probability sought is of the order of 0.01, the required sample size N_{MCS} is $1/(0.1 \times 0.01^2) =$ 100,000 simulations. Eq. (5.8) is a good indicator stating that we must have a sufficient number of "failed" simulations, or in other words the nominator N_H in Eq. (5.7) must be sufficiently large in order to have a reliable estimation of P_f.

For WDN problems, the reliability estimations were not found sensitive to the network size. However, when the dimension of the problem is large, depending on its complexity, the necessary number of simulations may vary and thus a more elaborate sampling scheme may be necessary. In all, significant computational effort may be required, depending on the order of the probability sought and the properties of the problem at hand.

The Monte Carlo calculation is based on reducing the network topology (i.e., neglecting network paths on which pipe segments are assumed as failing) and thus decreasing the total number of paths considered. Its basic steps are given below, while a more detailed algorithm is developed in subsequent sections. It should be noted that the method used for calculating the node-to-node probability of failure (and thus the corresponding reliability) is independent of the number of breaks per pipe segment.

- Map UWDN as a directed graph, with pipes represented as arcs and junctions represented as nodes
- For various earthquake scenarios (M_w, PGV, PGD), generate N_{MCS} simulations
  - Assign failure probabilities, p_f, to each arc
  - Based on a pipe's p_f determine if the pipe fails (1) or not (0)
  - Enumerate all ij node-to-node paths excluding the arcs that failed
  - If no path exists between input–output nodes, increase the "failed simulations" counter (N_U) by 1
- Calculate $P_{f,ij} = N_{failed}/N_{MCS}$

The flowchart of the Monte Carlo method is shown in Fig. 5.6. For every simulation a state vector is produced. In this vector, two states can be considered for every pipe: *0-state*, which refers to a failed state with probability of $\widehat{P}_{f,ij}(t)$ (Eq. (5.13)) and *1-state* that corresponds to success with probability $1 - P_{f,ij}(t)$. Once a state vector is obtained, the failed pipes are removed from the network. Using common graph algorithms we can determine whether a path between vertices i and j exists, thus allowing water flow delivery from node i to node j. In all our applications, a standard Dijkstra algorithm [65] gave quick and robust calculations. If at least one path exists, the simulation is successful, otherwise it has failed. The network reliability $\widehat{R}_{f,ij}$ can then be evaluated by dividing the number of successes with the total number of simulations performed. MCS can easily accommodate both pipe and node failures.

As a demonstration of the efficiency of the method, the MCS results for the sample network in Fig. 5.3 are shown in Table 5.3. As aforementioned in Section 5.2.2, if we assume that the reliability ($R_i = 1 - P_{f,i}$) of each pipe segment (i) is equal to that of the other segments (i.e., $R_1 = R_2 = R_3 = R_4 = R_5 \equiv R$), then the analytical solution for the reliability of the network with respect to node "d" is given by $\widehat{R}_{s,ad} = 1 - P_{f,ad} = 2R^2 + 2R^3 - 5R^4 + 2R^5 = 0.978$ [98]. Thus, if we further assume that $R = 90\%$ then $R_s = 0.978$. As Table 5.3 shows, the MCS results match the analytical solution and are obtained in a relatively small number of simulations (about 10,000 runs), without the need to develop explicit equations for the reliability of each path in the network. Notice that the MC simulations were repeated 30 times, the mean and the cov estimates of the 30 simulations are shown at the bottom lines of the Table.

Table 5.3 Computed network reliability by use of the Monte Carlo Method (sample network 1).

Simulation run	Simulations: 1000			Simulations: 10,000			Simulations: 100,000		
	Node ID			*Node ID*			*Node ID*		
	B	*C*	*D*	*B*	*C*	*D*	*B*	*C*	*D*
1	0.990	0.990	0.975	0.9892	0.9895	0.9795	0.98812	0.98813	0.97761
2	0.990	0.990	0.977	0.9885	0.9877	0.9775	0.98910	0.98872	0.97918
3	0.990	0.987	0.984	0.9862	0.9868	0.9761	0.98866	0.98858	0.97920
4	0.987	0.989	0.978	0.9875	0.9871	0.9773	0.98838	0.98810	0.97853
5	0.991	0.989	0.980	0.9876	0.9874	0.9779	0.98838	0.98850	0.97894
6	0.994	0.995	0.986	0.9874	0.9877	0.9766	0.98855	0.98861	0.97870
7	0.996	0.995	0.987	0.9882	0.9891	0.9803	0.98861	0.98899	0.97879
8	0.991	0.992	0.984	0.9861	0.9853	0.9757	0.98830	0.98847	0.97842
9	0.993	0.994	0.984	0.9892	0.9888	0.9790	0.98848	0.98841	0.97889
10	0.986	0.986	0.978	0.9884	0.9882	0.9801	0.98820	0.98821	0.97848
11	0.984	0.984	0.971	0.9881	0.9883	0.9790	0.98867	0.98853	0.97879
12	0.984	0.986	0.968	0.9878	0.9877	0.9774	0.98767	0.98776	0.97809
13	0.986	0.985	0.970	0.9878	0.9881	0.9794	0.98860	0.98814	0.97833
14	0.990	0.992	0.983	0.9882	0.9860	0.9770	0.98836	0.98842	0.97884
15	0.984	0.988	0.972	0.9879	0.9881	0.9771	0.98838	0.98837	0.97905
16	0.989	0.988	0.980	0.9886	0.9883	0.9791	0.98836	0.98829	0.97902
17	0.990	0.990	0.976	0.9882	0.9885	0.9769	0.98872	0.98859	0.97938
18	0.990	0.989	0.984	0.9888	0.9890	0.9790	0.98849	0.98847	0.97887
19	0.983	0.986	0.977	0.9885	0.9894	0.9796	0.98843	0.98829	0.97862
20	0.983	0.984	0.976	0.9883	0.9893	0.9791	0.98876	0.98849	0.97825
21	0.994	0.993	0.990	0.9884	0.9886	0.9791	0.98811	0.98826	0.97833
22	0.986	0.983	0.974	0.9883	0.9874	0.9792	0.98866	0.98847	0.97824
23	0.988	0.985	0.972	0.9877	0.9888	0.9791	0.98796	0.98825	0.97833
24	0.984	0.986	0.976	0.9870	0.9869	0.9777	0.98784	0.98803	0.97853
25	0.993	0.994	0.987	0.9889	0.9873	0.9801	0.98849	0.98822	0.97902
26	0.990	0.990	0.977	0.9891	0.9887	0.9801	0.98802	0.98780	0.97792
27	0.993	0.993	0.988	0.9888	0.9886	0.9796	0.98833	0.98815	0.97790
28	0.987	0.989	0.983	0.9880	0.9874	0.9786	0.98823	0.98822	0.97857
29	0.987	0.989	0.975	0.9876	0.9873	0.9763	0.98803	0.98806	0.97838
30	0.991	0.990	0.984	0.9875	0.9873	0.9771	0.98836	0.98849	0.97871
MEAN	0.989	0.989	0.979	0.9881	0.9880	0.9784	0.98838	0.98833	0.97860
COV	0.996	0.997	0.994	0.9992	0.9990	0.9987	0.99970	0.99974	0.99958
STD	0.678	0.689	0.718	0.9363	0.9176	0.9378	0.97417	0.97764	0.98050

5.3 SEISMIC VULNERABILITY ANALYSIS OF WATER DISTRIBUTION PIPES

5.3.1 Pipe Vulnerabilities According to the ALA Guidelines

The seismic vulnerability (or fragility) of buried pipelines is discussed in the American Lifelines Alliance (ALA) guidelines (2001), in which several failure parameters are identified and empirical vulnerability functions/curves are proposed by use of observations from past disruptive earthquakes. The vulnerability functions are related to the peak ground velocity (PGV) and the permanent ground deformation (PGD). PGV is related to strong ground shaking caused by seismic wave propagation, while PGD is used to measure factors that include landslides, liquefaction, ground settlement, and fault crossing. Other parameters identified are a pipe's diameter, age, and year of construction, as well as possible discontinuities along a pipe. The recommended by the ALA pipe vulnerability functions provide the repair rate (RR) per 1000 ft of pipe length and have the form:

$$\begin{aligned} RR_{PGV} &= K_1 \cdot 0.00187 \cdot PGV, \\ RR_{PGD} &= K_2 \cdot 1.06 \cdot PGD^{0.319} \end{aligned} \tag{5.9}$$

where PGV is measured in inches per second (in/s) and PGD is measured in inches (in). The above equations can be rewritten in SI units as:

$$\begin{aligned} RR_{PGV} &= K_1 \cdot 0.01425 \cdot PGV, \\ RR_{PGD} &= K_2 \cdot 4.281 \cdot PGD^{0.319} \end{aligned} \tag{5.10}$$

with the repair rate (RR) calculated per 100 m of pipe length, PGV measured in cm/s and PGD measured in cm. A flowchart of the ALA process is depicted in Fig. 5.4.

K_1 and K_2 are material-specific adjustment factors, for which the ALA guidelines provide tabulated values for various pipe materials. For example, in the case of cast iron (CI) or asbestos cement (AC) pipes, $K_1 = K_2 = 1$. The pipe repair rates of Eq. (5.9) can be due to a complete fracture, a leak or damage to an appurtenance of the pipe, or any other reason that requires the water agency to intervene. For typical water pipe networks, a rule-of-thumb is that for failure due to wave propagation approximately 15–20% of failures are breaks and the rest are leaks, while for failures due to PGD approximately 80–85% are breaks that result to the loss of pipeline hydraulic continuity [15].

It should be noted that even though earthquake intensity relates to both peak ground velocity (PGV) and peak ground acceleration (PGA), PGV has been shown to provide the best correlation with damage and to be a robust indicator of the potential of ground motion to cause structural damage. The primary reason for PGV being a better indicator of pipeline damage is its relation to ground strain, which is the main cause of pipeline damage due to seismic wave propagation [136]. Furthermore, PGV is sensitive to longer periods than PGA, thus making it potentially more predictable using

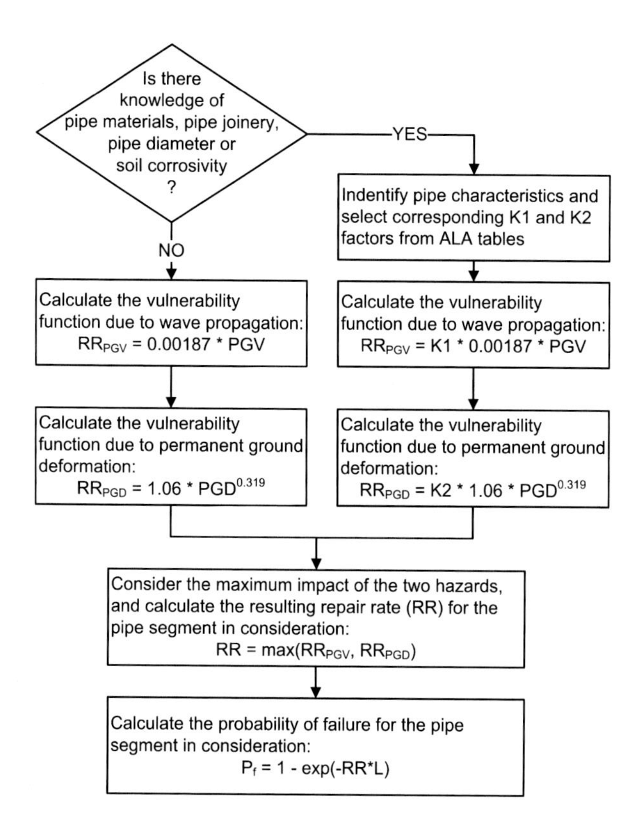

Figure 5.4 Pipe vulnerabilities according to the ALA guidelines (schematic).

deterministic models. For all the aforementioned reasons, PGV has been the most widely used seismic parameter for pipeline seismic fragility.

The value of PGV can be obtained from regional (or national) seismic hazard maps that spatially map the seismic risk into zones of similar seismic properties and provide PGV for a preassumed return period. An equivalent approach for getting valid PGV values is through using regional "ground motion prediction equations", formerly also known as "attenuation relationships." Such equations receive as input the magnitude and the distance from the epicenter of the earthquake.

Once the repair rate (*RR*) is known (i.e., the number of leaks or breaks, per pipe length), the failure probability of the pipe can be calculated, as one minus the probability of zero breaks along the pipe. Adopting Shamir and Howard's findings [152] on the breakage rate following an exponential distribution [152] and using the exponential distribution CDF formula, the pipe failure probability P_f is therefore calculated as [5]:

$$P_f = 1 - e^{-RR \cdot L} \tag{5.11}$$

where $RR = \max(RR_{PGV}, RR_{PGD})$ and RR_{PGV}, RR_{PGD} are calculated by use of either Eq. (5.9) (imperial units) or Eq. (5.10) (SI units). It should be noted that Eq. (5.11) is "memoryless", disregarding any failures that may have occurred along the pipe in the past. The symbology used should also be noted. $\widehat{P_f}$ refers to the failure probability at the network level, whereas P_f refers to the failure probability of a single pipe/edge.

5.3.2 Extending the ALA Guidelines – A Proposed Strategy for Pipe Vulnerability Assessment

The chapter's primary goal is to propose a seismic vulnerability assessment methodology for water pipe networks, while exploiting available data of everyday network failures due to sources other than seismic. Such methodology would be very useful for the analysis of networks under either normal or abnormal operating conditions (as is, for example, the case of intermittent water supply).

The proposed methodology takes into consideration the fragility that corresponds to pipe failures that occur frequently during the everyday operation of the water network and also more severe, but less often, failures due to earthquakes. To account for the vulnerability due to nonseismic loads we rely on survival analysis. To account for the vulnerability due to seismic hazard, we rely on the procedure described in the American Lifelines Alliance (ALA) [5] guidelines and we propose a simple and rational approach for combining it with the results of survival analysis in order to consider the effect of previously observed breaks in the network (Fig. 5.5).

As aforementioned, pipe vulnerability due to nonseismic causes is here assessed, among other methods, by use of survival analysis techniques on available everyday measurements. Survival analysis considers a number of parameters, e.g., number of observed previous breaks, pipe material, diameter, and age that affect the pipe survival curves [27] to develop survival/hazard rates and time-to-failure curves for system components based on a multitude of risk-of-failure factors and data stratifications.

Compared to failures caused by earthquakes, failures from nonseismic causes are more frequent and well distributed in time, while failures due to seismic effects occur intermittently and only when a major earthquake strikes. Thus, it is convenient to compile separately the data from the two

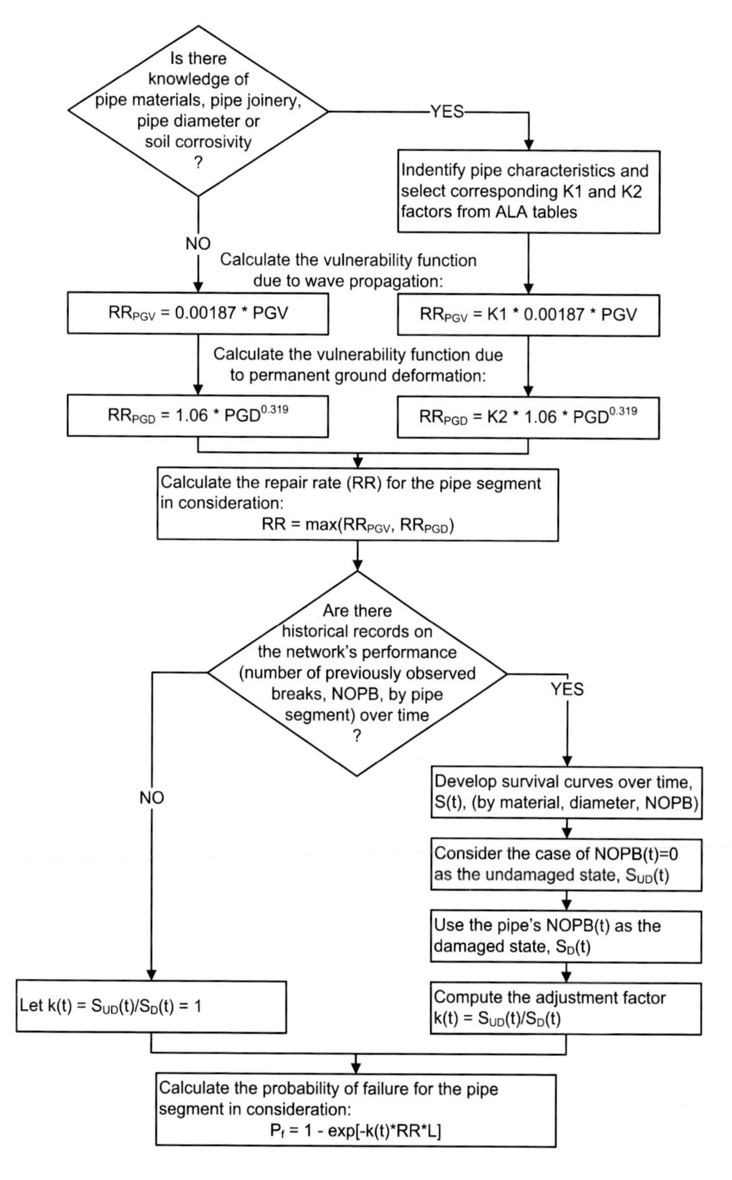

Figure 5.5 Proposed methodology (schematic).

failure causes. This approach is also close to the current practice, since usually it is the water agencies that maintain records of the everyday failure causes, while the seismic effects on the lifelines are usually given a more high-level attention by the civil protection agencies. Moreover, the approaches followed for seismic and nonseismic effects have distinct differences and therefore it is not always straightforward to postprocess the data in a manner that allows to combine consistently pipe survival curves and vulnerability curves.

In our study we adopt a simplified engineering approach that allows us to quickly combine data that are not similar. Having at our disposal the pipe survival curves (e.g., Fig. 5.2, p. 179) of $S(t)$ versus time (Eq. (5.1)) we can deduce the survival probability of a pipe, depending on the number of previous breaks (NOPB) and the pipe type (e.g., material, age, diameter), which we then use to adjust the seismic vulnerability curves.

In effect, we penalize the pipe vulnerability function of Eq. (5.9) by the ratio of the survival curve of the damaged case (NOPB $\neq 0$) over that of the undamaged pipe (NOPB $= 0$). Therefore, after t days, we define the ratio as

$$k(t) = S_{UD}(t) \big/ S_D(t) \geq 1 \tag{5.12}$$

where subscripts "UD" and "D" stand for "undamaged" and "damaged" state, respectively. The modified pipe failure probability that now includes memory of past nonseismic failures is obtained after modifying Eq. (5.11) as follows:

$$P_f(t) = 1 - e^{-k(t) \cdot RR \cdot L}. \tag{5.13}$$

Therefore, Eq. (5.13) allows the calculation of the failure probability P_f of the pipe after t days, given its NOPB metric obtained from historical records.

The Monte Carlo method implemented in this work (shown in Fig. 5.6) allows for the calculation of both the probability of failure $\widehat{P}_{f,ij}$ and of the network's servicability ratio $\widehat{eq:sr}$ (defined in the next section). During the applied MCS, a state vector is produced for every simulation run, holding one of two possible pipe states: *"0-state"*, which refers to a failed state with probability of $P_f(t) = 1 - \exp[-k(t) \cdot RR \cdot L]$ (from Eq. (5.13)) and *"1-state"* that corresponds to success with probability $1 - P_f(t)$.

5.4 RELIABILITY ASSESSMENT OF A WATER SUPPLY SYSTEM

5.4.1 Performance Indices

Once the failure probability, P_f, of every pipe is known, the performance of the network and its failure probability can be assessed. Depending on the problem at hand, different approaches may be preferable. Perhaps the

ASSUME
- Given network topology ($nn :=$ number of nodes, $ne :=$ number of edges)
- Given seismic hazard, e.g., magnitude–distance scenario (M_w, R), PGV, PGD
- Vector of inflow nodes, {denoted by i}
- Vector of outflow nodes, {denoted by j}
- Vector of failure probabilities for every pipe/edge ($\mathbf{p}_f$)

for all inflow nodes **do**
  for all outflow nodes **do**
    - generate a vector **r** of N_{MCS} numbers, so that $r_k \in [0, 1]$, $k = 1, \ldots, N_{MCS}$.
    for all pipes/edges **do**
      - according to $\mathbf{p}_f$, determine if the pipe survives (or fails) using a binomial distribution.
      - **if** the pipe survives (*1-state*) $\rightarrow x_s = 1$, **else** the pipe fails (*0-state*) $\rightarrow x_s = 0$.
    end for
    set $countFailedSimulations = 0$,
    for all samples with least one failed pipe **do**
      - REMOVE from the network all failed pipes {pipes with $x_s = 1$}
      - **if** no path exists between input-output nodes
        $countFailedSimulations = countFailedSimulations + 1$
    end for
    - calculate $P_{f,ij} = countFailedSimulations \big/ N_{MCS}$
  end for{inflow nodes loop}
  - obtain $\mathbf{X} = [1, 2, \ldots, X_k]$, $k = 1, \ldots, nn$.
end for{outflow nodes loop}
- calculate $SR = \sum_{k=1}^{nn} (\omega_k X_k) / \sum_{k=1}^{nn} \omega_k$ {serviceability ratio}

Figure 5.6 Outline of the Monte Carlo simulation algorithm adopted. The algorithm calculates both the failure probability ($\widehat{P}_{f,ij}$) and the serviceability ratio (SR) of the network.

most significant parameter that affects the selection of the strategy to follow is how the network performance is measured and thus how the failure probability of the network is defined. In the simplest case, the network fails when it is not able to deliver water from its sources (inflow vertices) to every house connection (outflow vertices). Another, approach would consider the number of customers that are left without water. If such, rather simplified, network performance definitions are adopted, the performance of the network can be quickly evaluated using methods based on Graph Theory [65]. Alternatively, if failure is defined with respect to hydraulic quantities, i.e., the hydraulic head in every house connection should not be less than a given minimum value, then hydraulic analysis of the network is required. Appropriate software is necessary in this later case.

We consider as failure of the network its inability to provide water to a consumer/house connection. Therefore, we define the failure probability as the probability of the network being unable to provide water from an inflow source vertex i to an outflow (e.g., house connection) vertex j. If the failure probability to deliver water between i and j is $\widehat{P}_{f,ij}$, the network reliability $\widehat{R}_{s,ij}$ is defined as

$$\widehat{R}_{s,ij} = 1 - \widehat{P}_{f,ij}. \tag{5.14}$$

For water networks with more than one inflow sources we can simply check if there is at least one path from all inflow sources if we use the Monte Carlo method. If analytical methods are chosen instead, we need to calculate the joint probability of water reaching the outflow vertex from at least one inflow source. In the latter case, one could consider as a proxy, of the failure probability the smallest probability of all possible sources and the outflow vertex examined. This definition refers to the probability that the outflow node j is left without water. To calculate failure probability we adopt an approach based on Monte Carlo simulation (MCS), which we compare against the path enumeration method [69].

The condition of a network can also be measured by adopting the concept of "Loss Of Connectivity (LOC)" [139], or by use of the "Serviceability Ratio (SR)" metric [1]. LOC and SR are metrics that allow establishing acceptable levels of network performance. For example, a LOC value equal to 20%, 50%, and 80% denotes minor, moderate, and major damage, respectively. This definition follows the principles of performance-based earthquake engineering as often done in structures. The loss of connectivity between an inflow source vertex i and an outflow vertex j, is simply measured as the ratio of available paths between the i and j ($N_{p,ij}$) of the damaged over the undamaged network:

$$LOC_{ij} = 1 - \left[N_{p,ij}^{\text{dam.}}\right] / \left[N_{p,ij}^{\text{orig.}}\right]. \tag{5.15}$$

The expression above holds for a single path ij, while the average value can be used to obtain a metric that measures the loss of connectivity of the whole network LOC_{net}.

The definition of the Serviceability Ratio (SR) is given in Eq. (5.16). If ω_j is the population or the number of consumers of house connection j, N is the number of nodes of the network, and X_j is a binary parameter that denotes whether node j is accessible or not, i.e., if water is able to go from the source to this node, then SR is defined as

$$SR = \frac{\sum_j^N \omega_j X_j}{\sum_j^N \omega_j}. \tag{5.16}$$

Both LOC and SR can be easily calculated using a Monte Carlo algorithm, as discussed in a following section. However, the calculation of LOC involves finding the number of paths available between nodes i and j. Computationally this can be obtained with exhaustive search, thus as the size of network grows, the computing cost of calculating this metric increases exponentially. In the language of computing this is an NP-hard problem and therefore the calculation of LOC may be impractical, or even impossible, for large networks. On the other hand, SR depends on the number of network nodes and therefore the cost of computing it grows almost linearly with the number of nodes. Given the above limitation, we chose the sensitivity ratio (SR) in order to define the overall state of the network.

5.5 CASE STUDIES

5.5.1 Simple Water Distribution Network

As a first example, we consider the case of the simple water distribution network of Fig. 5.7, reported by Christodoulou and Ellinas [34]. The network topology in study is a hypothetical simplified urban water distribution network with three inputs (could be thought of as water reservoirs or inflow pipes), six main valves, and 17 water distribution mains for a total length of 1375 m. Furthermore, unidirectional nodal connections are assumed (an acyclic graph). The pipe properties (e.g., material, diameter) are assumed to be identical throughout the network.

The peak ground velocity (PGV) is estimated based on the attenuation relationship proposed by [148], given by

$$\log_{10}(PGV) = -5.17 + 1.98 \cdot M_w - 0.14 \cdot M_w^2 - 0.10 \cdot \log\left(D^2 + 0.562\right) \tag{5.17}$$

where $M_w = \min(M_w, M_{sat})$, M_w is the moment magnitude, M_{sat} is a saturation moment magnitude value (taken as $M_{sat} = 7$) and D is the Joyner–Boore distance measured in km. Equations such as Eq. (5.17) are known as "ground motion prediction equations" (or "attenuation relationships")

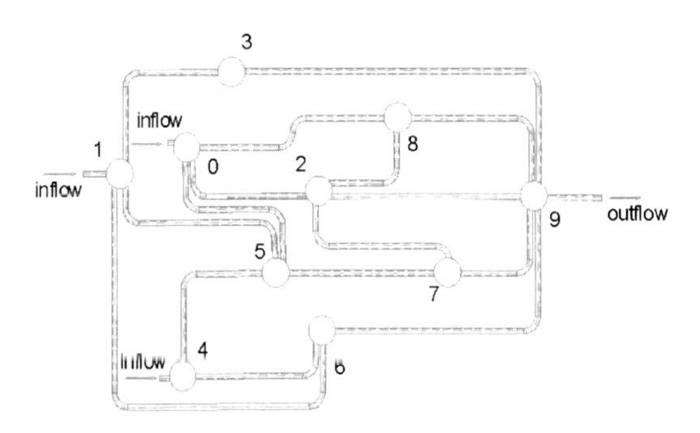

Figure 5.7 Topology of the case-study water pipe network (adopted from Christodoulou and Ellinas, 2010) (the numbers above the edges denote the length, in meters, of every pipe).

and are very common in earthquake engineering practice. These equations consider the tectonics and the seismicity of a region and given the seismic magnitude and the distance they provide the PGV, or other parameters that define the seismic intensity. For this limited example, we assume that the PGV is uniform throughout the water network, while an array of scenario events is assumed by assigning deterministic values to the magnitude M_w and distance D factors, in the range 5–7 and 5–50 km, respectively. Additionally, (i) we consider as seismic hazard only the wave propagation effects (thus neglecting the PGD term of Eq. (5.9)); (ii) we assume that all pipes in the network are cast-iron pipes with cement joints that lie on noncorrosive soil (thus the K_1 factor in Eq. (5.9) is equal to 0.70 (ALA 2001); and (iii) we assume no previous breaks (NOPB = 0) and thus the factor $k(t)$ in Eq. (5.12) is taken to be equal to 1.

Table 5.4 shows two cases of the $M_w - D$ scenarios examined and a comparison of the semianalytical path enumeration method with the numerical Monte Carlo method. For the Monte Carlo approach a total of 10^4 simulations are performed, with the system reliability, R_{ij}, computed each time for three pairs of inflow–outflow nodes due to the fact that the studied network has three inflow and one outflow node. The total system reliability, R_{tot}, is taken to be equal to the minimum of the reliability of the three paths and is also shown in the bottom row of Table 5.4. In general, the differences between the two methods are small (less than 1%) except from the case of R_{5-10}, where the error is 5% and 16% for the M_w being 5 and 7 earthquakes, respectively. For this small scale example, the path enumeration method was faster, but this is not the case as the size of the

Table 5.4 Comparison of the Path Enumeration and Monte Carlo methods (sample network 2).

	$M_w = 5,\ D = 20$ km		$M_w = 7,\ D = 20$ km	
	Path Enum.	*Monte Carlo*	*Path Enum.*	*Monte Carlo*
$R_{(1-10)}$	0.98440	0.99926	0.93530	0.93470
$R_{(2-10)}$	0.98480	0.99050	0.93610	0.92480
$R_{(5-10)}$	0.99090	0.94080	0.96060	0.80710
R_{tot}	0.98440	0.94080	0.93530	0.80710

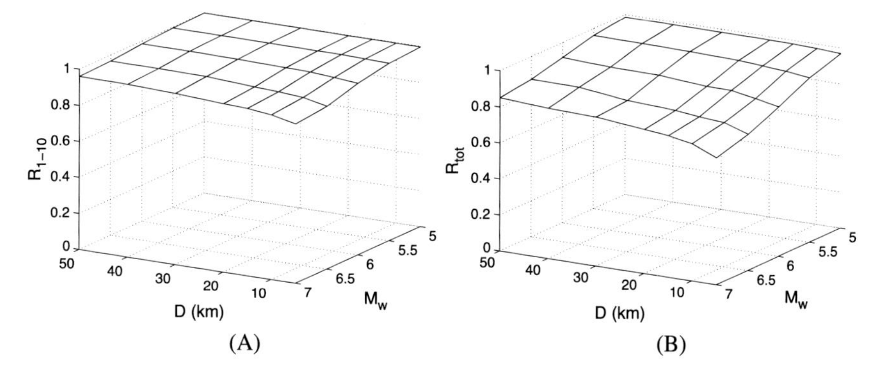

Figure 5.8 Fragility surfaces for the water pipe network assuming $k(t) = 1$: (A) reliability of path 1–10 (R_{1-10}), (B) total system reliability (R_{tot}).

network increases. The Monte Carlo method was shown to be quite stable when varying the number of simulations.

Fig. 5.8 shows the fragility surfaces of the network based on the Monte Carlo simulation, assuming that NOPB = 0 (i.e., $k(t) = 1$). Fig. 5.8A corresponds to the reliability of the "0–9" path, and Fig. 5.8B shows the total system reliability. As the figures show, the system becomes less reliable as the seismic event occurs closer to the network and as the earthquake magnitude increases. Also shown, for comparison purposes, is the case of $k(t) = 2$ (Fig. 5.9). The case corresponds to a network of increased damage due to past pipe breaks, which negatively affect the survival curves of the pipes in the network. By comparing Fig. 5.8 (undamaged network) with Fig. 5.9 (damaged network), it is evident that the system reliability quickly deteriorates as the $k(t)$ parameter increases. For example, for $M_w = 7$ and $D = 50$ the total system reliability, R_{tot}, for the undamaged state is approximately equal to 0.84 (Fig. 5.8B), while for the same earthquake parameters but, for the damaged state, the total system reliability drops to approximately 0.63 (Fig. 5.9B).

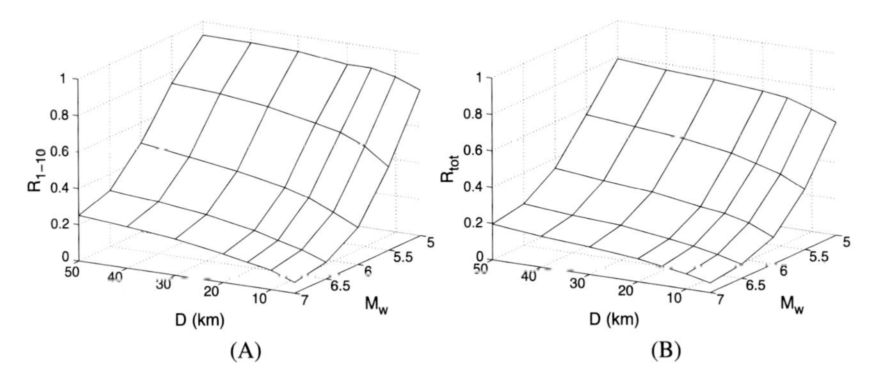

Figure 5.9 Fragility surfaces for the water pipe network assuming $k(t) = 2$: (A) reliability of path 1–10 (R_{1-10}), (B) total system reliability (R_{tot}).

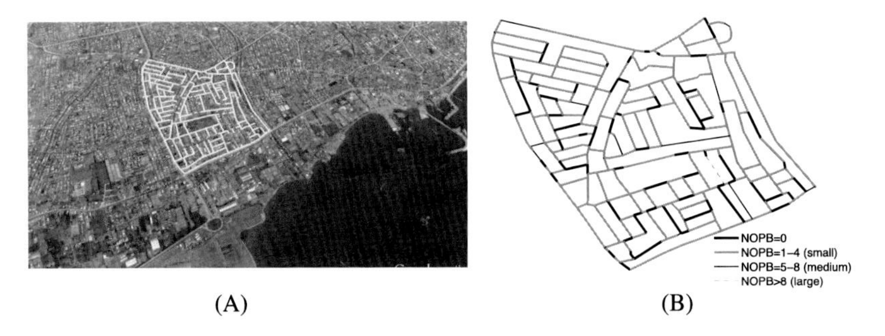

Figure 5.10 (A) Aerial view of the city of Limassol, Cyprus, and the DMA considered; (B) topology of the graph network and number of previous breaks (NOPB) of every pipe.

5.5.2 Complex Water Distribution Network (1)

The second case study is a district metered area (DMA) of the water network of the city of Limassol, Cyprus (Fig. 5.10). Fig. 5.10A shows an aerial view of the city together with the graph model used for simulating the network, which has been produced using available GIS data. GIS is a powerful tool for creating the graph network and obtaining details regarding the properties of the network, e.g., number of consumers in every house connection. However, it should be used carefully and the idealization made has to be as close as possible to the actual geometry of the network. Errors in this idealization may considerably affect the outcome of the analysis and introduce bias. For example, if the pipe length is not correctly modeled, the pipe failure probabilities $P_{f,ij}$ will vary considerably when calculated using Eqs. (5.9) and (5.11).

In total, the water network consists of 337 pipes/edges and 259 vertices/nodes and covers an area of 708 × 450 m. The total pipe length is 23,724 m and, according to the records of the Water Board, the number of consumers served by the DMA studied is 6585 people. On average, every node serves approximately 25–30 consumers, while the maximum number of consumers per node is 120. The pipe material is asbestos cement (AC) and is the same for every pipe. Since the elevation is practically constant throughout the network, we assume that the network is bidirectional. Fig. 5.10B shows the topology of the network and the number of previous breaks of every network pipe/edge. The pipe survival curves were those of Fig. 5.2 (p. 179), as derived from real data obtained from the Water Board of Limassol.

For buried pipelines, seismic hazards can be classified as either wave propagation hazards or permanent ground deformation (PGD) hazards, e.g., 1985 Michoacan earthquake in Mexico City. Typically pipeline damage is due to a combination of hazards. According to [124], roughly half of the pipe breaks in the 1906 San Francisco earthquake occurred within liquefaction-induced lateral spreading zones while the other half occurred over a somewhat larger area where wave propagation was apparently the dominant hazard. Thus, PGD damage typically occurs with high damage rates in isolated areas of ground failure, while wave propagation damage occurs over much larger areas, but with lower damage rates [124]. This is also evident from the repair rates of Eq. (5.9), where adopting typical values of PGV and PGD, the PGD equation will give rates of a different order of magnitude.

Based on the above observations, we consider two seismic scenarios. In the first scenario, damage is only due to wave propagation. Being consistent with the seismic hazard in the island of Cyprus which is mainly controlled by distant and moderate magnitude events, it is valid to assume uniform seismic intensity throughout the DMA. Here we measure seismic intensity with the aid of peak ground velocity (PGV). In the second scenario both PGV and PGD occur, but PGD is isolated in a small part of the network. For both scenarios, we produce fragility curves for every outflow node j.

Fig. 5.11 shows the fragility curve of every vertex with respect to peak ground velocity (PGV) for four time instances measured from the installation of the network. As PGV increases, we calculate the pipe failure probability $P_{f,ij}$ using Eq. (5.9) and the corresponding node probability $\widehat{P}_{f,ij}$ using Eq. (5.11). Therefore, the gray lines correspond to the probability of water being able to reach the corresponding valve, while the black

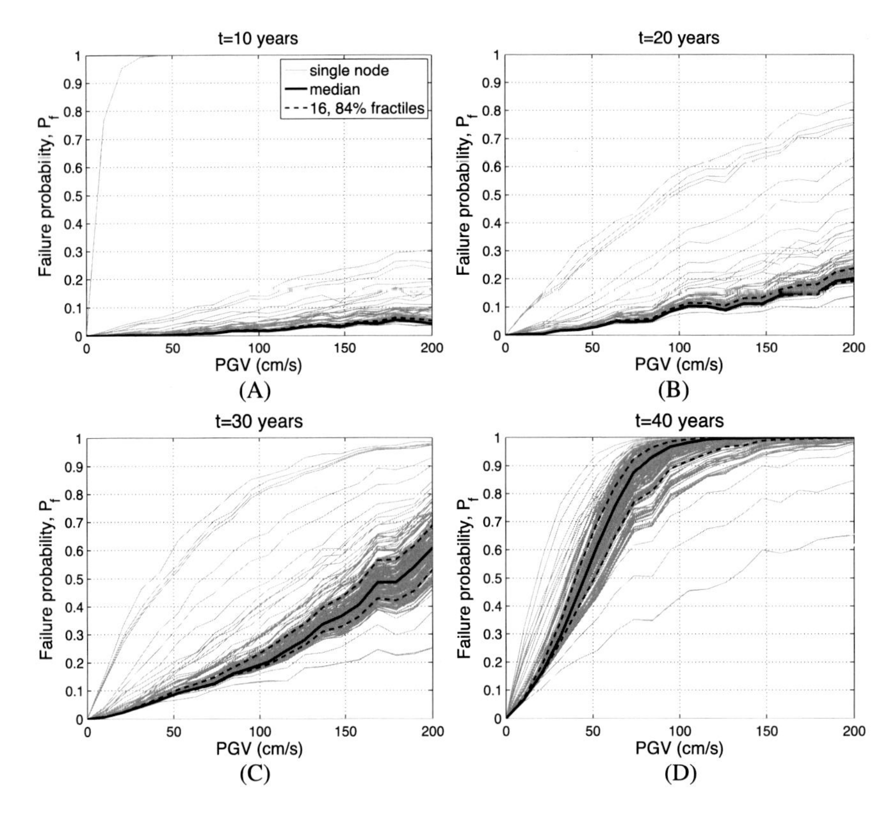

Figure 5.11 Fragility curves of every house connection versus the peak ground deformation (PGV) after: (A) $t = 10$, (B) $t = 20$, (C) $t = 30$, and (D) $t = 40$ years.

lines are the median (50% percentile) and the 16% and 85% percentile curves, which provide a measure of the overall condition of the network.

Figs. 5.11A–B show the fragility curves after 10 and 20 years of network operation. Since the network would be of "young age", the mean fragilities lie below 0.1, even for considerably high PGV values, e.g., PGV > 200 cm/s (Fig. 5.11A). Still, some house connections are vulnerable and their failure probability may exceed 20%. Moreover, there are nodes whose failure probability is very high, as also shown in the map of Fig. 5.12. This is due to the fact that these nodes are connected with the inflow source through pipes that are connected in series, thus if any of the connecting pipes fails the water will not be able to reach them. In this case, the remedy will be to create conditions of redundancy by forming alternative water paths.

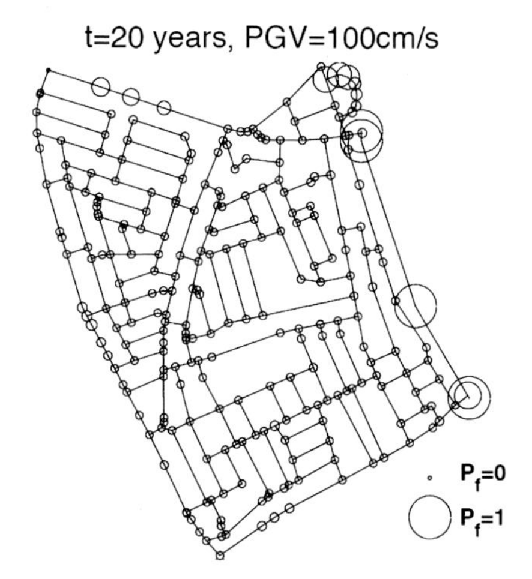

Figure 5.12 Geographical distribution of failure probabilities, for $t = 20$ years and PGV = 100 cm/s.

Figs. 5.11B–D have the well-known form of fragility curves, showing that the network vulnerability increased as the time passes and PGV increases. Note that since the construction of the pipes is made at a present time, the NOPB values are kept constant. Actually, NOPB will also vary as time passes, probably increasing the network vulnerability, but the prediction of survival analysis is based on the data available at the present time and therefore this effect can not be here considered unless a forecast model for NOPB is available. Again in Figs. 5.11B–C there are stray lines away from the average, indicating that the vulnerability of some house connections may considerably differ from the average and thus the interpretation of the analysis should also be done on a node-to-node basis and not rely purely on global metrics on the DMA level. After 40 years of operation (Fig. 5.11D), even a relatively moderate PGV ($\simeq 50$ cm/s) will lead to high failure probabilities and therefore extensive damage on the network.

Fig. 5.13 shows the network fragilities using the serviceability ratio (SR) of Eq. (5.16) as the damage indicator. In this figure we do not measure the network condition using percentiles, as in Fig. 5.11, but we explicitly define damage levels, e.g., minor, moderate or major, which are reached when SR exceeds some indicative thresholds. Here, we arbitrarily adopt as threshold values SR = 10, 50, and 90%, respectively. Contrary to the mean value of $\widehat{P}_{f,ij}$, SR also considers the population affected and its spatial variability with respect to the damage observed. Comparing, Figs. 5.11 and 5.13, the median curve of Fig. 5.11 has similar trends with the curves of

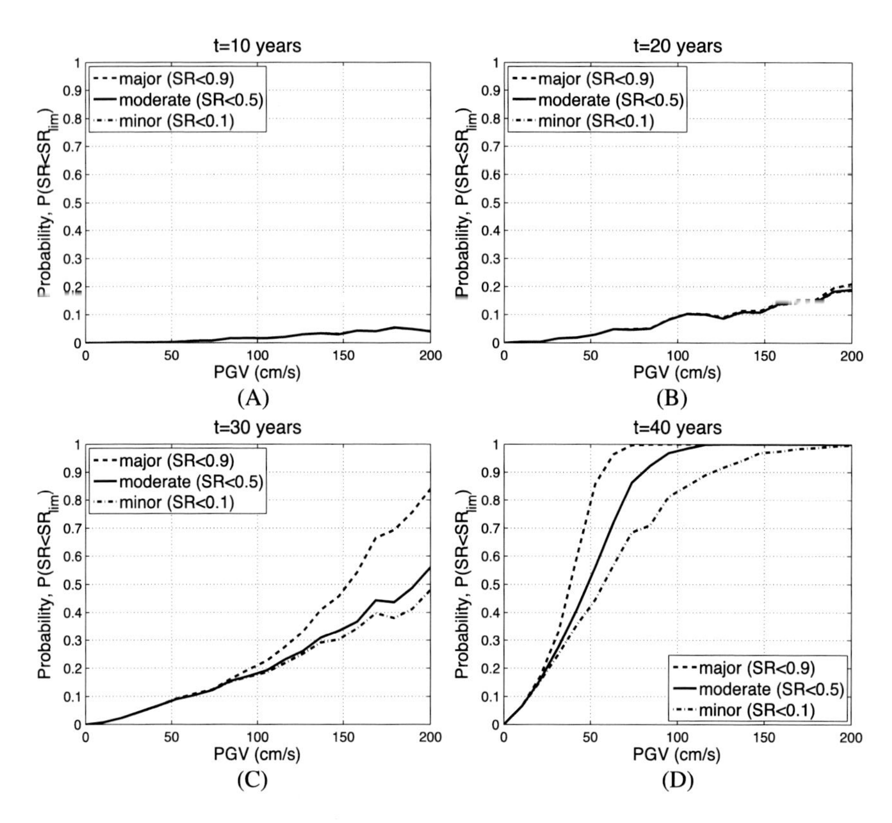

Figure 5.13 Network fragility curves using the Serviceability Ratio (SR) after: (A) $t = 10$, (B) $t = 20$, (C) $t = 30$, and (D) $t = 40$ years.

Fig. 5.13, especially the SR = 50% curve. It is also evident that the differences in the three damage levels considered, become apparent only after some initial damage has occurred, e.g., after 30 or 40 years (Fig. 5.13C–D), while for lower damage the difference is hardly noticeable (Fig. 5.13A–B). The geographical distribution of the damage for PGV = 50 cm/s is shown in Fig. 5.14.

In the second scenario considered, we assume a permanent ground deformation (PGD) due to a random cause (e.g. soil liquefaction). The PGD affects a wide area, of radius equal to 200 m, around a point shown in Fig. 5.15 with a red square. The pernanent ground motion deformation is assumed to be constant and equal to 12.6 cm (5 in). In Fig. 5.15 we also show with a thick red line the pipes that are affected by the imposed PGD. The fragility curves of every vertex for $t = 20$ and $t = 30$ years are shown in Fig. 5.16. It is evident from the plot that there are nodes whose

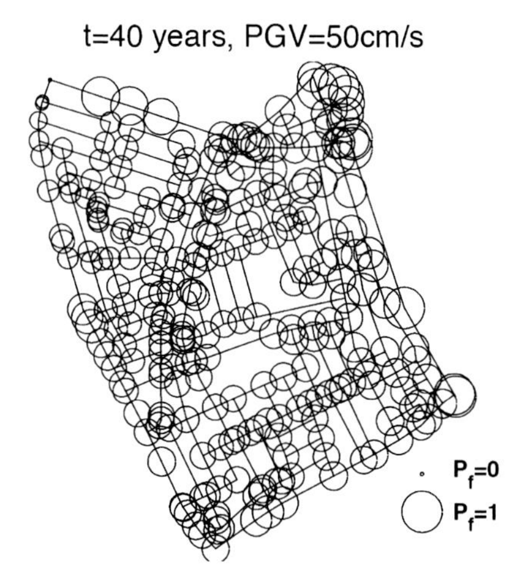

Figure 5.14 Geographical distribution of failure probabilities, for $t = 40$ years and PGV = 50 cm/s.

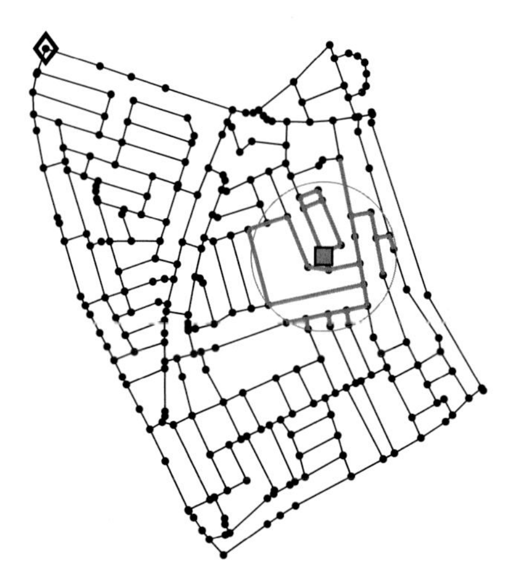

Figure 5.15 Water distribution network and the area where a PGD equal to 12.6 cm (5 in) is imposed. The circle shows the affected area and the square is its center.

vulnerability is considerably higher that the rest. More specifically, looking and PGV = 0 cm/s there are nodes whose probability $P_{f,ij}$ is larger than zero. The vulnerability of these nodes is governed by PGD and for visual purposes we show them with solid black lines, while in the legend of Fig. 5.16 they are denoted as "PGD-sensitive". A plot of the geographical distribution of risk after $t = 20$ and 30 years and for PGV values equal to

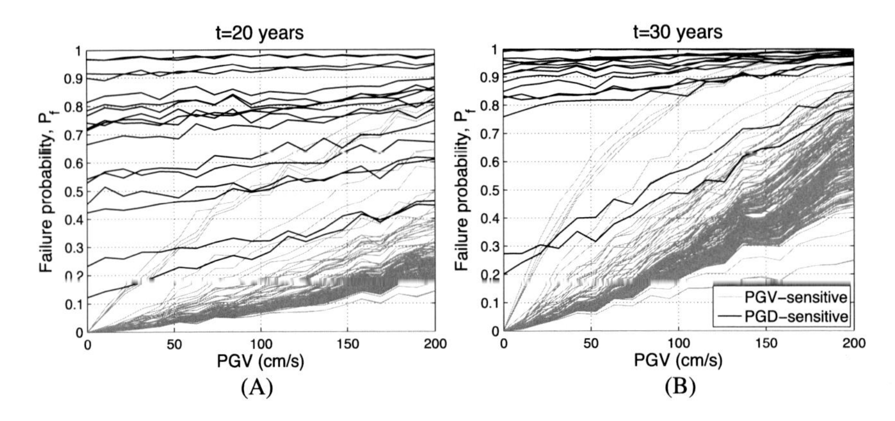

Figure 5.16 Fragility curves versus peak ground velocity (PGV) when PGD is imposed after (A) $t = 20$ and (B) $t = 40$ years.

50 and 100 cm/s (not included) shows that "PGD-sensitive" are the pipes in the vicinity of the PGD, while the risk in the rest of the network is not affected. The grey lines (Fig. 5.16) correspond to vertices whose vulnerability is "PGV-sensitive". For the $t = 30$ years case, the "PGD-sensitive" curves start from high probability values, and quickly approach one, while for $t = 20$ years a larger dispersion is observed. In any case, the "PGD-sensitive" curves are also affected by the increase of PGV (although with a smaller rate), since they operate within a network that combines PGV and PGD-sensitive components. Moreover, when considering both PGV and PGD the practice of producing average curves (e.g., Fig. 5.11) is not useful, since the probabilities vary considerably and depend on the location of the node with respect to where the permanent ground formation occurred.

5.5.3 Complex Water Distribution Network (2)

The third case study considered is also a district metered area (DMA) from the water network of the city of Limassol, Cyprus (Fig. 5.17). A DMA is a hydraulically discrete area with no unmonitored inflow or outflow (typically with one inflow vertex), with a defined and permanent boundary, supplied via a single source, having approximately equal pressure levels across its population of pipes, and with night flows regularly monitored. The city network is clustered into DMAs; a practice that allows the Water Board the isolation of any damage in the network within finite areas (the DMA or its sub-DMAs) and then the handling of any problem that

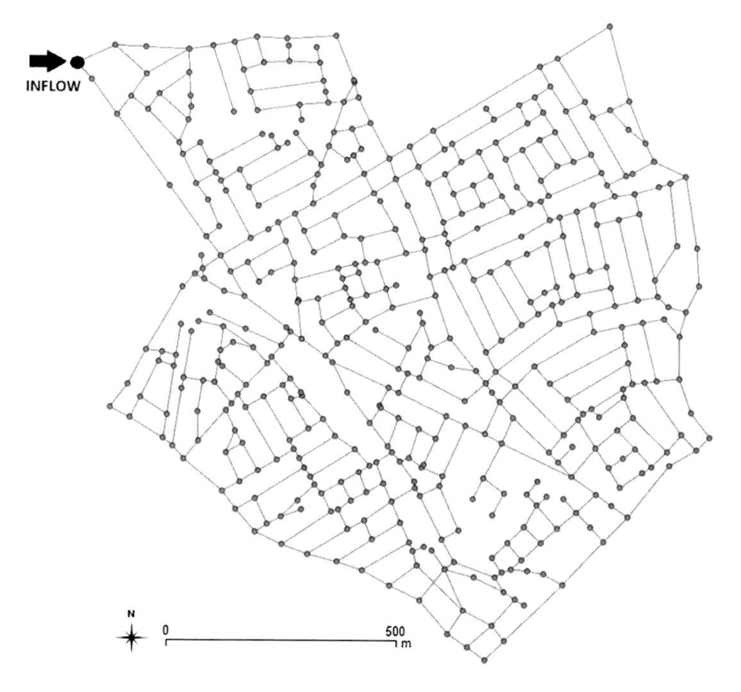

Figure 5.17 Topology of the DMA network considered, in an undisturbed state (*NOPB* = 0 for all pipes).

may occur within a DMA without the cascade of any effects on the overall network.

In total, the water network of the DMA studied consists of 633 pipes (graph edges) of approximate total pipe length of 35,754 m, 450 junctions (graph nodes) and serves a total of 9292 consumers. The material of all pipes in the network is asbestos cement (AC), and the corresponding pipe survival curves are those of Fig. 5.2. Fig. 5.17 shows the topology of the network in its original (undisturbed) state assuming that the NOPB for all pipes is zero, while Fig. 5.18 is a GIS depiction of the studied DMA network in its current (disturbed) state taking into consideration the NOPB values for each pipe, based on available operations and maintenance records from the Water Board.

Fig. 5.19 shows the fragility curves of the inflow vertex with respect to each outflow vertex, at various PGV and age values. In terms of seismic considerations, the seismic hazard in the island of Cyprus is mainly controlled by distant and moderate magnitude events and, thus, it is valid to assume that the seismic intensity will be uniform throughout the DMA.

Figure 5.18 Topology of the DMA network considered, in a disturbed state (*NOPB* ≥ 0 for every pipe).

The analysis is performed for $t = [0, 10, 20, 30]$ years, taking into consideration the condition of the network (NOPB values) and the corresponding survival curves. The light grey lines correspond to the probability of water being able to reach a particular network vertex as PGV increases, while the black lines are the median (50% percentile) and the 16% and 85% percentile curves. The median and percentile curves can be used to provide a measure of the overall condition of the network. As expected, the network's failure probability increases with increasing earthquake intensity (PGV), as well as with increasing network age (from 0 to 30 years of age). In effect, the analysis shows that the vulnerability of the network increases as its deterioration over time increases. As vulnerability one can refer to the expected impacts of either earthquake loadings, or abnormal operating conditions (such as intermittent water supply), and as deterioration one can refer to the survival curves of the network's piping components due to age, material, deterioration, previous breaks, operating pressure, etc.

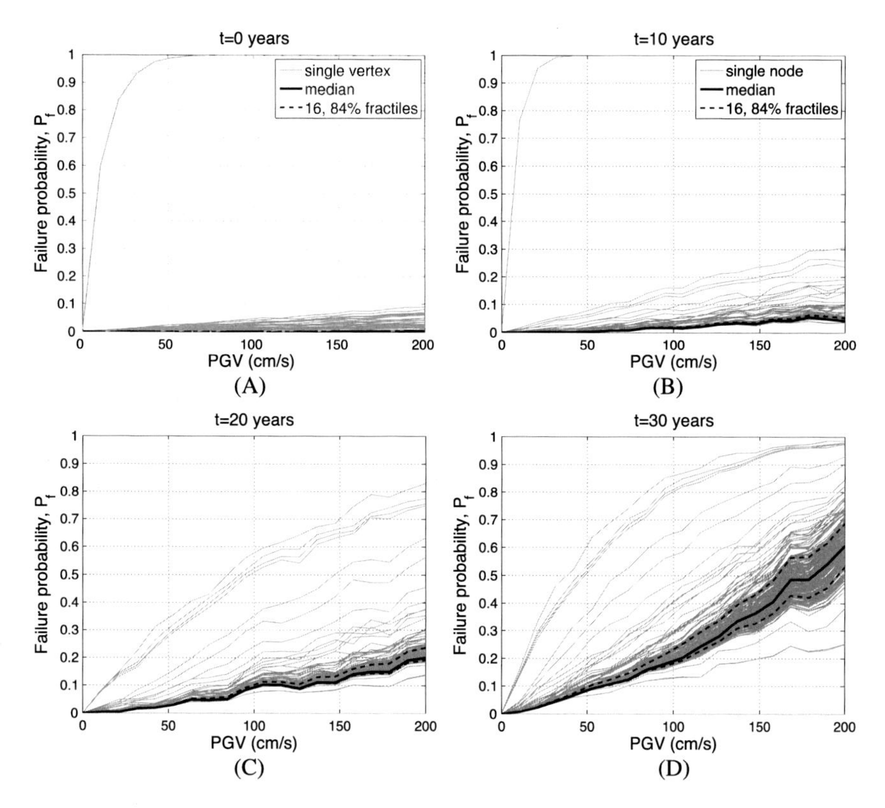

Figure 5.19 Fragility curves of the inflow node with respect to each outflow node, after: (A) $t = 0$, (B) $t = 10$, (C) $t = 20$, and (D) $t = 30$ years.

Similar conclusions can be deduced from Fig. 5.20, which depicts the network's serviceability ratio (SR) with respect to the network's age and the acting earthquake intensity. As seen, a network's serviceability ratio drops substantially as the network ages and as the seismic intensity increases. Fig. 5.20A (depicting SR at $t = 0$) corresponds to the expected serviceability ratio based on simply the topology (nodal connectivity) of the network and the ALA (2001) guidelines, without any adjustments to the network in study of the applicable survival curves. Figs. 5.20A–C depict the computed drop in the serviceability ratio of the network in study as the network ages. The computed drop is the direct outcome of the proposed adjustment factor to the ALA (2001) guidelines based on the applicable to the network survival curves (Fig. 5.2, p. 179).

A spatial analysis of the aforementioned case-study DMA network will be presented in Chapter 8.

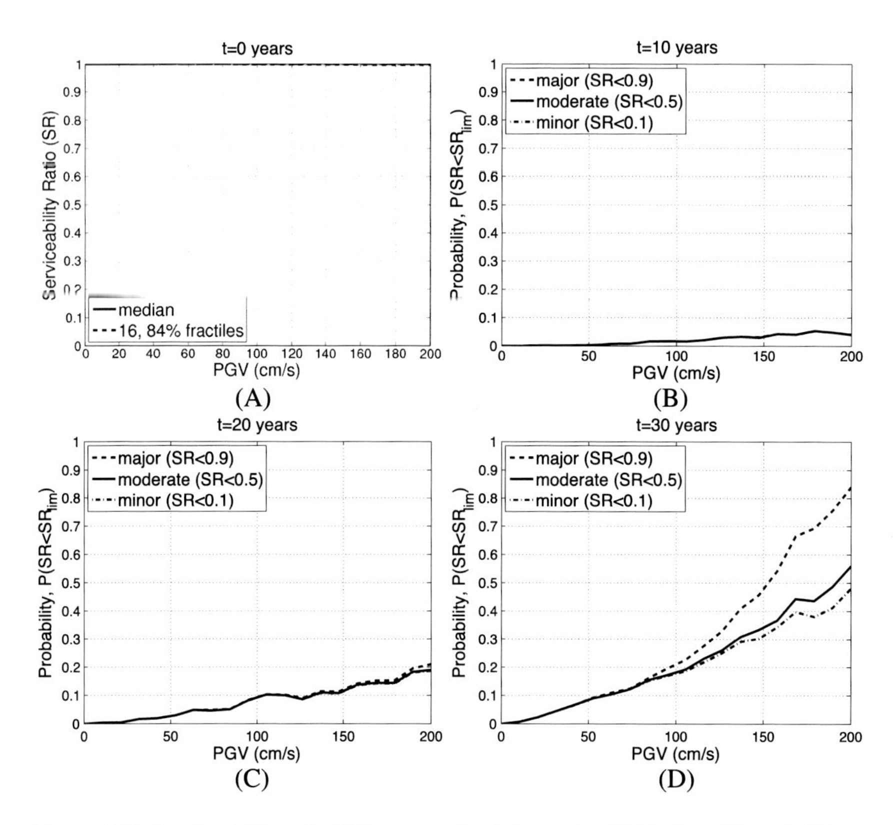

Figure 5.20 Serviceability ratio (SR) versus seismic intensity (PGV) after: (A) $t = 0$, (B) $t = 10$, (C) $t = 20$, and (D) $t = 30$ years.

5.6 CONCLUDING REMARKS

A general-purpose methodology for the reliability assessment of water pipe networks has been presented. The proposed approach is a rational, easy to apply, and will greatly assist taking proactive measures for the seismic risk mitigation of lifelines. The proposed methodology efficiently combines seismic fragility curves (e.g. those of the ALA) with localized information on failures caused by nonseismic sources. Since a network's vulnerability is negatively affected by periodic pipe breaks, assimilated information on historical pipe breaks is clustered based on various risk factors (e.g., material, age, diameter, etc.) and then postprocessed to develop survival curves for prevailing pipe clusters in a network and then utilized in adjusting the pipe fragilities to account for localized conditions.

Once the pipe failure probabilities are known, the reliability of the whole water network can be calculated using Monte Carlo simulation. The efficiency of the proposed Monte Carlo scheme is compared to reliability bounds obtained analytically with the aid of the path enumeration method. Even though the procedure proposed herein is presented based on performance data from urban water networks of the island of Cyprus, it is general in scope and applicable to any locale with historical records of pipe-break incidents on its water distribution network.

The proposed methodology arrives at fragility curves and serviceability ratios that take into consideration localized information on risk-of-failure factors pertaining to the particular network in study. Moreover, it also arrives at improved vulnerability analyses of water distribution networks acted upon by seismic and/or abnormal loadings over time. Future research work should focus on spatio-temporal comparisons of existing seismic fragility curves and the proposed method to examine spatio-temporal variations between the network vulnerabilities deduced by each method.

CHAPTER 6

Hydraulic Vulnerability Assessment of Water Distribution Networks

A network's reliability is a function of not only the reliability of the network components but also of its topology (connectivity) and hydraulic parameters.
Christodoulou, S.E., et al. (2014),
Journal of Infrastructure Systems.

6.1 INTRODUCTION

Since lifeline systems, such as water distribution networks, are of critical importance to the uninterrupted provision of services and thus to the resiliency of a city, their holistic vulnerability assessment and sustainable management is of paramount importance to society.

As noted in [6], "*when referring to vulnerability assessment and analysis of water distribution networks (WDN), one should note that the topic is multi-faceted and encompasses as a minimum the following three aspects: (1) component analysis (i.e. pipes, valves, reservoirs, etc.); (2) operations (i.e. the operating parameters of a network, such as water pressure) and (3) topology and connectivity (i.e. the number of arcs/nodes in a network, elevations, arc lengths, etc.). For every aspect, a number of possible analysis methods exist.*" Added to the above, one should also consider vulnerability aspects due to seismic events [35,59–61], as discussed in Chapter 5. These analysis facets and typical methods for evaluating them are shown in schematic form in Fig. 6.1, with the subject of this chapter highlighted in a different color shade.

A number of previous studies have assessed several of the said WDN vulnerability facets, but seldom have the nonseismic and seismic vulnerability, and the system-component interactions been considered in tandem with the hydraulic performance of such systems.

Most of the reported research work on the vulnerability of WDN, as previously pointed out, focuses on component analysis, i.e., the estimation of either the probability of failure or of the time-to-failure for the various network components, primarily the pipe segments. Commonly used

Urban Water Distribution Networks
DOI: http://dx.doi.org/10.1016/B978-0-12-813652-2.00006-2

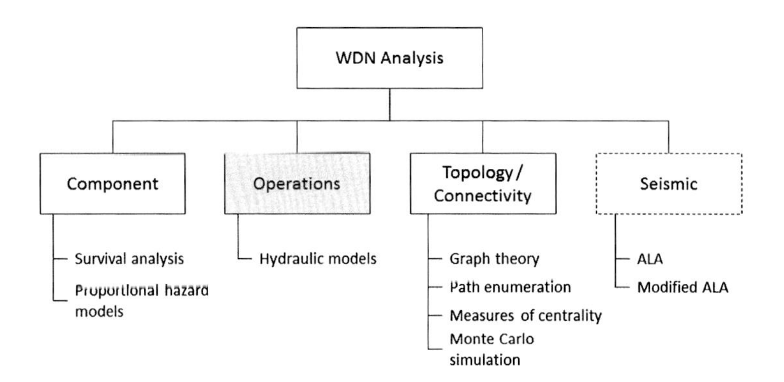

Figure 6.1 WDN vulnerability analysis (schematic).

methods are the proportional hazard model, survival analysis, neurofuzzy systems, and data mining techniques. The component-based approach, though, does not provide vulnerability metrics on a node-to-node basis since it does not enumerate the various paths in a network and it does not incorporate the hydraulic model of the network. A network's reliability, though, is a function of not only the reliability of the network's components and of its topology (connectivity) but also of its hydraulic operations and parameters.

6.2 WDN HYDRAULICS

As a prelude to the discussion on WDN hydraulic analysis at the network level, we first examine some basic ideas with regards to pipe failures and water dispersion in soils, as modeled by computational fluid dynamics (CFD). To assist in the analysis model, the soil in which a water leak disperses is modeled as a porous domain with certain characteristics, and the water is modeled as a fluid passing through this porous medium. The CFD numerical modeling results shown in Figs. 6.2–6.8 illustrate that the velocity distribution in the porous domain can be well simulated once the correct soil characteristics are known, and that in its passage through soil water seeks paths and zones of least resistance (i.e., zones of high soil permeability).

6.2.1 CFD Formulation of Water Loss Dispersion in Soils

A numerical model for flow in a porous medium was hereby adapted, in order to capture the effect of water loss in soils surrounding vulnerable underground WDN pipes. Such models, once calibrated and validated with in-situ data, can aid in the placement optimization of leak detection mechanisms and for a better overall water network management. For the analysis included herein, a commercially available computational fluid dynamics (CFD) software was used (ANSYS 14), in both two- and three-dimensional domains. The porous zone is treated as a fluid zone with the addition of a momentum source term to the standard turbulence flow equations ($\kappa - \epsilon$ in this case). The pressure drop is simulated by adding a sink term (S_i) as described in Eq. (6.1), which is the sum of the viscous loss term (Darcy's term) and the inertial loss term as discussed by Mossad and Aral [118]. The viscous and inertial resistance terms are empirically computed based on the assumed porosity and permeability of the soil materials that surround the vulnerable pipes or, in this case, the pipes with specific water leak rate:

$$S_i = -\left(\sum_{j=1}^{3} D_{ij}\mu v_j + \sum_{j=1}^{3} C_{ij}\frac{1}{2}\rho v_{mag} v_j\right) \tag{6.1}$$

S_i is the source term for the ith (x, y and/or z) momentum equation, D_{ij} and C_{ij} the respective matrices for the viscous and inertial loss terms, and v_{mag} is the magnitude of velocity while μ is the dynamic viscosity in (Ns/m^2). The viscous resistance matrix was estimated from the permeability and porosity of the different layers of the soil by the equation given by Vafai [164] (see later section on geometry decomposition) while the inertial resistance C matrix was ignored due to the laminar flow assumption within the porous region. This assumption/simplification effectively transports the inlet turbulence quantities through the region, but their effects on the fluid mixing and momentum are ignored. In addition, the generation of turbulence is set to zero in this region as well.

Fig. 6.2 depicts the first 2D numerical simulation geometry (XY plane) in which a 10 m long pipe of 10 cm in diameter (Internal Diameter, ID) experiences a 5 cm long longitudinal crack as it is typically shown in Fig. 6.3. The pipe is covered by a 5 cm layer of sand, then by a 12 cm layer of sand mixture, and then followed by a third layer of mixed dirt of approximately 1 m in height. The effects of the surrounding layers of soil and particularly fluidization of soil as a result of water leaks is not directly addressed purely by the addition of the extra momentum source via the Brinkman equation. Cassa et al. [22] and van Zyl et al. [166] provide results of the behavior of different types of leak openings on pressurized pipes assuming elastic behavior, along with a preliminary study of the effect of soil and particularly fluidization of soil surrounding leaking water pipes as those shown

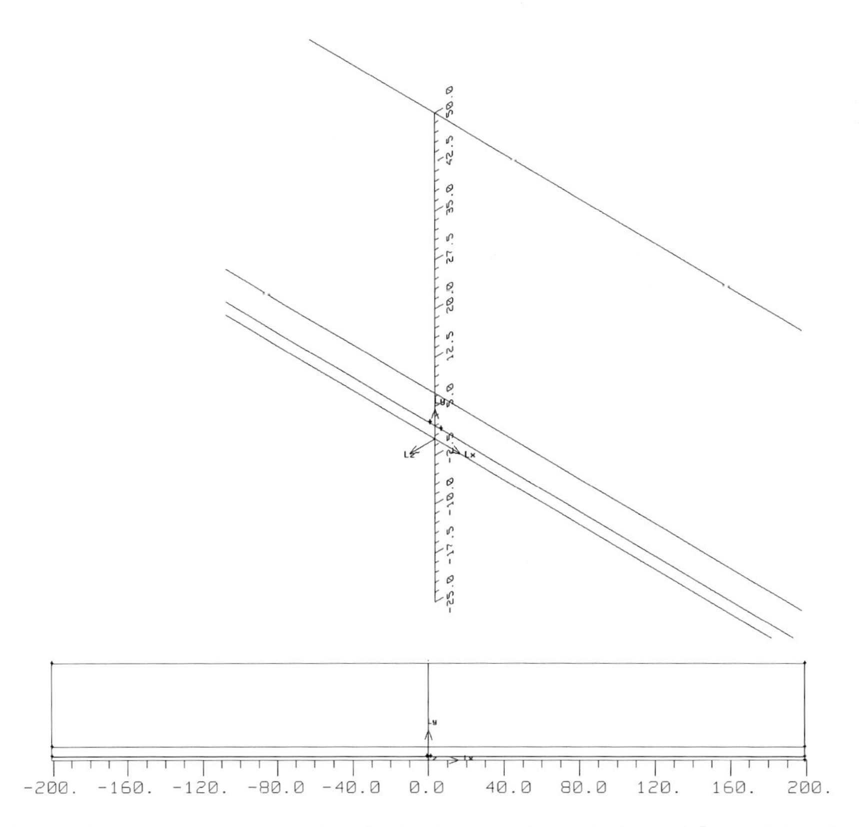

Figure 6.2 Geometry representation for 2D CFD water loss calculation (2D – XY (plane)).

in Fig. 6.3. It was observed that pipe stresses are highly affected by the leak opening and that significant head can be sustained in a shallow pipe without the fluidized ever breaking through the bed surface. Overall head loss can be attributed to losses through the orifice, in the fluid/mobile zones and through the static unaffected soil with the highest gradient experienced in the fluid/mobile zones.

Fig. 6.4 depicts the alternate 2D numerical simulation geometry (XZ plane) in which a the length of the pipe in normal to the plane the ID of the pipe is the same while suffering a circumferential (from 0 to 180 degrees from horizontal) crack as it is typically shown in Fig. 6.3. The pipe is covered by a 5 cm layer of sand, then by a 12 cm layer of sand mixture, and then followed by a third layer of mixed dirt of approximately 0.75 m in height. The imposed flow conditions are the same as in the other 2D CFD case.

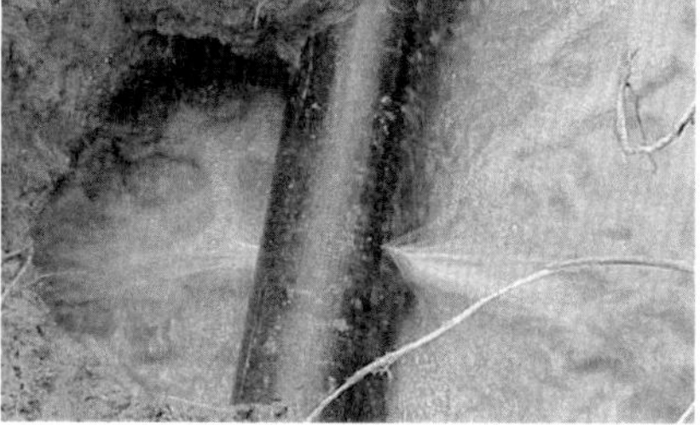

Figure 6.3 Typical longitudinal and circumferential cracks in water pipes. Image on left obtained from www.sheffield.ac.uk; image on the right from Alliance for Water Efficiency.

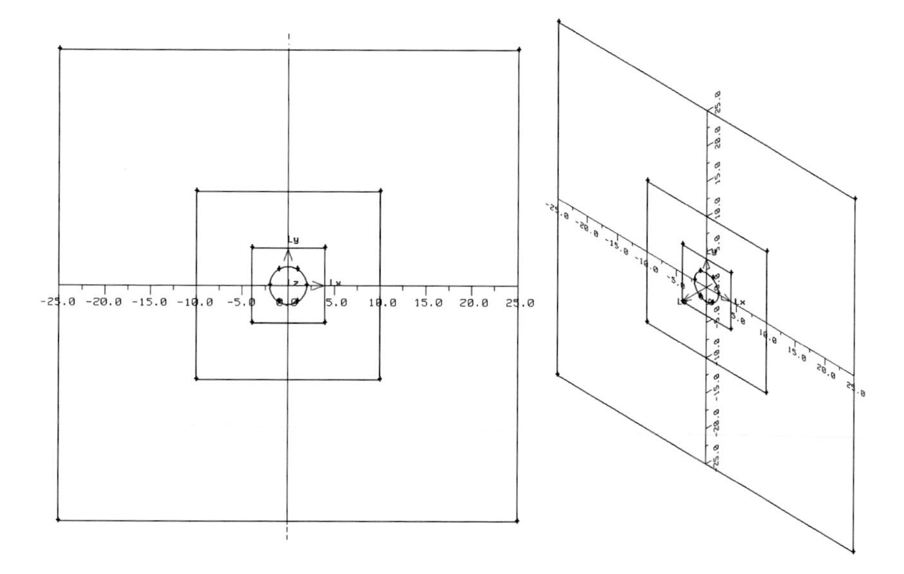

Figure 6.4 Geometry representation for 2D CFD water loss calculation (2D – XZ (plane)).

6.2.2 2D CFD Results and Comments

An initial 2D simulation on the geometry discussed in Fig. 6.2 is shown in Fig. 6.5, where the region is assumed to be the same as the fluid in the pipe without invoking the assumption of the porous region. The predicted flow field resembles one of a fluid leaking into the ambient air as in the case shown in Fig. 6.3. There should be no confusion here that the surrounding fluid is still water but not porous and flow is not assumed to be laminar.

Fig. 6.6 depicts the same domain with the porous medium active and hence with a much higher resistance to the flow from the porous approx-

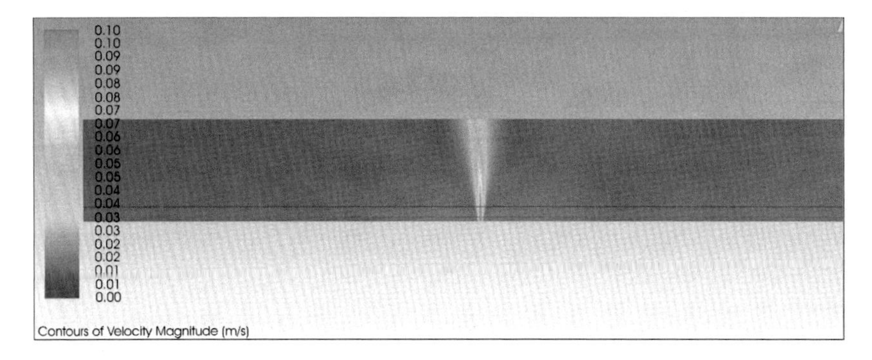

Figure 6.5 2D CFD water loss calculation with visible contours of velocity magnitude (2D – XY(plane)) (Control – Porous Region not activated).

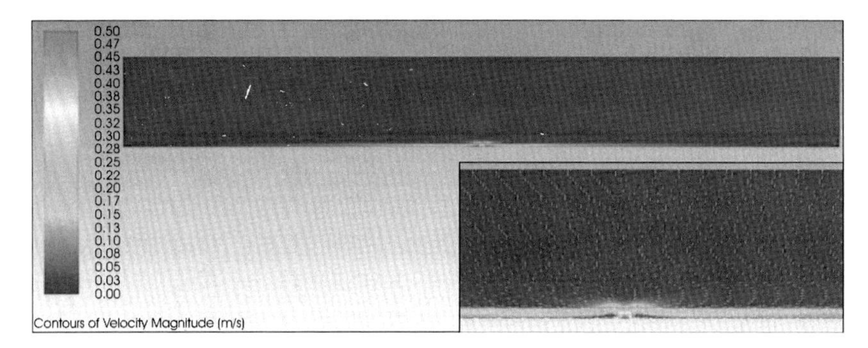

Figure 6.6 2D CFD water loss calculation with visible contours of velocity magnitude (2D – XY(plane)) (Control – Porous Region activated).

imation; zone A (Proximal to the pipe, sand) and zones B and C regular soil with different compaction values, hence different permeability values, thus imposing a anisotropic permeability field. In this case the water leak follows a preferential path along the length of the pipe and hence shows a low dispersion index or profile which makes it difficult for leakage sensors (soil moisture sensors) to identify. A close-up of Fig. 6.7 is shown on the right overlay image.

Fig. 6.8 illustrates similar results for the transverse section of the buried pipe (see Fig. 6.4), where the pipe cross-section is identified. This pipe is assumed to have suffered a circumferential crack with the water leak observed at the top section of the pipe (refer to Fig. 6.3). As shown, the imposed soil conditions have the same effect as in the XY plane case on the flow field, damping the flow velocity magnitude and allowing it to permeate in the surrounding region.

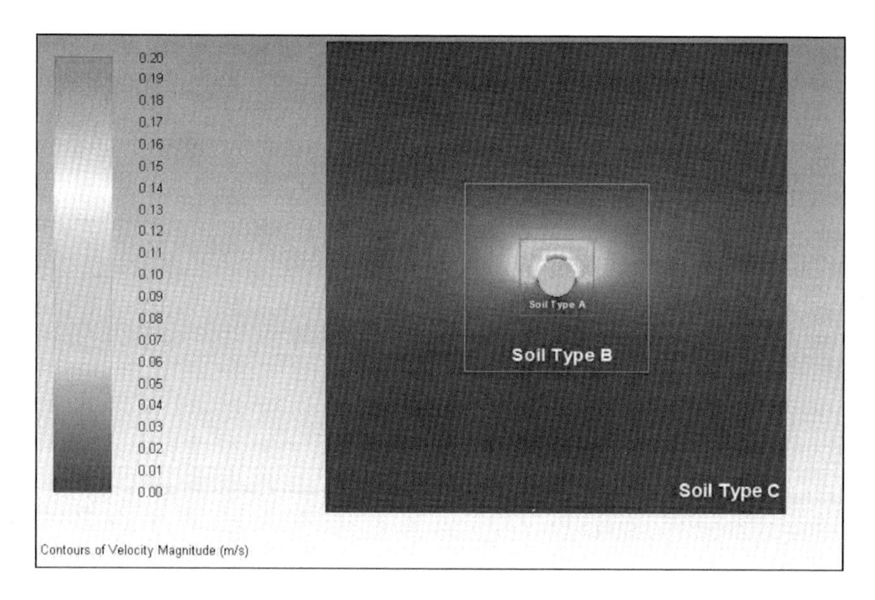

Figure 6.7 2D CFD water loss calculation with visible contours of velocity magnitude (2D – XZ(plane)) (Control – Porous Region activated).

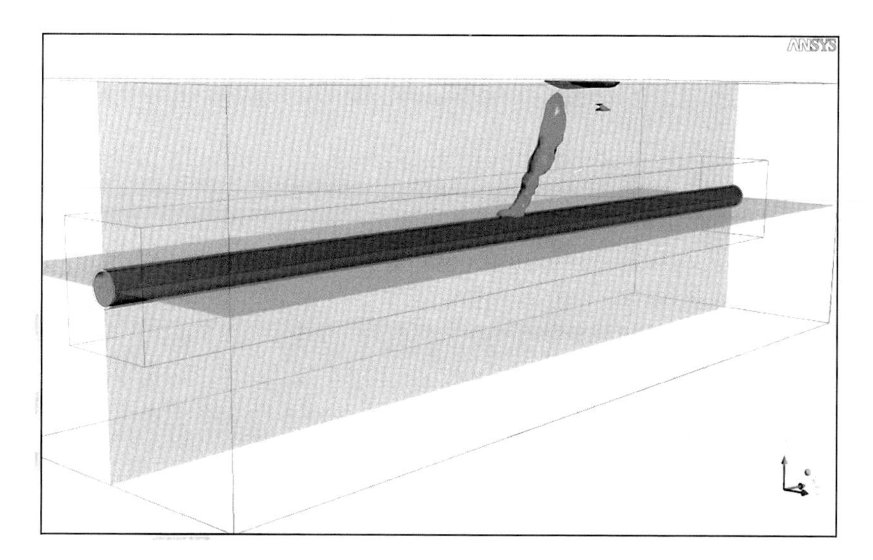

Figure 6.8 3D CFD water loss calculation water tracer conveyed into the porous medium (surrounding soil).

A 3D model was developed that simulated the conditions discussed in Fig. 6.2. Quantitative steady state results are presented via fluid particle tracers as they are conveyed in the porous medium shown in Fig. 6.8. As discussed by van Zyl et al. [166], the different zones of energy dissipation need to be modeled with alternate methods and results validated based on bench scale experiments. Empirically calculated modeling parameters that describe the three-dimensional nature of the surrounding soil need to be obtained for a well calibrated and validated simulation.

6.3 HYDRAULIC VULNERABILITY OF WATER DISTRIBUTION NETWORKS

6.3.1 Proposed Methodology

The performance of a network and its failure probability can be assessed based on the WDN's topology/connectivity and on the failure probability of every pipe in the network. In the simplest case, the network performance is measured by the probability that it fails to deliver water from its sources (inflow vertices) to every house connection (outflow vertices). If such, rather simplified, network performance definition is adopted, the performance of the network can be quickly evaluated using methods based on Graph Theory and Monte Carlo simulation (MCS). Alternatively, if failure is defined with respect to hydraulic quantities, then hydraulic analysis of the network is required.

In the case of nonseismic evaluations, the hydraulic vulnerability analysis is simpler in form and derived directly from a Monte Carlo simulation. In the case of seismic evaluations, the hydraulic vulnerability analysis could either be conducted after the dynamic/seismic investigation on selected network configurations, or coupled directly within the WDN reliability analysis as shown in Fig. 6.9, providing hydraulic results for all iteration steps of the MCS analysis step. The last step in the flowchart depicted in Fig. 6.9 relates to the WDN seismic vulnerability aspects presented in detail in Chapter 5 (p. 173), and briefly summarized in Section 6.3.3.

6.3.2 Hydraulic Vulnerability – Nonseismic Conditions

As aforementioned, in the case of nonseismic evaluations the hydraulic vulnerability can be derived directly from a hydraulic model (e.g., by use of the EPANET software), in conjunction with a Monte Carlo simulation (as depicted in Fig. 6.9). The hydraulic model is used in determining the water pressure/flows at each network arc (pipe segment) and junction (network node), while the Monte Carlo simulation is used in stochastically model-

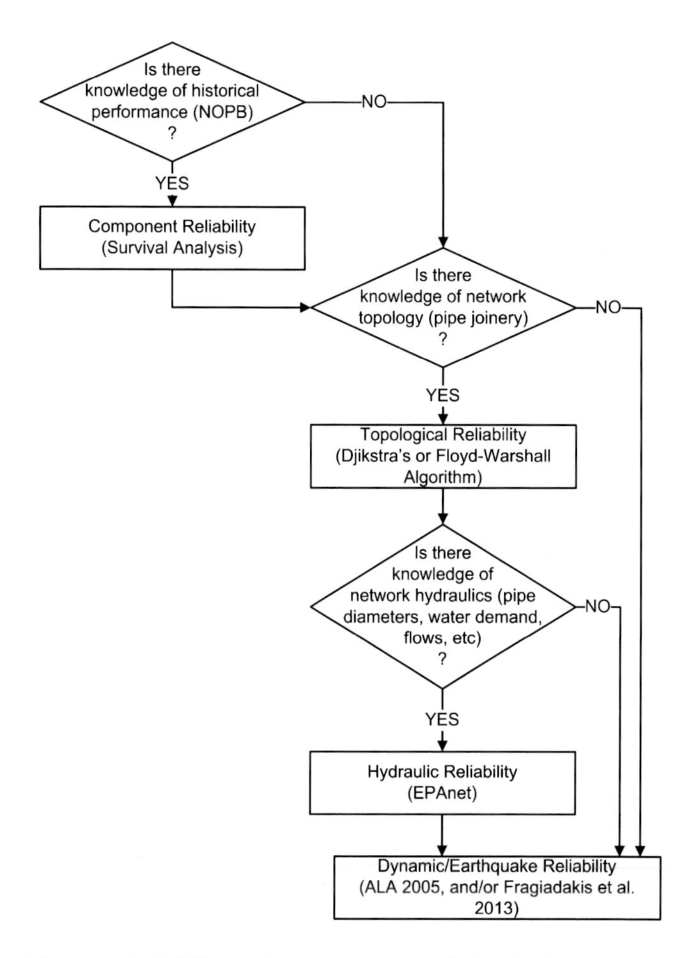

Figure 6.9 Proposed reliability analysis procedure, including hydraulics.

ing the water demand in the network and in dynamically modifying the network topology (to simulate random pipe breakages in the network).

Simulation of pipe breakages is performed on the basis of the probability of failure for each network component (as derived from the survival analysis for these components), and the network's topological reliability is computed by use of either the Dijkstra or the Floyd–Warshall algorithm.

6.3.3 Hydraulic Vulnerability – Seismic Conditions

Should one also consider seismic vulnerability, though, then as noted in [35, 59–61] and in Chapter 5 (p. 173), the American Lifelines Alliance guidelines [6] can be helpful. In the ALA guidelines, seismic vulnerability (or fragility) of buried pipelines is expressed in terms of several failure parameters, and empirical vulnerability functions/curves are proposed by use of observations from past disruptive earthquakes. The vulnerability functions are related to the peak ground velocity (PGV) and the permanent ground deformation (PGD). PGV is related to strong ground shaking caused by seismic wave propagation, while PGD is used to measure factors that include landslides, liquefaction, ground settlement, and fault crossing. Other parameters identified are a pipe's diameter, age, and year of construction, as well as possible discontinuities along a pipe. The reliability procedures proposed by [35,59–61] are shown schematically in Fig. 6.10.

As discussed in Chapter 5, the vulnerability curves suggested in [6] are combined with available and localized survival curves. The survival probability of a pipe is then deduced by use of clustering the underlying risk factors, such as the number of previous breaks factor (NOPB) (Fig. 6.11A) and the pipe type (e.g., material, age, diameter). These survival curves are then used to adjust the vulnerability curves proposed by the ALA.

Hydraulic Vulnerability – Case Study

Consider, for example, the real-life network shown in Fig. 6.11B consisting of multiple district metered areas (DMA). The network's number of normalized water main incidents and its nonseismic performance (expressed through the number of observed previous breaks, NOPB) are as shown in Fig. 6.12A and Fig. 6.12B, respectively.

The region in question is mapped as a graph and a Monte Carlo simulation, with an assumed seismic load of $M_w = 7$, is performed to produce the failure probabilities (and thus the network reliability) at every network node. The result is then mapped as a spatial risk plot (a "heatmap"), as shown in Fig. 6.13A. The produced heatmap identifies the areas in the network which are of high failure risk. In this case the areas of concern seem to be limited to three areas which, on closer inspection, are related to areas of open-ended pipe pathways (orphan nodes). The horizontal allocation of risk is actually shown to be highly sensitive to open-ended network topologies.

The situation changes drastically once the NOPB risk factor (Figs. 6.11A and 6.12B) is included in the analysis (Fig. 6.13B) by use of the algorithm proposed in [35,59,60]. The inclusion of a network's nonseismic performance increases the pipes' risk of failure and thus alters the network reliability. The increased network vulnerability is the direct outcome of the

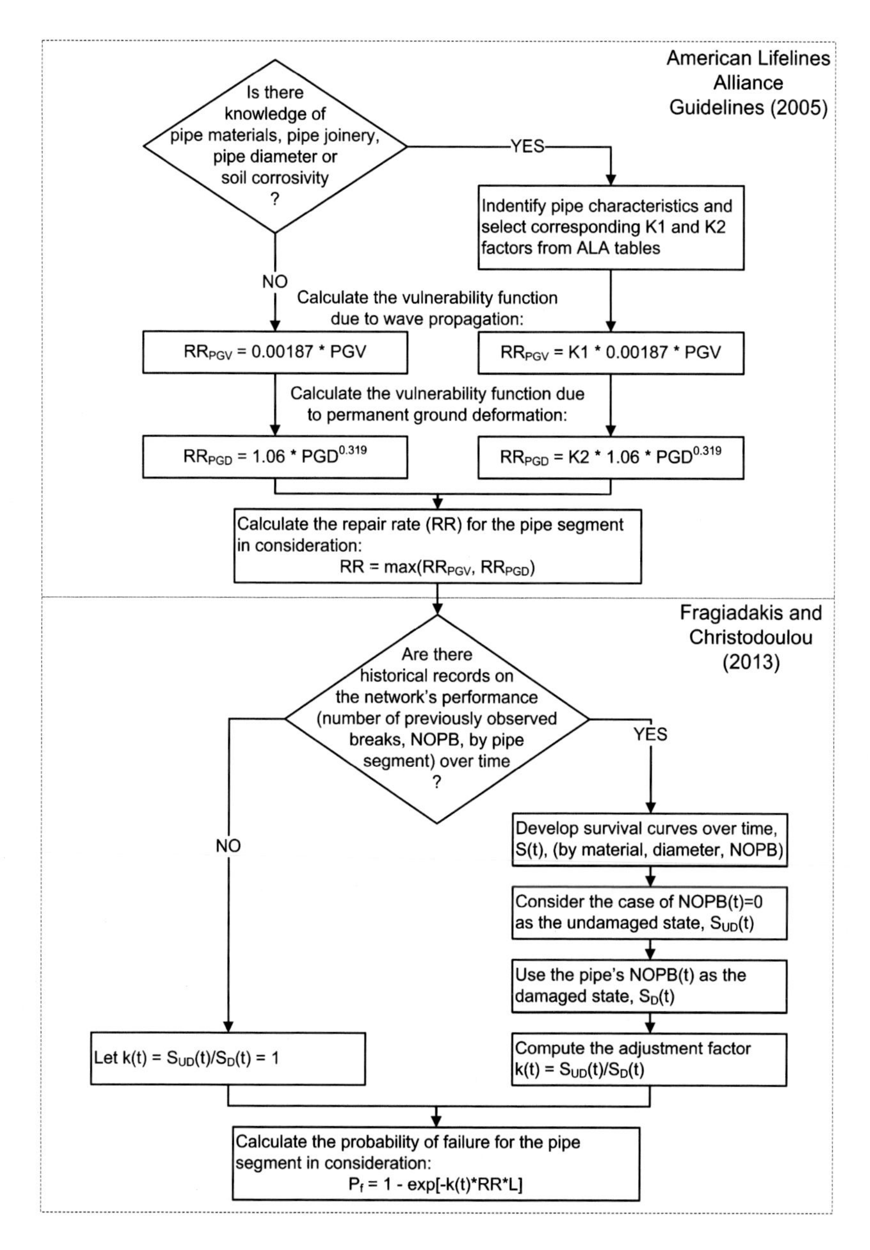

Figure 6.10 Reliability analysis under dynamic loading conditions.

increased pipe vulnerabilities, as impacted by their performance prior to the seismic event and their decreasing reliability. The probability of fail-

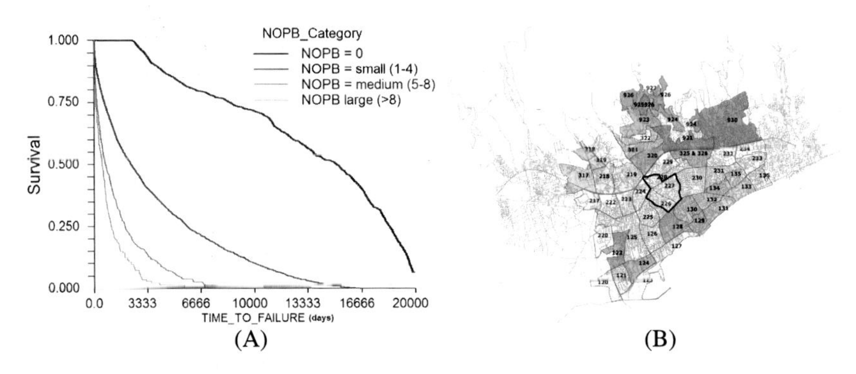

Figure 6.11 (A) Survival curves for asbestos cement (AC) pipes, as a function of the number of previous breaks (NOPB); (B) Real-life DMA as adapted from Limassol's (Cyprus) WDN's shown by dotted lines.

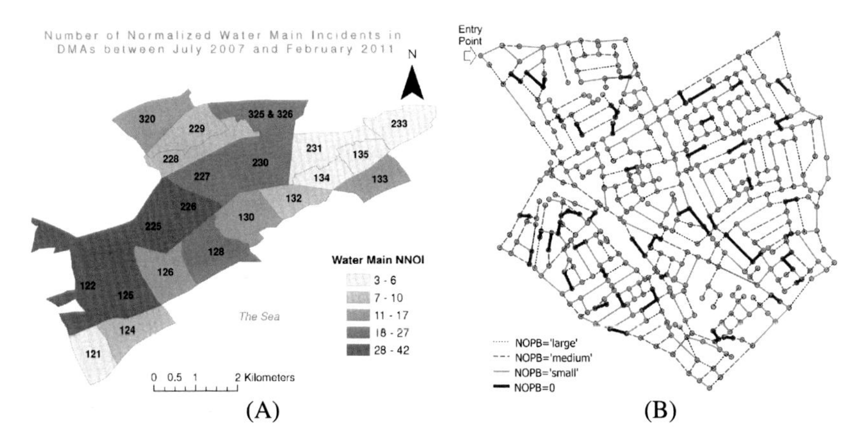

Figure 6.12 (A) The number of normalized water-main incidents, in DMAs of the Limassol network, between July 2007 and February 2011; (B) Topology of the DMA network considered, in a disturbed state ($NOPB \geq 0$ for every pipe).

ure is now highly sensitive to the condition of each network element, as manifested and influenced by their nonseismic performance over time (survival analysis), and introduced in the risk analysis by means of the proposed adjustments to the American Lifelines Alliance guidelines [6].

If the hydraulic parameters of the WDN in study are added to the vulnerability analysis, then the water pressure and flow values need to be considered. In the case-study example, the region of choice encompasses three DMA regions, it has a range of elevations from 15–65 m (50–215 ft) (Fig. 6.15A), a total consumer demand of approximately 10,000, an average

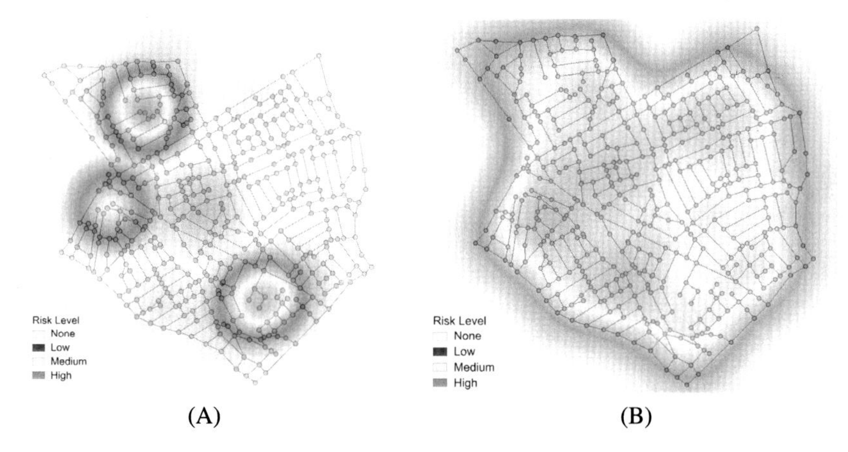

Figure 6.13 (A) Spatial analysis (heatmap) of studied network's reliability based on the ALA guidelines. Seismic effects are considered ($M_W = 7$); (B) Spatial analysis (heatmap) of studied network's reliability based on survival curves and considering the effects of the NOPB risk factor.

daily consumption of 200 L/day/capita (53 gpd/capita). The EPANET real time hydraulic simulation consists of 450 junctions (nodes), 621 pipes with diameter of 100 mm ($D = 4$ in), 1 reservoir with a total head of (90 m (295 ft)). All pipes are given a constant friction factor and the friction loss is calculated by the Hazen–Williams equation. Further details can be found from EPANET's manual [147].

The use of EPANET as a hydraulic analysis tool for seismic simulation analysis poses several challenges as reported by [25], especially when the dynamic link library is to be utilized in a multiple iteration vulnerability assessment application [35,59–61]. No-flow subnetworks often result in negative pressures at nodes where base demand is expected and hence the simulation stops. Moreover, whilst the MC simulation is performed, pipe failure could either signify that a selected pipe (link) would be completely decommissioned from the network (closed while experiencing catastrophic failure) or would still be connected but experiencing water leak (pressure loss) that is not known in advance based on the type of failure the pipe experienced. In both cases, the network description file would need to be adjusted and a new network file created for each iteration point without any assurance that the adjusted file would provide converged results. Hence, the steps proposed by the reliability analysis shown in Fig. 6.9 were followed, with the hydraulic analysis step completed selectively on a smaller

Figure 6.14 Real-life DMA as adapted from Limassol's (Cyprus) WDN: (A) Elevation map of simulated region; (B) Undisturbed analysis based on EPANET and initial conditions described in Section 3.3.2. (*Elevation of each node is shown in (B) in meters (m).*)

subgroup of scenarios as a stand-alone step on converged simulation results as those illustrated in Fig. 6.13. Initial simulation results for the undisturbed network are shown in Fig. 6.14 with an elevation map of the region and a flow path analysis for the boundary conditions described earlier.

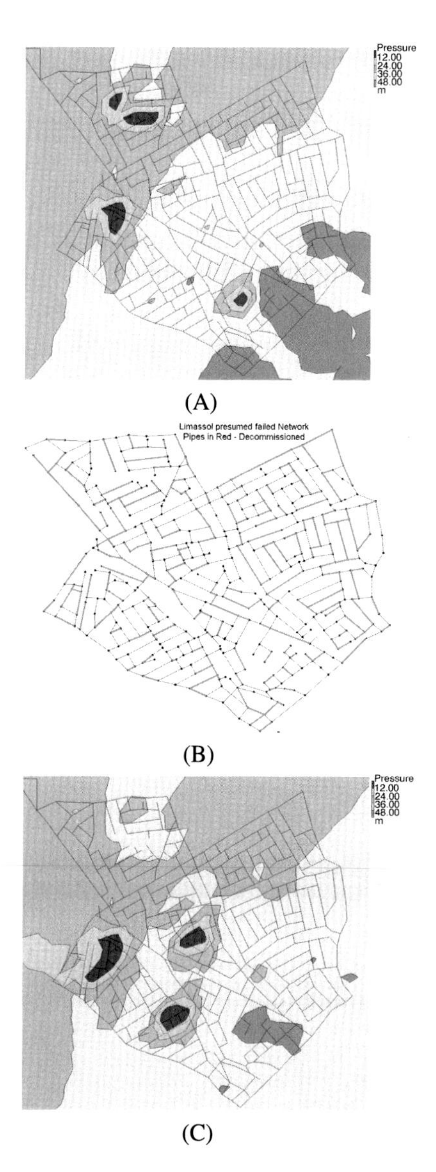

Figure 6.15 (A) Total Head of system (m) prediction by EPANET hydraulic simulation in Fig. 6.14A; (B) MC identification of failed pipes within the network; (C) Total Head of System (m) considering the effects of the NOPB risk factor as described in Fig. 6.13B.

Following the proposed outline of steps in Fig. 6.9, EPANET's dynamic link was incorporated within the existing code that performs the MC simulations and produces the resulting heatmaps as shown in Fig. 6.15. At each

MC step, several pipes are given a failure condition as shown in Fig. 6.15B and the EPANET library is then called and adjusted domain solved for all hydraulic quantities. Figs. 6.15A and 6.15C depict the preliminary results of total head of the system for the seismic scenario described in Fig. 6.13. Blue contours (zero demand availability) match well with the spatial results obtained from the MC simulations, identifying regions of high probability of failure and thus zero supply to the demand regions. It should be noted that the simulation stability was greatly improved by the alternate representation of the demand at the nodes. If demand was assigned for each node explicitly, a separate step in the iteration loop would require the identification of nodes that were unreachable either due to the assumption that their neighboring links were decommissioned, or if the only available flow path was at a much lower elevation. The subsequent step is a rather cumbersome and also complex process with high likelihood of divergence. The built-in capability of EPANET to model pressure-dependent flow issuing from emitters (i.e., pipe-leak model) with the outflow rate directly dependent on the local node pressure allows for a smoother and more stable simulation. The overall process and modeling set-up for the correct simulation and representation of the physical scenario needs to be further investigated. Factors that could be explored are the coefficients that describe the emitters behavior, the number of pipes that are decommissioned, the overall pressure drop and the assumption that water is readily available from the reservoir/tank at a specific head, etc. These results will be discussed in detail while a sensitivity and optimization matrix is run and validated against available data. Future work entails the use of richer datasets which will be used to validate the predictions along with a more appropriate hydraulic model, where the connectivity is not as redundant as the one here in which connectivity is always assumed at each node.

6.4 CONCLUDING REMARKS

A methodology is proposed by which hydraulic considerations can be utilized in the vulnerability analysis of WDNs, and by which the seismic performance of WDNs can be enhanced so as to enable the inclusion of a network's nonseismic and hydraulic performance in the calculation of a WDN's reliability. Available past network performance is processed through data-mining and survival analysis techniques, clustered by a number of different risk factors, and then used to calculate the generalized pipe vulnerability considering data that refer to the specific network. The network reliability is subsequently assessed using Graph Theory tools and Monte Carlo simulation. The proposed approach, which is demonstrated on a real-scale network with multiple district metered areas in the city of Limassol (Cyprus) allows for both the estimation of the probability that a network fails to provide the desired level of service (hydraulic vulnerability) and also the prioritization of retrofit interventions and of capacity-upgrade actions. A spatial analysis, by means of risk "heatmaps", of the proposed method's derived network hydraulic reliability is also included, showcasing the effects of past nonseismic performance of network components on a network's reliability.

A hydraulic model was developed to complement the results presented herein providing a robust methodology in the true identification of the reliability of a WDN. The complexities of the adaptation of such a hydraulic model, even though high, are not prohibitive against the implementation of such methodologies by WDN managers. Further, with regard to hydraulic vulnerability, the effects on the network of intermittent water supply operations which have been shown to cause an increase of leakage incidents in key WDN components (Chapter 3, p. 131), should also be considered, as IWS is related to hydraulic transients ("hammer effect") and thus to increased WDN vulnerability (typically higher number of pipe breaks).

CHAPTER 7

Real-Time Monitoring

The general goal in a time-domain anomaly analysis is the detection of events of interest which are anomalous, in the sense that they are out of the ordinary pattern and consecutively have a low frequency of occurrence, and which typically affect a subgroup of the data rather than an individual data point.

Christodoulou, S.E., et al. (2016),
Journal of Water Resources Management.

7.1 INTRODUCTION

As water distribution networks (WDNs) worldwide are rapidly degrading and damages in them are increasing, the need for efficient assessment and real-time monitoring is becoming increasingly pressing. The damages in WDNs are most often manifested as pipe-failure incidents and are leading to significant levels of nonrevenue water (typically in the range 20–30%). While data on pipe failures can, post-event, be analyzed and provide useful knowledge on the causalities of the failures, records of these incidents are usually hard to produce, are inconsistent and poorly recorded, and most often only span short periods of time. Thus, the need for timely detection of any anomalies in the WDNs and the correlation of such anomalies to water loss, has become of utmost importance to the efficient operations of WDNs.

Chapter 7 presents an approach to the timely identification of anomalies in WDN by use of a change-point anomaly detection algorithm and of streaming water consumption data from consumers and from district meter areas (DMA), by use of "automatic meter reading" devices (AMR). The proposed anomaly detection approach is then coupled with spatio-temporal analysis to arrive at spatial decision support systems (DSS) that automate the process of water loss detection, help reduce water loss, and increase the efficiency in the management WDN, and they eventually positively contribute to the sustainability of such networks. The proposed framework (Fig. 7.1), adopted from Gagatsis et al. [64], is an integrated

Urban Water Distribution Networks
DOI: http://dx.doi.org/10.1016/B978-0-12-813652-2.00007-4

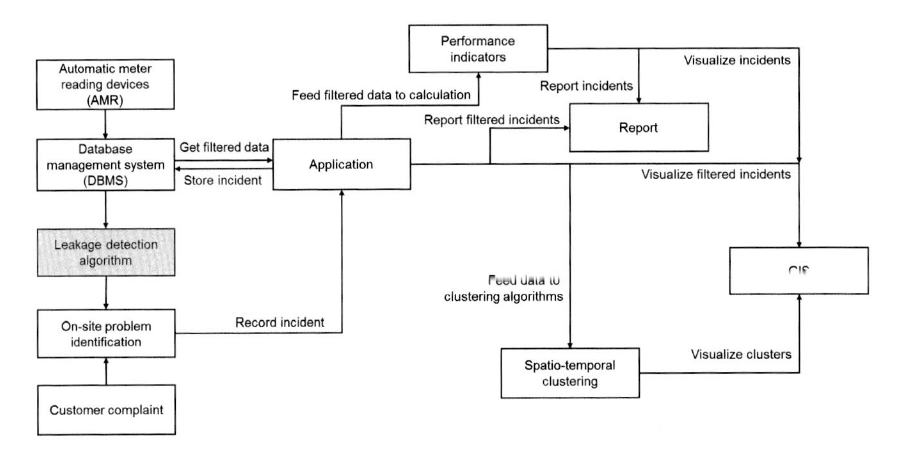

Figure 7.1 Schematic of the WDN monitoring, anomaly detection, and visualization solution proposed (adopted from Gagatsis et al., 2012).

monitoring solution in the form of an application software integrating a relational database management system, automatic metered reading (AMR) devices, geographic information systems (GIS) and spatio-temporal clustering techniques.

The aforementioned system provides real-time performance monitoring of water distribution networks and near-real-time anomaly detection in water consumption, at either an aggregated (DMA) or an individual (household) consumer level. The anomaly detection component is highlighted with a darker shade, for quick reference.

Automatic meter readers are devices embedded in house connection meters which can provide readings in frequent intervals and transmittal of such readings to a remote server, without the need for physical presence at the water meter. By examining unusual fluctuations in the water consumption, water board technicians can be made aware of potential anomalies in the WDN (such as water leaks) and investigate them, thus resulting in an automated, faster and more efficient leak detection method. Furthermore, the use of AMR can provide information on the water consumption within a district metered area (DMA) in near-real time, and compare this information with the total water input in that DMA. Since a DMA is an area of the WDN where water is distributed to the network from a single point of entry and where the network's water pressure is approximately uniform, large differences between periodic readings often indicate leaks within the network.

7.2 STATE OF KNOWLEDGE ON WATER LOSS DETECTION

Anomalies in WDN operations usually manifest themselves in the form of pipe-failure incidents (pipe breaks) and in the form of water loss. While pipe-failures are not a perfect indicator of the condition state of a WDN, they are amongst the primary symptoms of network degradation and one of the few sources of information on the condition of WDNs. However, recording pipe-failure incidents is a relatively recent practice. Furthermore, knowledge of the type of pipes and their condition over an area is often unknown prior to a pipe-failure incident which requires repair.

Past efforts had led to the identification of several risk factors influencing the fragility of WDNs and to the devising of decision support systems for sustainably managing them. For example, Christodoulou and Deligianni [33] used data mining techniques and neurofuzzy systems in order to identify pipe-failure data clusters and convert these into decision support rules for water loss management, with survival analysis curves provided for several pipe clusters by Christodoulou [27]. Subsequently, Christodoulou and Agathokleous [28] studied the effects of intermittent water supply on the vulnerability of WDNs, reporting that intermittent water supply policies resulted in increased stresses in the network they were applied to that eventually lead to an abnormally high number of pipe-failure incidents and higher volumes of water losses. The spatial distribution of pipe-failure incidents was studied by Gagatsis et al. [64] who first used spatial analysis techniques and a spatial clustering method to study the extend of pipe-failure clustering in the Limassol WDN and then transformed the deduced knowledge into a decision support system.

Records on pipe-failure incidents are usually the result of customer complaints (e.g., low pressure, no water, high bills, etc.) or of technical observations, which subsequently lead to field investigation by a Water Board's technical staff. However, these sources of information often require a certain amount of time before the source of a water leak is discovered, thus resulting in significant water losses until the damage is repaired. In order to resolve this problem, the use of automatic meter reading (AMR) technologies is being advocated and is currently widely implemented.

In principle, the use of AMR devices permits Water Boards to detect water leaks before they are reported by consumers; a feature that is particularly useful in cases when consumers are away on vacation or in cases when the leak is not detectable by the consumers. By examining unusual rises or drops in water consumption, water board technicians can be made aware of potential leaks and investigate them, thus resulting in a faster, more efficient leak detection method. Further, the use of AMR can provide near real-time information on the water consumption within a district metered area (DMA), and a possibility to compare it with the total water input in that DMA.

The AMR devices send consumption information in frequent intervals to the central database. This data is then analyzed for detecting abnormal increases in water consumption readings sent from the AMR units. When such an abnormal increase is detected, an alert is sent to the WDN operators of possible WDN malfunction and then the WDN operators are tasked with linking the event with a possible water leakage.

This process is in many aspects subjective (in terms of how the anomaly thresholds are set) and manual (thus time consuming) with regards to the analysis of indicators (outliers) of water consumption anomalies. Further, the architecture shown in Fig. 7.1 relies heavily on the wide use of AMR devices in the WDN (one for each house connection) and thus it is not cost-efficient.

A solution could be the selective location of AMR devices (or sensors) within the WDN, for which several sensor placement optimization techniques exist (Ostfeld et al., [125]). The problem, though, of analyzing the sensed operational data of a WDN still remains, in terms of processing time-related data and sifting through this data for possible anomalies in the WDN's behavior.

As Laucelli et al. [101] state, *"recent approaches implementing data mining techniques for analyzing pressure/flow data appear very promising, because they can automate mundane tasks involved in data analysis process and efficiently deal with sensor data collected. Furthermore, they rely on empirical observations of a WDN behaviour over time, allowing reproducing/predicting possible future behaviour of the network."*

Related recent work on water loss detection by means of data mining and informatics can be found in [2,24,77,85,101]. Adnan et al. [2] discussed gas leak detection in pipeline systems by means of an acoustic method and the wave propagation in the pipeline, using signal processing to decompose the raw signal. Even though the study focused on gas pipelines, the methods used could be extended to water pipelines. Jung and Lansey [85] used a nonlinear Kalman filter (NKF) method to identify a WDN's system condition, to estimate the system state, and to detect bursts, while Hutton and Kapelan [77] presented a preliminary investigation into the application of a probabilistic demand forecasting approach to identify pipe bursts, comparing time-series data to mean daily night flows and quantifying the probability that a future observation is abnormal. Also, Choi et al. [24] investigated the effect of sampling interval when an adaptive Kalman filter is used for detecting bursts in a WDN, presented a sampling algorithm that adjusts the sampling interval depending on the normalized residuals of flow after filtering and applied their algorithm to a virtual sinusoidal flow curve and real DMA flow data, showing a reasonable accuracy in burst detection. Finally, in the work of Laucelli et al. [101] the effectiveness of the evolutionary polynomial regression (EPR) paradigm is investigated in reproducing the behavior of a WDN using online data recorded by low-cost pressure/flow devices. Data from a real DMA is then used to show that by using the EPR paradigm a model can be built which enables the accurate reproduction and prediction of the WDN behavior over time and detection of flow anomalies. The authors then suggest that such an EPR model might be integrated into an early warning system for when anomalies are detected.

In terms of near-real-time detection of pipe bursts, recent related work can be found in Yazdekhasti et al. [179], Romano et al. [145], Kim et al. [92], and Loureiro et al. [108], with brief summaries given below.

Yazdekhasti et al. [179] presented a noninvasive approach for leak detection involving continuous monitoring of the changes in the correlation between surface acceleration measured at discrete locations along the pipeline length, and the use of a leak detection index based on the cross-spectral density of measured pipe surface accelerations, for detecting and assessing the severity of leaks.

Romano et al. [145] presented a methodology for the automated near-real-time detection of pipe bursts and other events that induce abnormal pressure/flow variations at the (DMA) level. The methodology makes synergistic use of several self-learning artificial intelligence (AI) techniques and statistical data analysis tools, including wavelets for denoising of the recorded pressure/flow signals and artificial neural networks (ANNs) for the short-term forecasting of pressure/flow signal values. The methodology was tested on a case study involving several DMAs in the UK with both real-life and simulated pipe burst/other events.

A similar method was also presented by Kim et al. [92], who used a cumulative integral of shifted pressure data and localization based on statistical estimation to detect leakages. Verification was performed using two different field leakage data sets and normal data sets, with the researchers stating that the proposed detection scheme exhibited fewer false alarms than other previous methods.

Loureiro et al. [108] suggested a four-step methodology in detecting water leakages (data acquisition; data validation and normalization; anomalous observation detection; anomalous event detection and characterization), and an approach based on outlier regions. The researchers conclude that their approach is flexible and applicable to the detection of different types of events (e.g., pipe burst, unusual consumption) and to different flow time series (e.g., instantaneous, minimum night flow).

A recent work on the dynamic analysis of time series related to water consumption was recently reported upon by Christodoulou et al. [36], in which wavelet change-point detection classifiers were utilized for identifying anomalies in the consumption patterns. The wavelet change-point method utilizes the continuous wavelet transform (CWT) of time-series (signals) to analyze how the frequency content of a signal changes over time. In the case of water distribution networks the time-series relates to streaming water consumption data from automatic meter reading (AMR) devices, at either the individual consumers' level or at an aggregated district meter area (DMA) level. The wavelet change-point detection method analyzes the provided time-series to acquire inherent knowledge on water consumption under normal conditions at household or area-wide levels, to then make inferences about water consumption under abnormal conditions.

7.3 TIME SERIES ANALYSIS FOR ANOMALY DETECTION

The general goal in a time-domain anomaly analysis is the detection of events of interest which are anomalous, in the sense that they are out of the ordinary pattern and consecutively have a low frequency of occurrence, and which typically affect a subgroup of the data rather than an individual data point. Mathematically, if the general form of the time series of a dataset D is given by $D = x_1, x_2, \ldots, x_n$, where x_i are the observations over time, then the goal is to detect a time t when an abnormal event occurs, to characterize the event (identify the affected data subset and its features), and to measure the event's severity. Further, the goal is to detect the event as accurately and as early as possible. The detection may either be temporal, or spatio-temporal depending on the problem at hand and the detection needs.

In terms of temporal detection methods, the general detection framework is based on developing a model to predict expected signal values, to measure the difference between actual and expected values, and to compute an alarm value that relates to the observed data anomaly. The most common univariate temporal methods are the following:

- Moving Average

  If we let W be the window size, then a moving average window predicts the following:

$$X_{t+1} = \frac{1}{W}\left(X_t + X_{t-1} + \cdots + X_{t-W-1}\right). \tag{7.1}$$

  If a Gaussian distribution, $N(\hat{\mu}, \hat{\sigma})$, is then set so that it fits the W observations within the window, the alarm level can be defined as

$$AlarmLevel = \Phi\left(\frac{\max(0, X_i - \hat{\mu})}{\hat{\sigma}}\right) \tag{7.2}$$

  where Φ is the cumulative distribution function (CDF) for $N(0, 1)$. The method performs better on seasonal effects but has drawbacks if recent history is anomalous.

- Control Chart

  The mean ($\hat{\mu}$) and standard deviation ($\hat{\sigma}$) of the data up to current time is estimated and an upper control limit is set, equal to $\hat{\mu} + 3\hat{\sigma}$. If the set upper control limit is exceeded then an alarm is issued. Alternately, a threshold may be set and used in conjunction with

$$AlarmLevel = \Phi\left(\frac{\max(0, X_i - \hat{\mu})}{\hat{\sigma}}\right) \tag{7.3}$$

  where Φ is the CDF for $N(0, 1)$. An alarm is issued when the *AlarmLevel* is greater than the set threshold. The method raises too many false alerts because of the seasonal effects and is not sensitive to small shifts.

- Exponentially Weighted Moving Average (EWMA)

  The method, which is a variation of the moving average, uses a statistic, Z_i, defined as

$$Z_i = \lambda X_i + (1 - \lambda)Z_i - 1, \quad 0 < \lambda \leq 1 \tag{7.4}$$

  with observations in the past $[(1 - \lambda)Z_i - 1]$ receiving a decreasing amount of weight compared to current observations.

- Cumulative Sum (CUSUM) statistics
  The CUSUM method, which is good at detecting shifts from the mean more quickly than the control chart method, keeps a running sum of excesses in each time period over the mean and when this sum exceeds a set threshold H an alarm is issued and the sum is reset. The shift is defined by

$$S_h = \sum (X_k - r) + S_k - 1 \tag{7.5}$$

  where X_i is the ith observation, r is the reference value (for example, the mean) and S_i is the ith cumulative sum. The method is better at detecting small shifts in the mean.
- Regression
  The time series is fitted to a regression equation by use of

$$Y = \beta_0 + \beta_1(X_{1,t}) + \beta_2(X_2) + \epsilon_i \tag{7.6}$$

  where $X_{1,t}$ is the time variable and X_2 is a second, Boolean, variable used for accounting for certain data of interest and for assigning modified weights to such data points. The time series variable, $X_{1,t}$, can be modeled by a sinusoidal signal as

$$X_{1,t} \sin\left(\frac{2\pi n_p}{365.25} - \frac{\pi}{2}\right) \tag{7.7}$$

  where n_p is number of time periods since start of time series.
  Regression learns the equation factors (β) from the time series to minimize the residual sum of squares, with the error, ϵ_i, defined as a normally distributed noise with mean 0 and variance σ^2.

A graphical comparison of the performance of some of the above methods on a synthetic time series of 400 hours in duration can be seen in Fig. 7.2.

The synthetic time series, which will be discussed in more detail in subsequent sections, contains an induced anomaly (increased water consumption values for the $t \in [150, 250]$ hours of the time series). The simple, exponential, triangular, weighted, and modified moving averages of a synthetic water consumption hourly time series were produced, using a 48-hour time window. As observed, the best-fit moving average curves for the studied time series are the weighted and the exponential moving average curves.

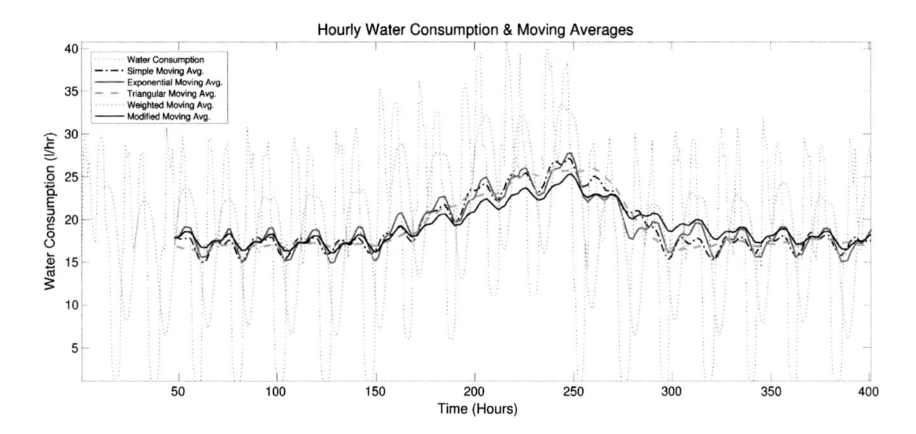

Figure 7.2 Moving averages of sample synthetic time series.

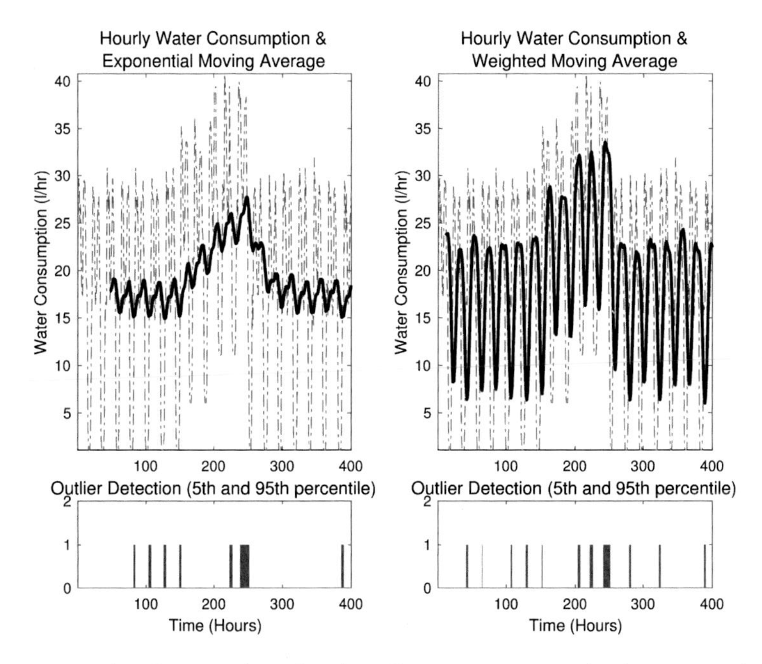

Figure 7.3 Outlier detection based on best-fit moving averages for the case-study synthetic time series.

A further analysis of the case-study time series using the two aforementioned moving average curves for outlier detection, produces the indicators shown in Fig. 7.3. The top two graphs in Fig. 7.3 show the moving average curves, whereas the bottom two barcharts show the time periods where the values are outliers (lower and upper 5% outliers). The outliers are hence used as indicators of anomalies in the time series.

As can be observed from Fig. 7.3, the moving average methods do detect anomalies in a time series, but the time reference for these anomalies differs depending on the method used. Further, the methods produce several false-positive indications, thus making them ineffective for problems such as water loss detection.

Other univariate methods include Box–Jenkins models, wavelets, change-point detection, Kalman filters, and hidden Markov models.

Similarly, multivariate temporal methods may be applied to time series where each data point is a multivariate vector. In such case there are two classes of detection methods, depending on the level of detection required: multivariate change-point detection and multivariate event detection. Both methods detect the occurrence of a change, but while the former does not identify the subgroup of data that has changed the most, the latter method does identify it.

7.4 WATER LOSS DETECTION IN STREAMING WATER FLOW TIME SERIES USING CHANGE-POINT ANOMALY METHODS

Change-point detection refers to the identification of both whether or not a change (or several changes) has occurred in a time series (signal), and of the times of any such changes. Common change-point detection methods relate to changes in the mean, variance, correlation, density, or slope of the signal in study.

Among the several change-point detection methods found in literature, notable is the approach proposed by Yamada et al. [173] and Liu et al. [106], who presented a statistical change-point detection algorithm based on nonparametric divergence estimation between time series samples from two retrospective segments. The algorithm uses the relative Pearson (PE) divergence as a divergence measure and a method of direct density-ratio estimation. Mathematically, the PE divergence from $p(x)$ to $p'(x)$ is defined as

$$PE\left[p(x), p'(x)\right] := \frac{1}{2}\int\left[\left(\frac{p(x)}{p'(x)} - 1\right)^2 p'(x)dx\right] \tag{7.8}$$

and, in their work, Yamada et al. [173] and Liu et al. [106] proposed an approach to distribution comparison called α-relative divergence estimation, by which they estimated the α-relative divergence (defined as the divergence from $p(x)$ to the α-mixture density $\alpha p(x) + [1 - \alpha p'(x)]$ for $0 \leq \alpha < 1$). By use of Eq. (7.8), the α-relative Pearson (PE) divergence is given by

$$PE_\alpha\left[p(x), p'(x)\right] = \frac{1}{2}\int\left[\left(\frac{p(x)}{\alpha p(x) + (1-\alpha)p'(x)} - 1\right)^2 \left[\alpha p(x) + (1-\alpha)p'(x)\right] dx\right] \tag{7.9}$$

and the α-relative divergence was estimated by the researchers by direct approximation of the α-relative density ratio given by

$$r_\alpha(x) := \frac{p(x)}{\alpha p(x) + (1-\alpha)p'(x)}. \tag{7.10}$$

Yamada et al. [173] also showed that using an estimator of the α-relative density-ratio $r_\alpha(x)$, we can construct several estimators of the α-relative PE divergence, of which the following one is easy to compute:

$$\tilde{PE}_\alpha := \frac{1}{2n}\sum_{i=1}^{n}\left[\hat{r}(x_i) - \frac{1}{2}\right]. \tag{7.11}$$

In its final form, and by using an estimator $\hat{g}(Y)$ of the α-relative density ratio, the RuLSIF method approximates the α-relative PE divergence by use of the following equation [173]:

$$\tilde{PE}_\alpha := -\frac{\alpha}{2n}\sum_{i=1}^{n}\hat{g}(Y_i)^2 - \frac{1-\alpha}{2n}\sum_{i=1}^{n}\hat{g}(Y_i')^2 + \frac{1}{n}\sum_{i=1}^{n}\hat{g}(Y_i') - \frac{1}{2} \tag{7.12}$$

where $y(t)$ is the time series sample at time t, $Y(t)$ is the subsequence of k time series samples at time t, and n is the number of points in the analysis window.

7.5 CASE-STUDY: OVERVIEW

The applicability of change-point methods to anomaly detection in the operations of water distribution networks is case-studied on a synthetic hourly time series of about two months in duration (approximate 1500 hourly data readings), the signal of which is as shown in Fig. 7.4. The signal includes

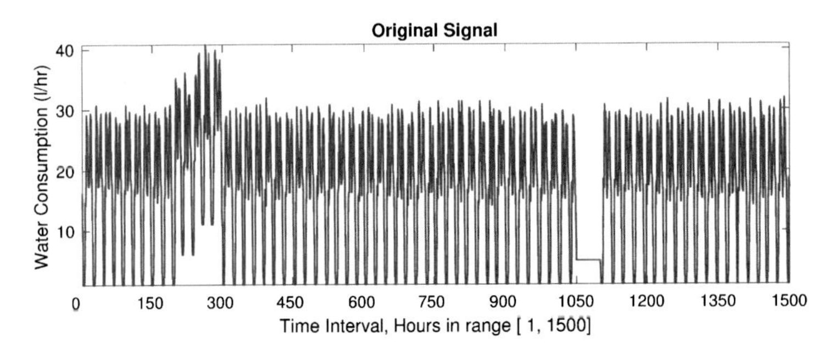

Figure 7.4 Two-month-long hourly water consumption signal for case-study household.

two induced water-flow anomalies at $t \in [200, 300]$ and $t \in [1050, 1100]$ hours, corresponding to a water loss incident and to an abnormal drop in water consumption, respectively.

The synthetic data has been generated based on studies reported in the literature (Athuraliya et al., [13]), subsequently adjusted to observations from real-life water flows from Nicosia, Cyprus (Christodoulou et al., [36]). The basis of the time-series in study is an hourly water consumption profile (indoor and outdoor usage) for a general household, as depicted in Fig. 7.5. Water consumption starts in the early morning hours (around 06:00), increases peaking up at around 10:00, then drops until the early afternoon hours (16:00), peaking up again in the late afternoon and early evening hours (18:00–21:00), before dying down at night (21:00–06:00). It should be also noted that the household's daily water consumption does not, at any point in time, vanish. This is in agreement with the "minimum night flow (MNF)" concept, commonly used in WDN operations. MNF is a common method used to evaluate water loss in a water network, and refers to the water volume flowing through the network even when all true water demand is zero (typically in the time band of 02:00–04:30).

The average daily water use, in liters, per person and household is given in Table 7.1 [13], along with the two regression models fitted to the water consumption data [13] relating water use to the household size (HH size).

Further to the profiled daily water consumption, three induced anomalies in the household's consumption are recorded (as shown in Fig. 7.4): (1) a two-day-long drop in consumption ($t \in [1050, 1100]$ hours); (2) a 5-L/hour water loss for a duration of 48 hours ($t \in [200, 250]$ hours); and (3) a 10-L/hour water loss for a duration of 50 hours ($t \in [1050, 1100]$ hours). The water consumption's time series (Fig. 7.4) is first processed

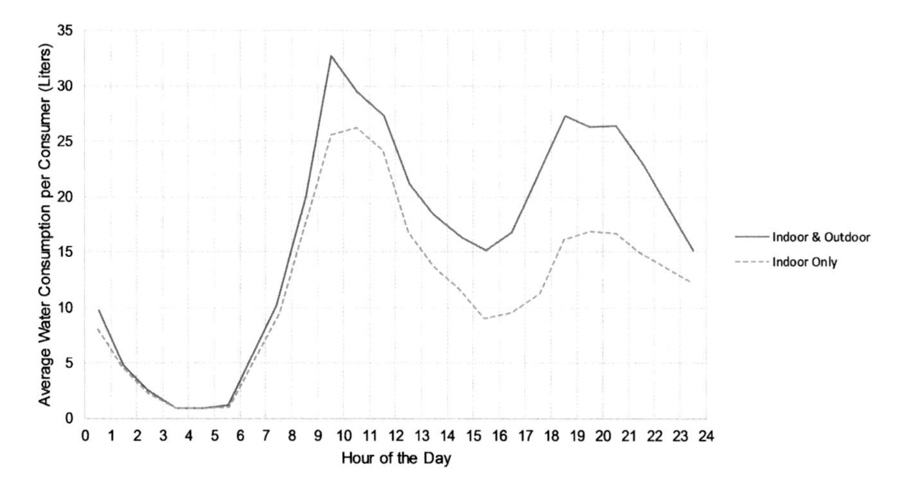

Figure 7.5 General daily water consumption profile (indoor vs. indoor & outdoor usage).

Table 7.1 Average daily water use per person and per household.

Household size	Daily per person[1] water use (L)	Daily per household[2] water consumption (L)
1	141	147
2	152	319
3	149	419
4	132	490
5	100	546
Weighted average	137	336

[1] Per person use $= -6.99 \times (HHsize)^2 + 31.685 \times (HHsize) + 116.72$.
[2] Per household use $= 247.82 \times \ln(HHsize) + 146.85$.

macroscopically to identify the time periods of concern, and then microscopically to zoom in on possible consumption anomalies. As a point of reference and subsequent assessment of the RuLSIF approach, the water consumption signal is first analyzed for anomalies by use of outlier detection and the cumulative sum (CUSUM) methods.

7.6 CASE-STUDY: PRELIMINARY ANALYSIS

Let us first examine the efficiency of the outlier and CUSUM methods (Section 7.3, p. 232), by use of the case-study time series of Section 7.5 depicting the water consumption of a 3-person household following the consumer profile of Fig. 7.5. The average total daily water consumption

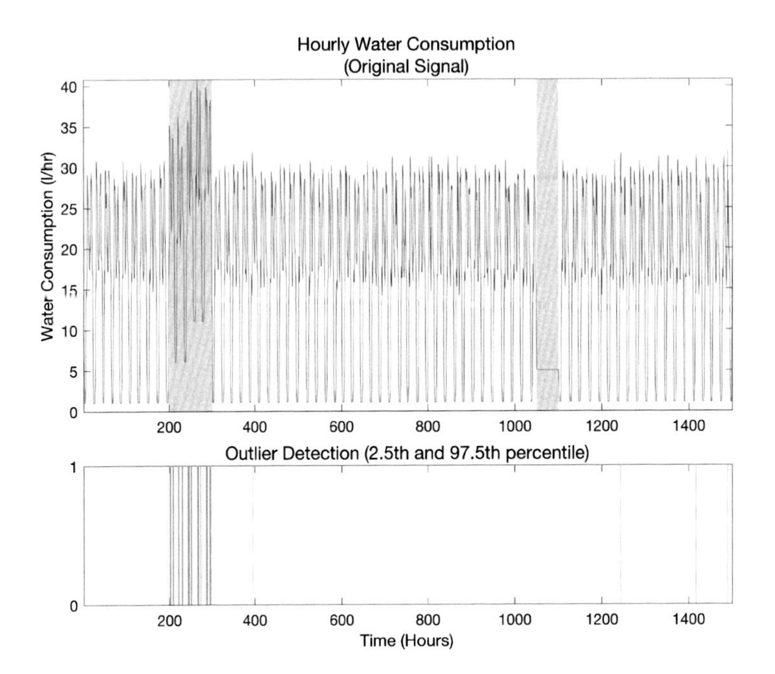

Figure 7.6 Water consumption signal and outlier analysis (disturbed signal, hours 1–1500).

of the household is presumed to be 419 liters (L) (Table 7.1), distributed hourly across the day as per Fig. 7.5.

The water consumption's time series is first processed macroscopically (the analysis's range is set to the total signal duration of 1500 hours) by use of outlier detection (Fig. 7.6) to identify the time periods of concern, and then by use of the CUSUM method (Fig. 7.7) to assess the suitability of the method in detecting signal discontinuities. In order to simplify the computations, for the macroscopic analysis a True/False anomaly indicator was used instead of an anomaly score, and in order to assist in the ensuing discussion the time periods of known abnormalities in the signal are highlighted (Fig. 7.6).

As Fig. 7.6 shows, the outlier detection method identifies several outliers in the water-flow data time series, but not all of them are necessarily related to WDN abnormalities. Rather, the outlier detection approach simply detects deviations from the norm (the outliers) and it performs better at picking up unusual highs while it is less efficient in picking up unusual lows in the signal. For example, as shown in Fig. 7.6, the outlier detection method correctly identifies the period of increased water

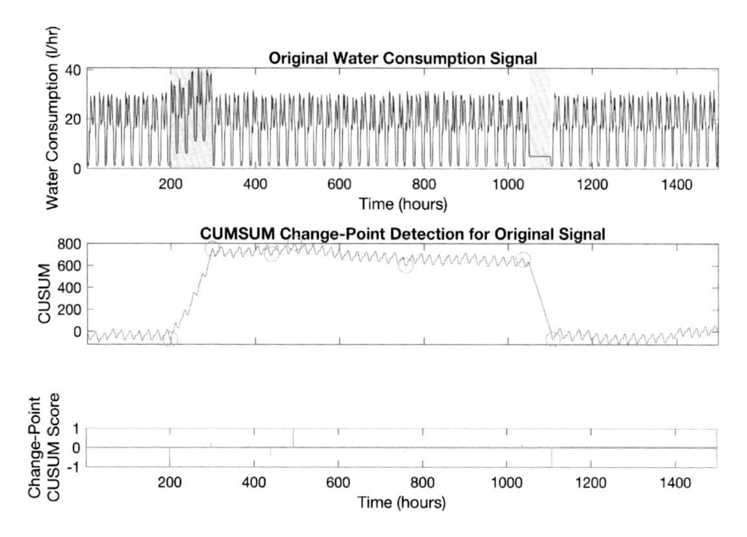

Figure 7.7 Water consumption signal and CUSUM analysis (disturbed signal, hours 1–1500).

flow (water loss; $t \in [200, 300]$ hours), whilst it fails to detect the period of decreased water flow (broken meter, or absence/vacation of consumer; $t \in [1050, 1100]$ hours). Further, several false-positives are produced ($t \approx 400, 1250, 1410, 1480$ hours).

In the case of the CUSUM method (Fig. 7.7), the identified signal inflection points are related to the abnormalities in the water flow data, and used to pinpoint such abnormalities. These inflection points, which are points of the curve at which a change in the direction of curvature occurs, are better suited for detecting sudden changes in the signal. The CUSUM method is actually a change-point detection method, and it is closer in philosophy to the RuLSIF method (to be discussed in the following sections) compared to the outlier detection methods.

As depicted in Fig. 7.7, the CUSUM method correctly identifies the two time periods of known anomalies in the consumption signal ($t \in [200, 300]$ and $t \in [1050, 1100]$ hours), since these two anomalies are related to sudden changes in the direction of the signal's curvature. At the same time, the method detects discontinuities in the signal which were not otherwise noticeable (at $t \approx 440$, 490, and 760 hours). However, these discontinuities are not directly related to abnormal conditions, as they may simply be a continuation of a progressively increasing signal and not of an

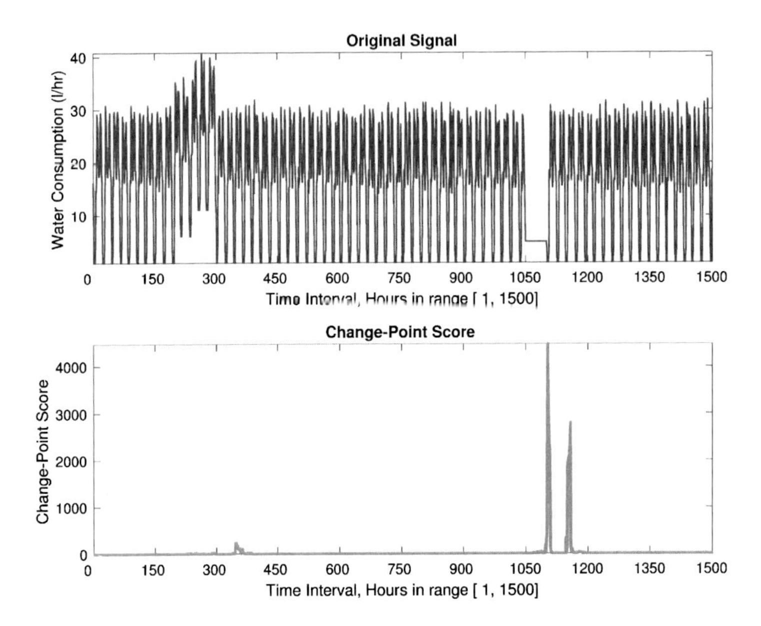

Figure 7.8 Water consumption signal and change-point score (Hours 1–1500).

abrupt change in it, and they are thus false-positive indications of known abnormalities in the water consumption.

7.7 CASE STUDY: RULSIF-BASED ANOMALY DETECTION

As aforementioned, the case-study time series contains two induced anomalies, in the time regions of $t \in [200, 300]$ and $t \in [1050, 1100]$ hours. This case-study signal is first processed macroscopically (the signal duration is set to 1500 hours) to identify the time periods of concern, and then microscopically to zoom in on possible water-flow anomalies.

The data stream was segmented in a sliding window manner with window width $n = 50$ and sliding step $k = 10$ time-periods. The value of parameter n affects the resolution (time interval) of the analysis and the value of k affects the immediacy (time shift) of the anomaly detection.

The results of the macroscopic analysis are depicted in Fig. 7.8, with the top half of the figure showing the signal in study, and the bottom half of the figure showing the RuLSIF-computed anomaly scores. The water consumption signal oscillates over time, with an escalation in water consumption in the region $t \in [200, 300]$ and a drop in consumption in the region $t \in [1050, 1100]$ hours (Fig. 7.8).

As initial observations, the following statements can be made:

- Both these signal anomalies are picked up by the RuLSIF change-point detection method,
- A slight time-shift in the anomaly detection is observed (this relates to the RuLSIF parameters being used),
- The detection is accompanied with anomaly scores, alluding to the severity of the detected anomaly,
- One of the detected anomalies (in period $t \in [1050, 1100]$ hours) has an anomaly score that outweighs any other anomalies in the time series.

A closer, microscopic, analysis of the detected anomalies is warranted.

7.7.1 Abnormal Drop in Water Consumption

At first, let us investigate the applicability of the RuLSIF change-point method to the detection of an anomaly in the water-flow time series related to an abnormal drop in water consumption. Such abnormal drop may stem, for example, from either a malfunction in the household's water meter, or from a loss of connection with the water meter (in the case of AMR devices) and thus intermittent data transfer, or from an absence of the household's occupants for a period of time (e.g., a summer vacation).

In order to get a closer look at the underlying time series, a microscopic analysis is performed for the specific time interval of concern (Fig. 7.9). The RuLSIF parameters used in the microscopic analysis are now $n = 25$ and $k = 1$, aiming for a finer resolution and a smaller time shift in the analysis and prediction.

The change-point detection method identifies both the start and the end points of the anomaly (with a small time-shift in the computed points), assigning high anomaly scores to them.

7.7.2 Water Loss Detection

We now turn our attention to the suitability of the RuLSIF change-point method to water loss detection, reexamining the aforementioned case-study time series and zooming in on time period $t \in [200, 300]$, in which an escalating water leak is present. The water leak is presumed to be 5 L/hr for the first 50 hours and then 10 L/hr for the next 50 hours.

As a prelude to the analysis let us first consider a period of no known water losses, such as the time interval $t \in [600, 1000]$, and perform a microscopic analysis of the water consumption signal for this time period. The

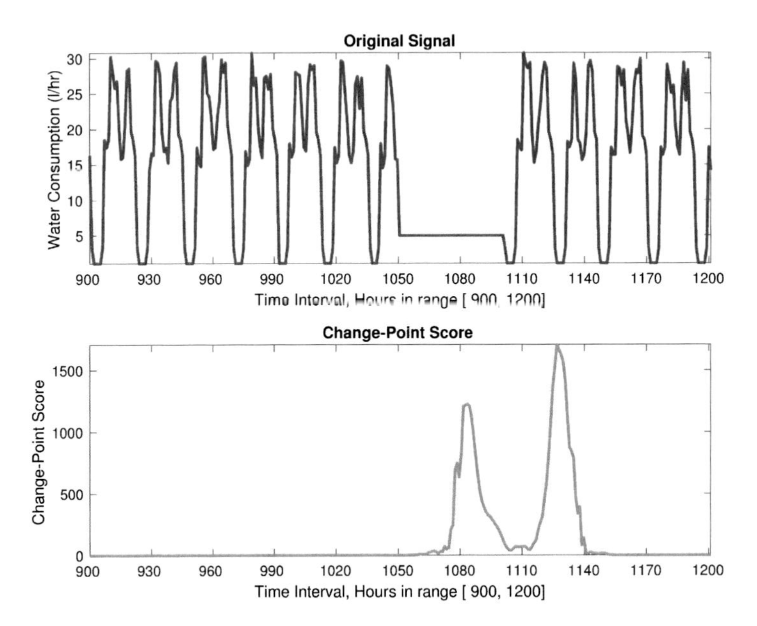

Figure 7.9 Water consumption signal and change-point score (Scenario 1, Abnormal drop in water consumption; Hours 1050–1100).

computed anomaly scores for this section of the signal (which corresponds to no abnormalities in water consumption) are shown to be oscillating in the range $[0, 0.5]$ (Fig. 7.10).

With this in mind, we shift our attention to the water loss section of the signal ($t \in [200, 300]$ hours). The macroscopic analysis (Fig. 7.8) showed the increase in water consumption during the water leak period ($t \in [200, 400]$) and identified this as an anomaly in the general water consumption signal, albeit overweighed by the computed anomaly score for the period of $t \in [200, 300]$ hours.

A microscopic analysis (Fig. 7.11) confirms the findings of the macroscopic review (Fig. 7.8). A spike in the change-point score is first recorded at approximately $t \approx 220$ hours, followed by a second spike at $t \approx 270$ hours, and then by a third and bigger spike at $t \approx 320$ hours, highlighting the anomaly in the water consumption pattern. The spikes in the anomaly score indicate the approximate start of the two phases of the recorded water leak (5 L/hr in range $t \in [200, 250]$ and 10 L/hr in range $t \in [250, 300]$). As with the previous case, the time shift in the anomaly score spikes are due to the RuLSIF parameters used in the analysis.

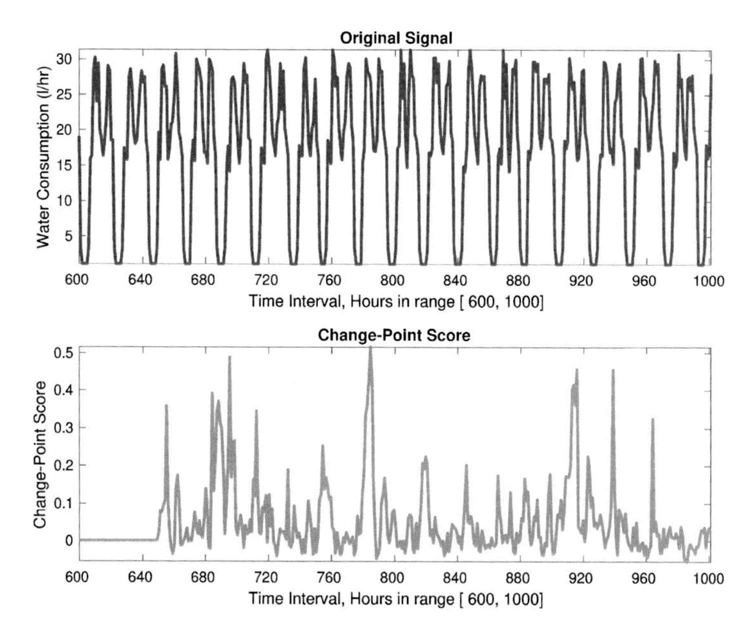

Figure 7.10 Water consumption signal and change-point score (Regular water consumption; Hours 600–1000).

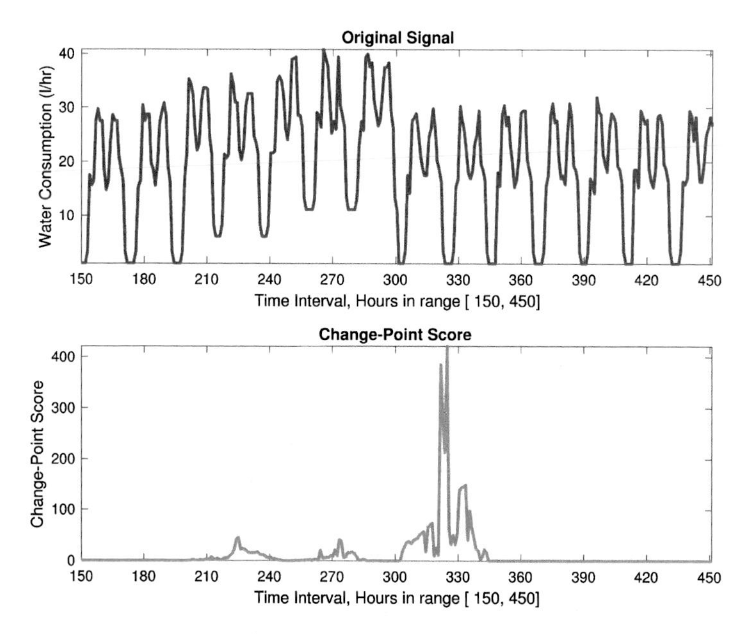

Figure 7.11 Water consumption signal and change-point score (Scenario 2, Escalated water loss; Hours 200–300).

Further, the utilized change-point detection method scores these water loss related anomalies with a higher score, compared to other signal anomalies such as those observed during normal conditions (Fig. 7.10), indicating the high probability that the detected deviation in water consumption is highly unusual and that it is not attributed to "signal noise" (i.e., small variations in water consumption patterns).

7.8 CONCLUDING REMARKS

The work described in the proceeding sections evaluated the utilization of the relative unconstrained least-squares importance fitting (RuLSIF) change-point method for the detection of anomalies in water consumption time series, and its applicability to water loss detection in water distribution networks. The RuLSIF change-point detection method was implemented on an hourly water consumption signal of about 1500 hours in duration, successfully detecting unusual reductions and increases in the water consumption patterns, and classifying them as anomalies. The first water consumption anomaly type related to a discontinuity in the signal (a break in the consumer's water consumption patterns), whereas the second type related to an unusual increase in the signal (two water loss incidents).

Unlike traditional approaches that monitor the average current periodic consumption and compare this average to the one from a corresponding past period, the RuLSIF change-point detection approach is dynamic and sensitive to any change in the time series in study. Even though the proposed analysis does not predict future anomalies, it is suitable for past and near-real-time anomaly detection; an attribute that is sufficient for water loss management as it allows for a timely detection of anomalies is streaming water flow data. Further, the method dynamically assigns anomaly scores to the detected changes in the signal, thus easing water loss detection and appraising the severity of each detected incident. The change-point detection method is able to extract knowledge from past consumption patterns (the water flow signal over time) to detect anomalies in the signal, both in the past and in near-real-time, and for both physical and apparent water losses.

Ongoing research work on wavelet change-point detection addresses signal identification, spatial mapping and heatmap generation for the identification of areas of concern in a distribution network, as well as the blending of automatic meter reading (AMR) technologies and water auditing at a district meter area (DMA) level, for the dynamic (and near-real-time) detection of water leaks in an urban water distribution network.

CHAPTER 8

From Historical and Seismic Performance to City-Wide Risk Maps

A vulnerability map gives the precise location of sites where people, the natural environment or property are at risk due to a potentially catastrophic event that could result in death, injury, pollution or other destruction. Such maps are made in conjunction with information about different types of risks.

Handbook for Vulnerability Mapping,
United Nations Environment Programme (UNEP).

8.1 INTRODUCTION

The chapter combines the work presented in Chapters 2–5, on the historical nonseismic vulnerability assessment of urban water distribution networks (WDN) and of their components by use of survival analysis, and on seismic vulnerability of WDNs by use of simulation and a graph-based shortest-path algorithm, to holistically evaluate WDN vulnerability. Further, the chapter discusses spatial aspects of WDN vulnerability, arriving at spatio-temporal depictions of it.

The intent is to first show how WDN vulnerability is spatially allocated, and then to depict how this spatial distribution is affected by the various sources of vulnerability and their estimation.

8.2 ESTIMATION OF THE PROBABILITY OF FAILURE – A RECAP

8.2.1 Estimating the Probability of Failure, P_f, of Every Network Component

As discussed in Chapter 2, there are several ways to estimate the nonseismic probability of failure for a WDN component. For example, one could simply use statistical probability based on the component's past performance (e.g., the number of breaks per unit length), or use heuristic/empirical

Urban Water Distribution Networks
DOI: http://dx.doi.org/10.1016/B978-0-12-813652-2.00008-6

methods. Other possible approaches rely on analytical (stochastic) models, or on survival analysis.

Survival analysis is probably the most powerful way to estimate the probability of failure of a component, for it formally accounts for missing data (past or future) and it allows for data clustering based on chosen risk-of-failure metrics. Survival analysis is a branch of statistics dealing with deterioration and failure over time and involves the modeling of the elapsed time between an initiating and a terminal event [70]. In the case of piping networks, an initiating event can be the installation of a pipe, a water leak observation or the start of a pipe treatment. Cases of terminal events can be a relapse of a previous leak, a fix or a failure. The method is based on estimating the reliability of a component and its lifetime, subject to multiple risk factors and the aim is to provide answers on the population fraction (e.g., pipes) that survives past an expected lifetime subject to various risk factors [70,74]. The data values used in the analysis are a mixture of both complete and censored observations. In the former case, a terminal event is thought to have occurred, while in the latter case a terminal event has not occurred yet. A terminal event is assumed to occur just once for every subject. A more detailed description of the method, its mathematical framework and a small numerical example can be found in section on pp. 175–179.

As aforementioned (please refer to section on p. 175), in the case of piping networks the survival function has been shown to be dependent on several factors, the most important of which have been shown to be the 'number of observed previous breaks' (NOPB), the age and the material of the pipes [33]. These risk factors have been studied extensively [27,33], both when acting separately or when acting in tandem. For example, a survival analysis reported by [27] of an urban water distribution network under abnormal operating conditions, revealed almost identical survival curves for the network mains and its house connections, but when clustered by the NOPB, the survival curves varied substantially. The NOPB value for a pipe is typically increased by a unit when a leak is observed on the pipe, and zeroed when the pipe is replaced.

A typical set of survival curves is shown in Figs. 8.1–8.2. The curves have been derived from a real network's performance data reported upon in literature [27]. The former figure refers to pipes of different material, while the latter figure specifically refers to asbestos cement (AC) pipes. Additionally, the data in Fig. 8.2 have been clustered according to the "number of observed previous breaks" (NOPB), with four survival curves derived based

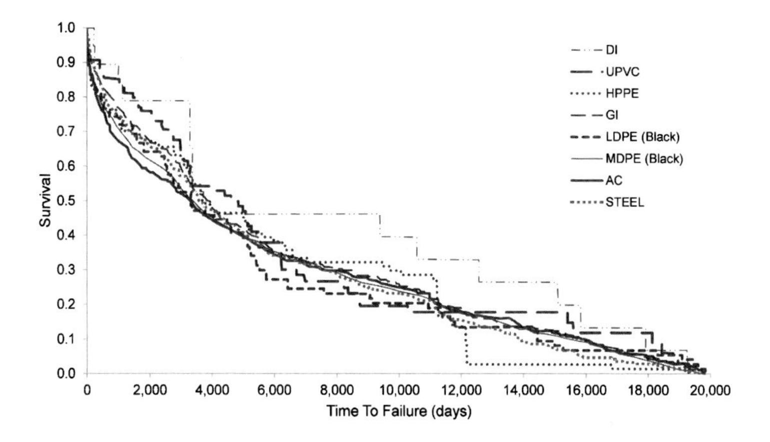

Figure 8.1 Survival curves for pipes, as a function of material type.

on NOPB clusters of size "zero", "small" ($1 \leq NOPB \leq 4$), "medium" ($5 \leq NOPB \leq 8$) and "large" ($9 \leq NOPB$).

It can be seen from Fig. 8.1 that pipes follow approximately the same survival curve over time, regardless of material type. By contrast, as Fig. 8.2 depicts, the survival curves differ dramatically when the "NOPB" clustering is examined. Even in the "NOPB = 0" case, a pipe will have to be eventually replaced after approximately 20,000 days (≈ 55 years), while a pipe that has already broken more than 8 times[1] is not expected to survive more than 18 years. Moreover, a pipe that has broken at least once is very vulnerable, since its survival curve reduces rapidly compared to a pipe that has never been damaged.

It should be noted that the survival curves depicted in Figs. 8.1 and 8.2 may be used to calculate a pipe's nonseismic fragility, and thus its forecasted nonseismic repair rate, at a time of interest t. For example, by use of Fig. 8.2, a pipe of length $L = 100$ feet belonging to the "NOPB = small" cluster of the network in study has at time $t = 3333$ days (≈ 9 years) a survival rate of 0.45, thus $P_f = 1 - 0.45 = 0.55$. Assuming that the risk of failure is uniformly distributed across the length of the pipe, the expected repair rate (RR) for the pipe becomes P_f/L ($= 0.55/100 = 0.0055$) and, by extension, approximately 6 repairs per 1000 feet of network length.

Thus, given a pipe, its length and the cluster in which the pipe belongs to based on its previous nonseismic performance, one can first obtain the

[1] The NOPB metric in this case refers to street-length pipe runs; hence the high values for it.

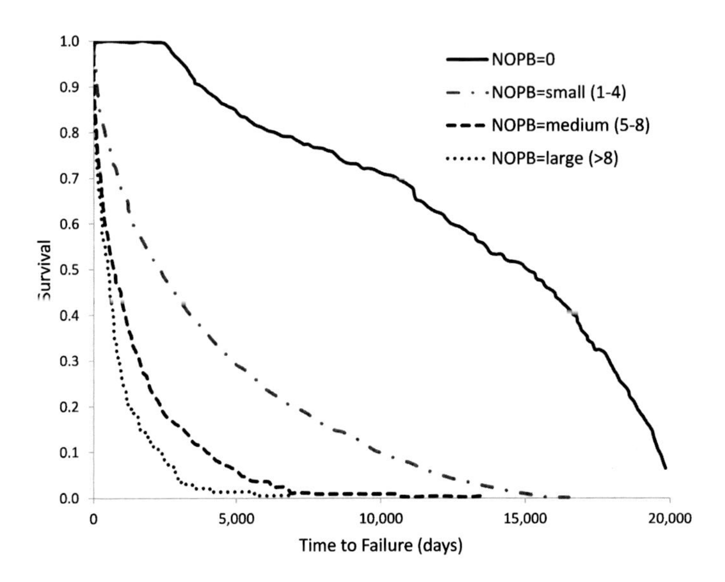

Figure 8.2 Survival curves for asbestos cement (AC) pipes, as a function of the number of observed previous breaks (NOPB).

pipe's survival rate (from Fig. 8.2) and then its nonseismic repair rate. In the case of seismic risk, the probability of failure is deduced by use of the "repair rate" metric, which is a function of the peak ground velocity (PGV) and the permanent ground deformation (PGD). PGV and PGD are metrics which depend on seismic intensity and the local conditions.

8.2.2 Estimating the Reliability of the Network – Monte Carlo Analysis

The performance of the network and its failure probability P_f can be assessed based on the network's topology/connectivity and on the failure probability, $\bar{P}_f$, of every pipe. In the simplest case, the network performance is measured by the probability that it fails to deliver water from its sources (inflow vertices) to every house connection (outflow vertices). If such, rather simplified, network performance definition is adopted, the performance of the network can be quickly evaluated using methods based on Graph Theory and Monte Carlo simulation (MCS). Alternatively, if failure is defined with respect to hydraulic quantities, then hydraulic analysis of the network is required. In our study we adopt the former approach and therefore we calculate the probability that the network is not able to deliver water from an inflow source vertex i to an outflow (e.g., house connection) vertex j (symbolized by $P_{f,ij}$). The corresponding reliability $R_{f,ij}$ is equal to $1 - P_{f,ij}$.

For water networks with more than one inflow sources, the failure probability is the smallest probability of all possible sources and the outflow vertex examined. The failure probability $\widehat{P}_{f,ij}$ depends on the available paths between i and j, and, in general, does not provide information of the overall condition of the network.

For the estimation of network reliability (topological or hydraulic) a combination of Graph Theory and Monte Carlo simulation (MCS) is employed. The MCS of pipe networks is based on reducing the network topology. This is achieved by removing the components which are assumed as failed and creating different realizations of the network. This is repeated N_{MSC} times, thus N_{MSC} network realizations are created to obtain $\widehat{P}_{f,ij}$. MCS can easily accommodate both pipe and node failures, although in this work only pipe failures are considered.

In order to determine whether it is possible for water to reach node j starting from vertex i we create a graph model of the network and we subsequently apply a standard Dijkstra algorithm [65]. The Dijkstra algorithm is a very common mathematical algorithm of Graph Theory that gives quick and robust calculations in all our applications. Alternatively, the connectivity matrix $\mathbf{B}$ can be calculated using the adjacency matrix $\mathbf{A}_{ij}$ [177]. The elements of the adjacency matrix are either zeros or ones depending on whether there is a pipe connecting node i with node j. Therefore, if n is the number of the nodes of the graph, $\mathbf{B}$ is calculated as:

$$\mathbf{B} = \mathbf{A}^1 + \mathbf{A}^2 + \cdots + \mathbf{A}^{n-1}; \tag{8.1}$$

nodes i, j are connected only if $\mathbf{B}(i, j)$ is not zero.

Monte Carlo simulation (MCS) is employed when the analytical solution is not attainable and/or the failure domain cannot be expressed or approximated analytically. This is mainly the case in problems of complex nature with a large number of basic variables where other reliability methods are not applicable. Therefore, in order to calculate the failure probability $\widehat{P}_{f,ij}$, we perform a large number (N_{MSC}) of simulations. In every simulation, depending on the failure probability of every pipe we randomly create different realizations of the water network and subsequently check whether it possible to deliver water from the source vertex i to the outflow node j. If N_H is the number of simulations that the network failed to deliver water between the two vertices, the network failure probability is equal to

$$\widehat{P}_{f,ij} \approx \frac{N_H}{N_{\text{MCS}}}. \tag{8.2}$$

N_{MSC} and N_H should be sufficiently large numbers in order to have a robust estimation of $\widehat{P}_{f,ij}$. In other words, the nominator should be sufficiently large and depending on the order of the probability sought N_{MSC} has to be increased accordingly. The flowchart of the Monte Carlo method implemented in our analysis is shown in Fig. 5.6 (p. 191).

The calculation of the failure probability $P_{f,ij}$ is a useful indicator of the condition of the network. However, it does not account for the condition of the whole network, while it neglects some key information of the problem, such as the number of customers that every node serves. Therefore, other indices of the network are often useful. [1] proposed the concept of "Serviceability Ratio (*SR*)", which is defined as

$$SR = \frac{\sum_j^N \omega_j X_j}{\sum_j^N \omega_j} \tag{8.3}$$

where ω_j is the population or the number of consumers of house connection j, N is the number of nodes of the network and X_j is a binary parameter that denotes whether node j is accessible or not. *SR* depends on the number of network nodes and therefore its computing cost grows almost linearly with the number of nodes in a network.

Within a Monte Carlo algorithm, *SR* can be calculated N_{MCS} times and its response statistics (mean and standard deviation) can be used to identify the overall condition of the network for the given set of $\widehat{P}_{f,ij}$ values. Such metrics when appropriately calibrated, e.g., $SR \geq 0.8$ signifies "good" performance, while $SR \leq 0.3$ is "poor" performance, can further assist the network performance assessment.

8.3 CASE STUDY – PART 1

8.3.1 Overview

The network studied is a district metered area (DMA) from a real-life UWDN (Fig. 8.3), first presented in Section 5.5.3 (p. 202). As a reminder, a UWDN is typically clustered into DMAs; a practice that allows the UWDN operator to isolate any damage in the network within finite areas (the DMA or its sub-DMAs) and then to take action without the risk

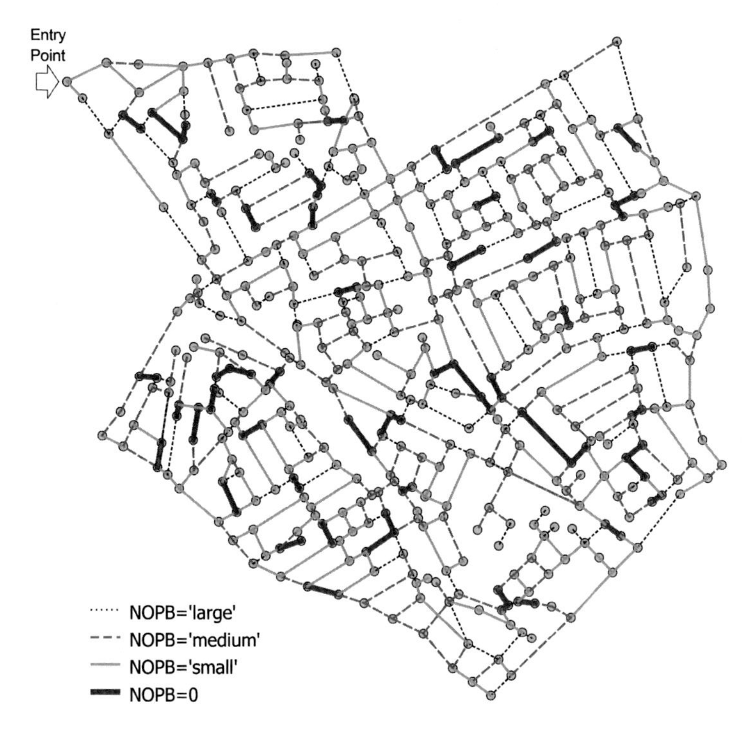

Figure 8.3 Topology of the DMA network considered, in a disturbed state (*NOPB* $\geq$ 0 for every pipe).

of cascading any effects on the overall network. Therefore, a DMA is a hydraulically discrete area with no unmonitored inflow or outflow, and having a defined and permanent boundary. A DMA is supplied via a single source and has approximately equal pressure levels across its population of pipes, while the night flows are regularly monitored.

In total, the water network of the DMA studied consists of 633 pipes (graph edges) of approximate total pipe length of 35,754 m, 450 junctions (graph nodes) and serves a total of 9292 consumers. The material of all pipes in the network is asbestos cement (AC), and the corresponding pipe survival curves are those of Figs. 8.1 and 8.2. The curves have been derived using real historical data obtained from the network operator. Fig. 8.3 is a GIS depiction of the studied DMA network in its current (disturbed) state taking into consideration the NOPB values for each pipe, based on available operations and maintenance records. Different line types are used to denote the NOPB of each pipe.

8.3.2 Preliminary Analysis – Topological Reliability

As a preliminary analysis, let us consider only the effects on network reliability of (i) the nodal density, and (ii) of the topology (connectivity) of the network. The analysis enables us to get a first glimpse at the topological reliability of the network, which depends on the density of the nodes (pipe junctions) and the arcs (pipes) and also on the "open ends" in the network.

Prior to moving to a detailed analysis, let us first precede the sections on topological reliability with stating that in graph theory two well-known methods for evaluating the importance of a node and the robustness of a network are the betweenness centrality and the algebraic connectivity metrics, respectively. *Betweenness centrality* is an indicator of a node's centrality in a network, and it is equal to the number of shortest paths from all vertices to all others that pass through that node. A node with high betweenness centrality has a large influence on the transfer of items through the network, under the assumption that item transfer follows the shortest paths. The *algebraic connectivity* of a graph G is the second-smallest eigenvalue of the Laplacian matrix of G. This eigenvalue is greater than 0 if and only if G is a connected graph, and the magnitude of this value reflects how well connected the overall graph is (it has been used in analyzing the robustness of a network). Both the betweenness centrality and algebraic connectivity metrics are a good starting point, but not sufficient indicators as water distribution networks are weighted graphs with specific input and output vertices.

In the case of nodal density, no consideration is given to pipe vulnerabilities. Rather, the network is spatially analyzed for point clusters by use of kernels based on bivariate distributions positioned at the network nodes. The results, shown in the spatial heatmap of Fig. 8.4, indicate the areas in which nodal density is higher and thus the likelihood of a nodal failure is higher than other areas (assuming that all nodes are of the same condition/state at the time of a catastrophic event). A heatmap is a graphical representation of data where numeric values are represented as colors. In this case, the heatmap maps the probability of failure at each network node based on the network reliability algorithm of Fig. 5.6.

Fig. 8.5, on the other hand, shows the spatial heatmap of network reliability based solely on network connectivity and ignoring the effects of historical nonseismic performance, of the material type, and of seismic risk. The failure probabilities are taken to be equal for all pipes and of minimal value of $\widehat{P}_{f,ij}$ equal to 0.01 for each and every pipe. The network reliability is calculated based on Graph Theory and Monte Carlo simulation. As

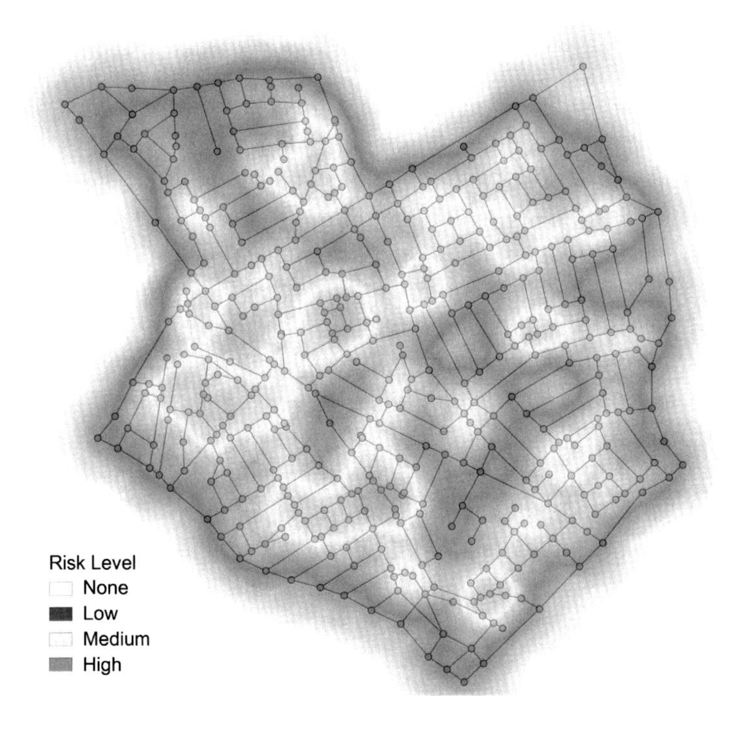

Figure 8.4 Spatial analysis (heatmap) of studied network's nodal density.

Fig. 8.5 shows, the probability of failure is higher in "open-ended" areas because the paths to these areas are not fail-safe.

This preliminary analysis showcases some of the aspects related to network reliability; namely the importance of nodal density and of network connectivity, how these issues affect the overall network reliability, and how risk varies spatially ("heatmaps") based on what aspect of network topology is included in the analysis. In effect, the computed network reliability values can be thought of as baseline values (based solely on the network's topology). The pipe vulnerabilities constitute an additional risk factor which will be examined in subsequent sections.

8.4 SEISMIC ASSESSMENT OF A WDN – A RECAP

8.4.1 Pipe Vulnerabilities According to the ALA Guidelines

As presented in Section 5.3.1 (p. 185), the seismic vulnerability (or fragility) of buried pipelines is discussed in the American Lifelines Alliance (ALA)

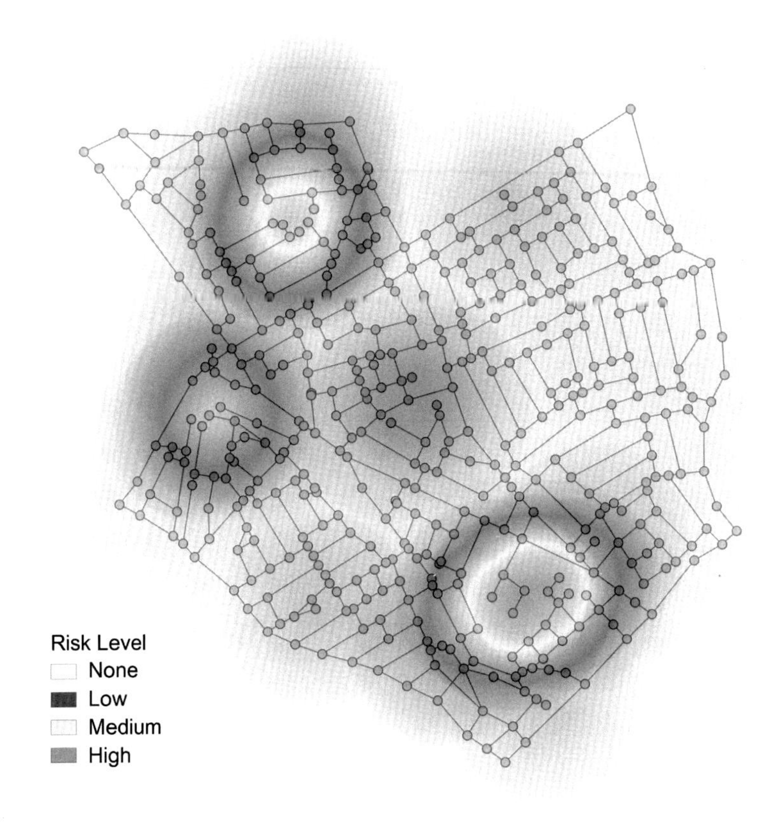

Figure 8.5 Spatial analysis (heatmap) of studied network's topological reliability.

guidelines [5,6], in which several failure parameters are identified and empirical vulnerability functions/curves are proposed by use of observations from past disruptive earthquakes. The vulnerability functions are related to the peak ground velocity (PGV) and the permanent ground deformation (PGD). PGV is related to strong ground shaking caused by seismic wave propagation, while PGD is used to measure factors that include landslides, liquefaction, ground settlement and fault crossing. Other parameters identified are a pipe's diameter, age and year of construction, as well as possible discontinuities along a pipe. The recommended by the ALA pipe vulnerability functions provide the repair rate (*RR*) per 1000 ft of pipe length and have the form:

$$RR_{PGV} = K_1 \cdot 0.00187 \cdot PGV,$$
$$RR_{PGD} = K_2 \cdot 1.06 \cdot PGD^{0.319} \quad (8.4)$$

where PGV is measured in inches per second (in/s) and PGD is measured in inches (in). K_1 and K_2 are material-specific adjustment factors, for which the ALA guidelines provide tabulated values for various pipe materials, assuming that for cast iron (CI) or asbestos cement (AC) pipes, $K_1 = K_2 = 1$. The pipe repair rates of Eq. (8.4) can be due to a complete fracture, a leak or damage to an appurtenance of the pipe, or any other reason that requires the water agency to intervene. For typical water pipe networks, a rule-of-thumb is that for failure due to wave propagation approximately 15–20% of failures are breaks and the rest are leaks, while for failures due to PGD approximately 80–85% are breaks that result to the loss of pipeline hydraulic continuity.

Once the repair rate (RR) is known (i.e., the number of leaks or breaks, per pipe length), the failure probability of the pipe can be calculated, as one minus the probability of zero breaks along the pipe. Adopting Shamir and Howard's findings on the breakage rate following an exponential distribution [152] and using the exponential distribution CDF formula, the pipe failure probability P_f is therefore calculated as [5]

$$P_f = 1 - e^{-\max(RR_{PGV}, RR_{PGD}) \cdot L} \tag{8.5}$$

where RR_{PGV}, RR_{PGD} are calculated by use of Eq. (8.4). It should be noted that Eq. (8.5) is "memoryless", i.e., the time until a certain event does not depend on how much time has elapsed already. In essence, the probability of a future pipe failure disregards any failures that may have occurred along the pipe in the past. The symbology used should also be noted. $\widehat{P_f}$ refers to the failure probability at the network level, whereas P_f refers to the failure probability of a single pipe/edge.

8.4.2 Limitation of the ALA Guidelines

Despite their power in terms of providing guidance on the expected seismic effects in a water distribution network, the ALA guidelines have a number of inherent limitations:

- They are based on posterior statistical analysis of a small number of earthquakes and their observed effects on networks.
- They do not consider a network's past nonseismic performance when estimating the seismic performance (risk factors such as pipe age and NOPB are not considered in the analysis).
- They do not consider the network topology (they arrive at an estimate of the repair rate and not at a network reliability metric).

- They are in essence generic in nature ("horizontal risk assessment") and nonlocalized ("vertical risk assessment").

The issue of the NOPB risk factor, its significance and its differentiation from the repair rate (RR) factor used in the ALA guideline is worthy of a closer look. The ALA guideline furnishes us with an estimation for the repair rate (RR) (Eq. (8.4)) based on the PGV and PGD values, and on the material of the pipes. This metric is in essence postulated statistically and it does not relate to any pipe condition prior to the seismic event. In contrast, the NOPB factor is a metric of the historical performance of a pipe (how many breaks the pipe experienced in the past, under nonseismic operation) and relates to us a pipe's operational condition. Furthermore, the NOPB is a risk factor which, by use of survival analysis, can account for incomplete data. In the following section we discuss a rational approach for combining the NOPB factor with the RR metric.

8.4.3 Proposed Strategy for Pipe Seismic Vulnerability Assessment

The ALA guidelines present procedures and fragility relationships that can be used to evaluate the probability of earthquake damage to water transmission systems and to make informed decisions on how to mitigate risks. However, the generic form of the pipe fragility curves obtained through the ALA method does not take into consideration a network's past performance and its effects when calculating the pipe repair rates due to seismic loading. A friendly amendment to the ALA guidelines, discussed previously in Section 5.3.2 (p. 188), aimed at proposing a seismic vulnerability assessment methodology for water pipe networks, while exploiting available data of everyday network failures due to sources other than seismic.

The proposed methodology takes into consideration the fragility that corresponds to pipe failures that occur frequently during the everyday operation of the water network (such as pipe breaks and water loss incidents), and also more severe, but less often, failures due to earthquakes. To account for the vulnerability due to nonseismic loads the proposed method relies on survival analysis. To account for the vulnerability due to seismic hazard, the method relies on the procedure described in the American Lifelines Alliance (ALA) 2001 guidelines, adjusted by the added vulnerability due to past performance (the observed previous breaks in the network).

Having at our disposal the pipe survival curves (e.g., Figs. 8.1–8.2) of $S(t)$ versus time, we can deduce the survival probability of a pipe, depending on the number of observed previous breaks (NOPB) and the pipe type

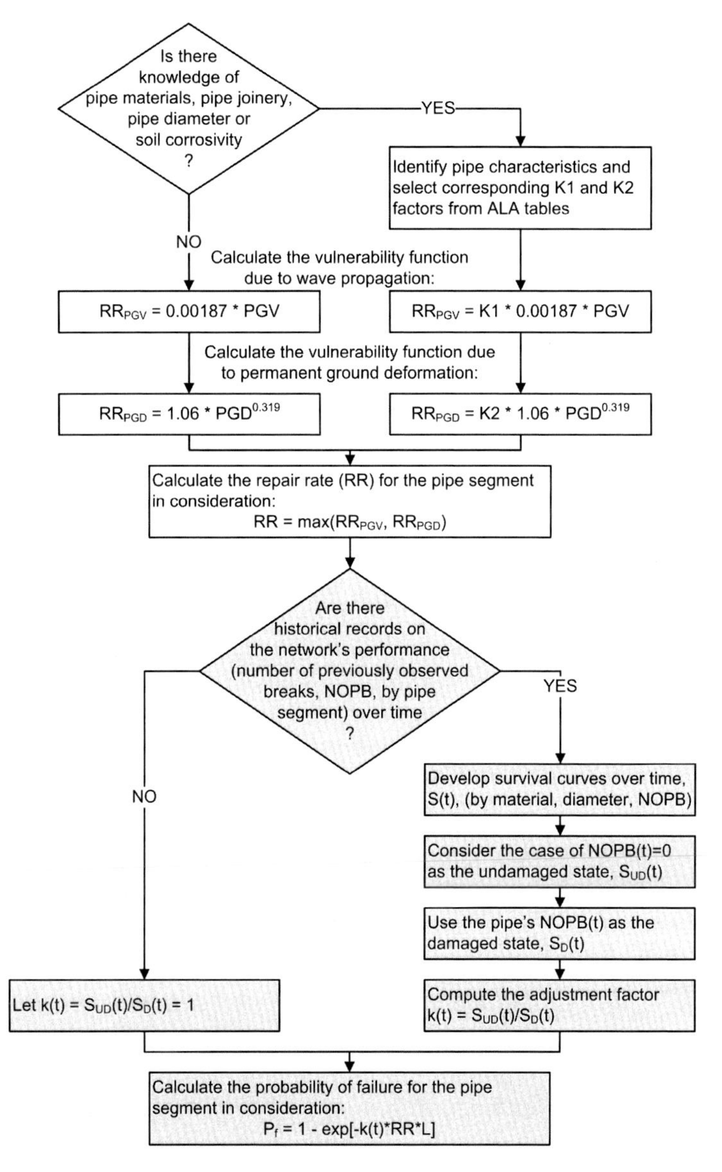

Figure 8.6 Proposed methodology (shaded shapes are proposed modifications to the ALA (2001) guideline).

(e.g., material, age, diameter). This survival probability is then used to adjust the vulnerability curves proposed by the ALA, as per the flowchart steps shown in Fig. 8.6 and the expected repair rate (RR) for varying values of $k(t)$ (Eq. (8.6); Fig. 8.7).

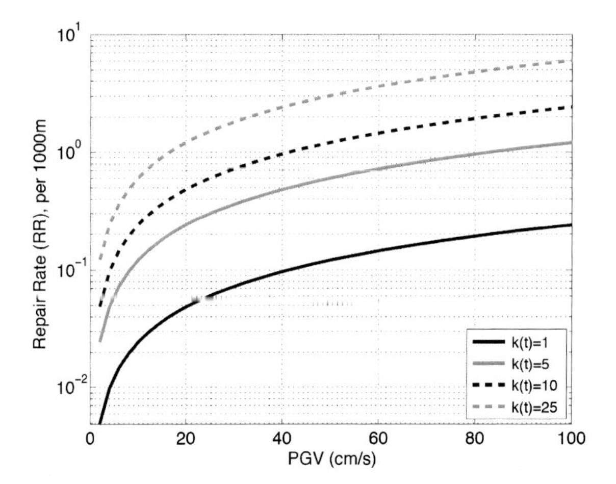

Figure 8.7 Expected repair rate (RR) for varying values of $k(t) = S_{UD}(t)/S_D(t)$.

In effect, we penalize the pipe vulnerability function of Eq. (8.4) by the ratio of the survival curve of the damaged case (NOPB $\neq 0$) over that of the undamaged pipe (NOPB $= 0$). Therefore, after t days, we define the ratio

$$k(t) = S_{UD}(t) \big/ S_D(t) \geq 1 \tag{8.6}$$

where subscripts "UD" and "D" stand for "undamaged" and "damaged" state, respectively. The modified pipe failure probability that now includes memory of past nonseismic failures is obtained after modifying Eq. (8.5) as follows:

$$P_f(t) = 1 - e^{-k(t) \cdot RR \cdot L}. \tag{8.7}$$

Therefore, Eq. (8.7) allows the calculation of the failure probability P_f of the pipe after t days, given its NOPB metric obtained from historical records.

A spatial analysis is subsequently employed to demonstrate the effects of such nonseismic performance data on the seismic network reliability.

8.5 CASE STUDY – PART 2

In order to demonstrate the impacts of nonseismic performance on a network's reliability and how the ALA guideline could be improved, let

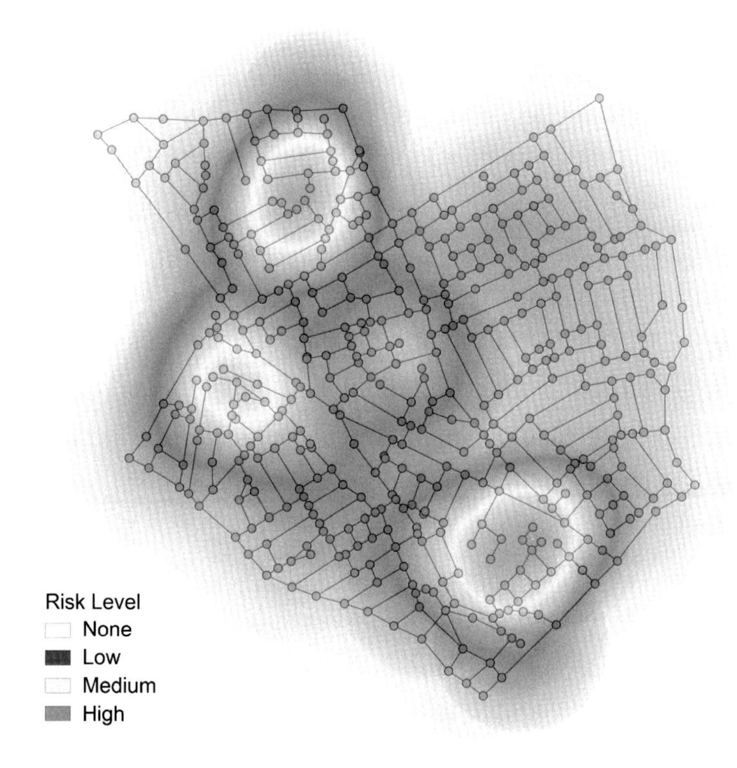

Figure 8.8 Spatial analysis (heatmap) of studied network's reliability based on statistical analysis (NOPB per unit length).

us continue the discussion on the case-study network from Section 8.3 (p. 252) and consider various scenarios and the resulting network reliability.

8.5.1 Network Reliability Based on Nonseismic Performance Data

At first let us consider pipe vulnerability as a posterior statistical metric, based on the historical performance of the pipes. The metric is simply defined (per pipe) as the pipe's number of breaks over its length ($P_{f,i} \equiv NOPB_i/L_i$), and for the network in study the values range from 0 to 0.9. As in the previous cases of nodal density and of topological reliability, the analysis does not take into consideration other risk factors such as the material type and the seismic risk. The network vulnerability is shown spatially in Fig. 8.8.

A comparison of Fig. 8.8 with Fig. 8.5 (showing the topological reliability of the network in study) indicates that there is a great level of similarity between the computed risk levels resulting from the two scenarios. It can be seen that the statistical approach to pipe vulnerability based on the metric *NOPB*/*L* results to the same risk areas in the network as in the case of topological reliability but with slightly more acute risk level. In other words, the uniform allocation of risk (based on the metric *NOPB*/*L*) has the same impact on the network reliability as the topology of the network where the open ends were found to be the weak points in the network.

8.5.2 Network Reliability Based on Survival Curves, Clustered by "Material Type" (Seismic Effects Not Included)

If we incorporate the variation in pipe vulnerabilities due to the pipes's material type, stemming from the impacts of survival analysis on the pipe population (clustered by material type), the risk levels increase and the network reliability map changes dramatically (Fig. 8.9). The network reliability is no longer directly related to the network topology and the material-clustered survival curves (Fig. 8.1) compound the risk levels and spread them spatially across a wider area in the network.

Each pipe's probability of failure is taken to be in the range of $[0.25, 0.35]$ depending on the "material-type cluster" the pipe belongs to, as deduced from Fig. 8.1 at a time horizon of 5 years ($t \approx 1825$ days).

8.5.3 Network Reliability Based on the ALA Guideline (Seismic Effects Are Included)

We subsequently study the DMA including seismic effects. In terms of seismic considerations, the seismic hazard in the island of Cyprus is mainly controlled by distant and moderate magnitude events and, thus, it is valid to assume that the seismic intensity will be uniform throughout the DMA.

Therefore, we turn back to the ALA guideline and we use the repair rate suggested by the guideline (Eqs. (8.4) and (8.5)) as a metric of a pipe's vulnerability, then the obtained risk map is again similar to the risk maps produced by the topological (Fig. 8.5) and the statistical (Fig. 8.8) vulnerabilities. The resulting heatmap (Fig. 8.10) shows an increase in the risk levels compared to Fig. 8.5, but the areas of concern in the network remain the same. This is an indication that the horizontal treatment of risk (i.e., its uniform distribution across the pipe segments based on the repair rate) has the same impact to network reliability as the network topology and that,

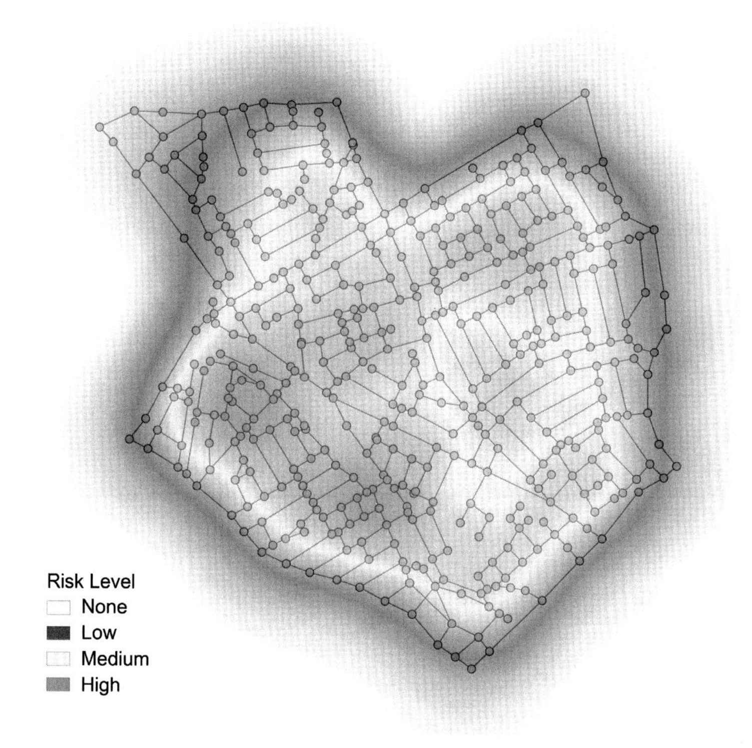

Figure 8.9 Spatial analysis (heatmap) of studied network's reliability based on survival curves (material, time). Seismic effects are not considered.

in essence, the seismic effects are directly related to the topology of the network (nonseismic condition).

The material of each and every pipe is taken to be asbestos cement ($K_1 = K_2 = 1$) and a pipe's probability of failure is computed based on Eq. (8.5). Seismic effects are included in the analysis, with $M_w = 7$.

8.5.4 Network Reliability Based on Survival Curves, Clustered by "NOPB" (Seismic Effects Not Included)

The situation changes drastically once the NOPB risk factor is included in the analysis (Fig. 8.11). The inclusion of a network's nonseismic performance, as expressed by the survival curves clustered by NOPB (Fig. 8.2), increases the pipes's risk of failure and thus alters the network reliability, while at the same time distributing their values over space. The increased network vulnerability is the direct outcome of the increased pipe vulner-

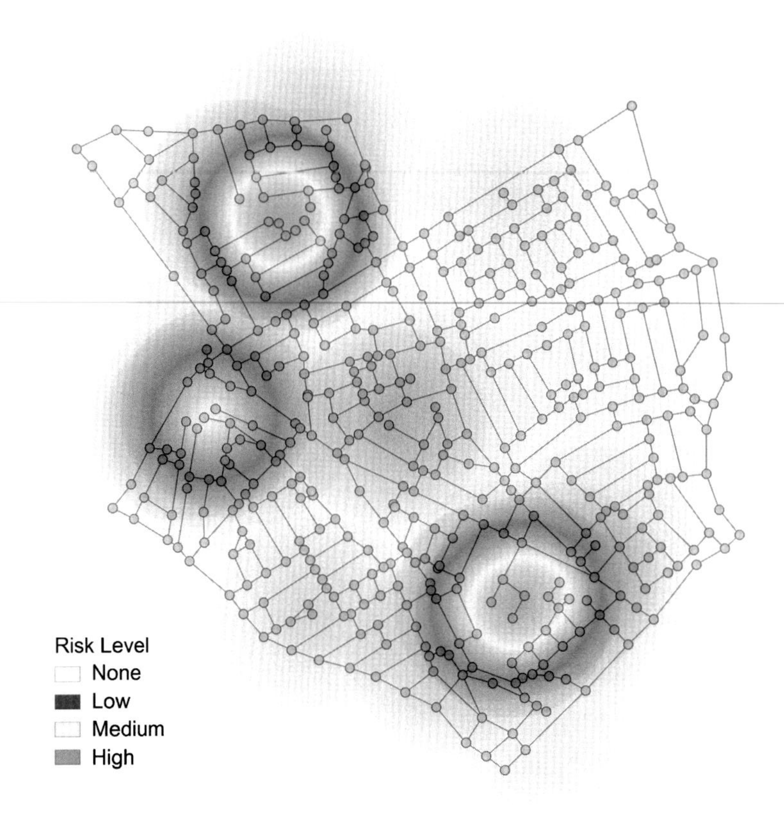

Figure 8.10 Spatial analysis (heatmap) of studied network's reliability based on the ALA guidelines. Seismic effects are considered ($M_W = 7$).

abilities, as impacted by their performance prior to the seismic event and their decreasing reliability. The severity and impact of risk are increased compared to the risk level deduced by utilizing the survival curves of varying material types (Figs. 8.1 and 8.9).

Each pipe's probability of failure is taken to be in the range of $[0.01, 0.80]$ depending on the NOPB cluster the pipe belongs to ("0", "small", "medium", "high"), as deduced from Fig. 8.2 at a time horizon of 5 years ($t \approx 1825$ days). For a pipe with $NOPB = 0$ its probability of failure (P_f) is taken to be in the range $[0.01, 0.05]$ so as to account for possible defects; for $NOPB =$ "low", $P_f \in [0.1, 0.3]$; for $NOPB =$ "medium", $P_f \in [0.3, 0.5]$; and for $NOPB =$ "high", $P_f \in [0.5, 0.8]$. Equivalently, should one consider Eq. (8.7), these failure probabilities correspond to adjusted $k(t)$

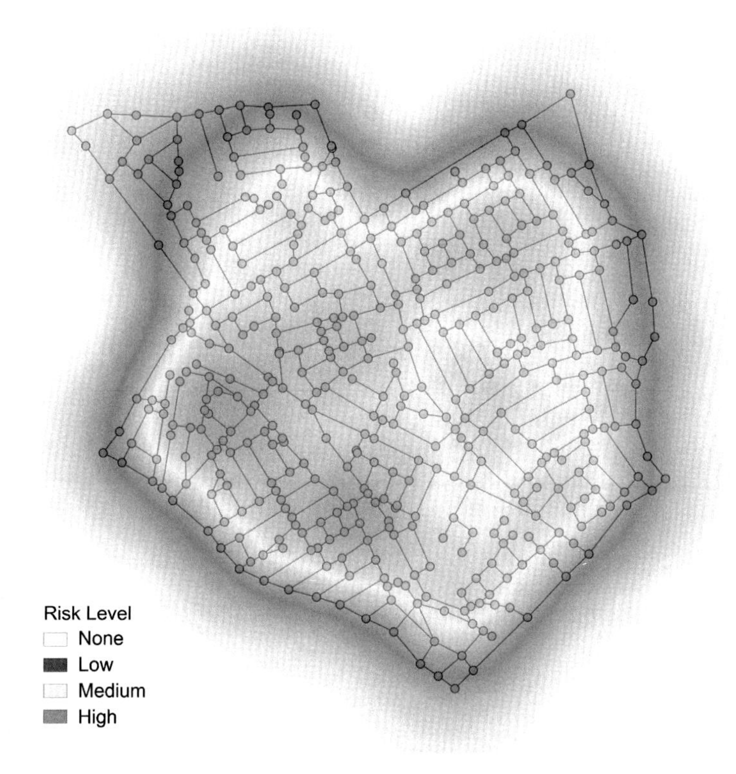

Figure 8.11 Spatial analysis (heatmap) of studied network's reliability based on survival curves (material, NOPB, time). Seismic effects are not considered.

factors of approximately 1.00, 0.60, 0.35, and 0.20 for the "0", "small", "medium", and "high" NOPB clusters, respectively.

8.5.5 Spatial Analysis (Heatmap) of Studied Network's Reliability, Based on a Combination of the NOPB-Clustered Survival Curves and the ALA Guideline

If we now consider the combined effects of nonseismic performance, as differentiated for each pipe by the NOPB risk factor and subsequently distributed over space (Fig. 8.11), and the expected vulnerability due to seismic effects (as described in the ALA guideline), then the network reliability can be calculated based on the proposed new methodology (Fig. 8.6) and mapped spatially as shown in Fig. 8.12.

The spread of the revised network vulnerability is shown to be in sharp contrast to the point-like vulnerability obtained by the ALA guideline

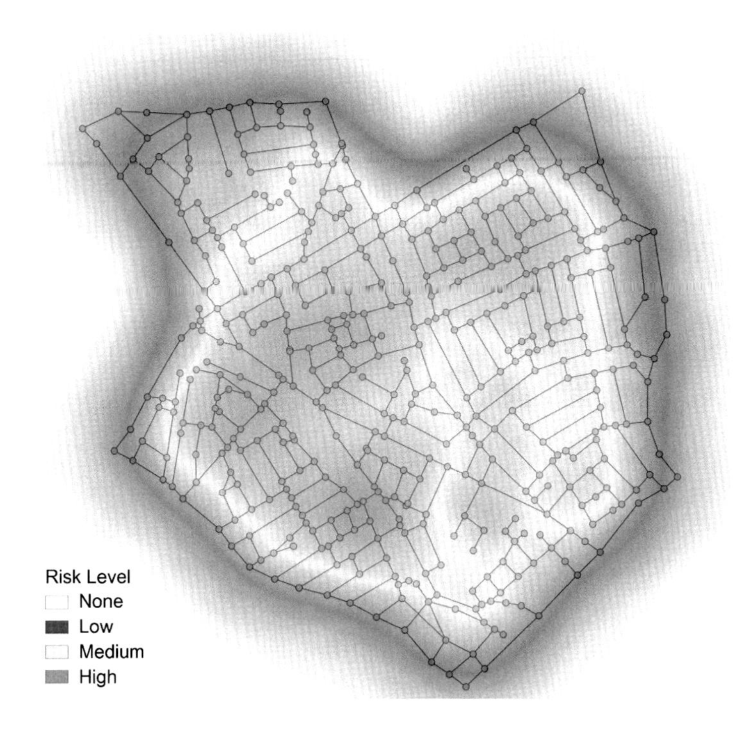

Figure 8.12 Spatial analysis (heatmap) of studied network's reliability based on survival curves and considering the effects of the NOPB risk factor. Seismic effects are considered ($M_W = 7$).

(Fig. 8.10), for it takes into consideration the localization and allocation of risk due to the nonseismic performance of the network. The variation compared to Fig. 8.11 is slighter, indicating that the resulting risk-map has heavy influences from the NOPB-clustered risk-map.

As with the previous scenario examined, each pipe's probability of failure is taken to be in the range of $[0.01, 0.80]$ depending on the NOPB cluster the pipe belongs to ("0", "small", "medium", "high"), as deduced from Fig. 8.2 at a time horizon of 5 years ($t \approx 1825$ days). For a pipe with $NOPB = 0$ its probability of failure (P_f) is taken to be in the range $[0.01, 0.05]$ so as to account for possible defects;, for $NOPB$="low", $P_f \in [0.1, 0.3]$; for $NOPB$="medium", $P_f \in [0.3, 0.5]$; and for $NOPB$="high", $P_f \in [0.5, 0.8]$. Equivalently, should one consider Eq. (8.7), these failure probabilities correspond to adjusted $k(t)$ factors of approximately 1.00, 0.60, 0.35, and 0.20 for the "0", "small", "medium", and "high" NOPB clusters, respectively.

8.6 CONCLUDING REMARKS

The study demonstrates that a network's historical nonseismic performance (i.e., the number of observed previous breaks, NOPB) has a direct impact on the network's seismic performance and its overall vulnerability, and demonstrates this impact by means of spatial vulnerability maps of a case-study network of a real-scale DMA.

Available past network performance is processed through data-mining and survival analysis techniques, clustered by a number of different risk factors, and then used to calculate the generalized pipe vulnerability considering data that refer to the specific network. The network reliability is subsequently assessed using Graph Theory tools and Monte Carlo simulation. When seismic effects are considered, an enhancement to the methodology of the ALA guidelines for the piping components is presented. This approach allows calculating the pipe fragility considering available performance records of critical risk-of-failure metrics pertaining to the specific WDN. This allows enhancing the ALA-derived seismic pipe reliabilities with knowledge on the network's past nonseismic performance. The proposed approach allows for both the estimation of the probability that a network fails to provide the desired level of service and also the prioritization of retrofit interventions and of capacity-upgrade actions. A spatial analysis, by means of risk "heatmaps", of the proposed method's derived network reliability is also included and compared to the network reliability based on the ALA guidelines, showcasing the effects of past nonseismic performance of network components on its reliability.

The study finally concludes that the ALA guideline's "*Repair Rate (RR)*" metric underestimates the seismic effects on the vulnerability of a network and recommends how the damaged ("prior performance") and undamaged network states can be included in the calculation of a pipe's probability of failure.

In essence, the proposed methodology improves on the statistically-based nonlocalized metrics proposed by the ALA guideline by introducing the utilization of localized nonseismic performance data through survival analysis.

CHAPTER 9

Disaster Resilience of Water Distribution Networks

The resilience of a system is a manifestation of the states of the system. Perhaps most critically, it is a vector that is time dependent. Resilience in this article is defined as the ability of the system to withstand a major disruption within acceptable degradation parameters and to recover within an acceptable time and composite costs and risks.

Haimes, Y.Y. (2006), "Systems Engineering".

9.1 RESILIENCE

9.1.1 The Concept of Resilience

The term *resilience* comes from the Latin word "resilio" which means "to leap back." Resilience denotes the ability of a system to recover from events that disrupt its operation and usually cause damage [181]. Disaster resilience of infrastructure during and after natural or man-triggered disasters is vital for the recovery of the community. Resilience-based design refers to planning for hazardous conditions and prioritizing a series of actions in order to: (*i*) deal with emergencies immediately after an event, and (*ii*) gradually restore the services offered by the network [20]. A network with a high level of resilience is expected to recover quickly, whereas systems with low resilience would experience a slow recovery after a disaster. Resilience planning and designing requires prioritizing the services that should be operational after an earthquake, e.g., access to critical infrastructure such as hospitals should be a top priority, while a resilient system should demonstrate reduced failure probabilities, reduced consequences from failures, and reduced time to recovery.

The seminal work of Bruneau et al. [20] provides the conceptual basis of resilience. The study points out that resilience is a multidimensional problem that includes technical, organizational, social and economical facets. Latter studies (e.g., [41]) attempt to propose consistent frameworks that enable quantifying resilience and allow resilience-based design in engineering

Urban Water Distribution Networks
DOI: http://dx.doi.org/10.1016/B978-0-12-813652-2.00009-8

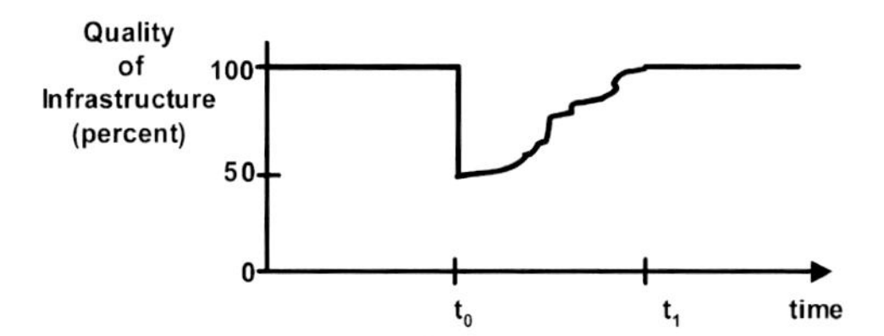

Figure 9.1 Schematic representation of system resilience (adopted from [20]).

practice. Resilience-based design requires meaningful resilience objectives, i.e., prescribing the level of service and a series of actions at different time instants following an event [23,111]. For example, the resilience of water distribution networks is examined in [48] using as a testbed the WDN of the city of Los Angeles, while Cimellaro et al. [43] study the resilience of the water distribution network of a town in Sicily.

The concept of resilience can be better understood with the aid of Fig. 9.1. The plot shows the "functionality" of the network versus time. At t_0 a damaging event (e.g., an earthquake) occurs and suddenly the functionality of the network, i.e., its capacity to provide a service, is dramatically reduced. The time required for restoring the system's "functionality" is the *recovery time* T_{RE}. If the system is restored to a stable condition at time t_1, then $T_{RE} = t_1 - t_0$. The system's resilience is measured as the area of the plot of Fig. 9.1 [41]:

$$R_{es} = \frac{1}{T_{RE}} \int_{t_0}^{t_0+T_{RE}} Q(t)\, dt \tag{9.1}$$

where $Q(t)$ is a parameter that measures the functionality that may represent different aspects of a lifeline's performance, such as serviceability, economic losses, number of deaths, etc. Note that we divide with T_{RE} so that R_{es} is dimensionless. The recovery time T_{RE} provides the downtime or, in other words, the overall time that the system is not in operation. T_{RE} includes the time necessary to repair the system but in the general case it may include other factors such an idle time period immediately after the event, evacuation times, or the time necessary to isolate the damaged area. The recovery time can be thus seen as a random variable since it depends on many factors that are difficult to predict or determine.

A system is considered resilient if it can be restored within a given time period into acceptable levels of functionality. Depending on the system at hand, the functionality following a disaster may be increased compared to

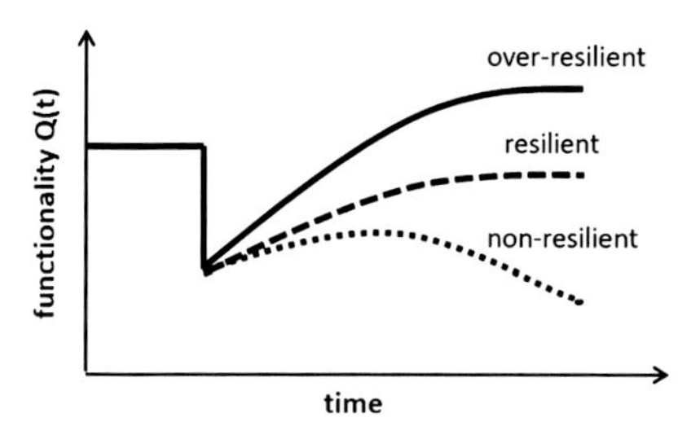

Figure 9.2 Systems with different resilience capacity.

its functionality prior the disaster, or it may be restored to a stable condition but at a lower service level. Such stably restored systems are shown in Fig. 9.2. A stable system that returns to a condition with functionality equal or less than its functionality before the event it is simply "resilient", while if it is restored to a higher level it is "over-resilient." On several occasions, the system may be "nonresilient", or simply it cannot be restored to a stable functionality level.

9.1.2 Properties of Resilience

Bruneau et al. [20] suggest that resilience has four main properties: robustness, rapidity, redundancy, and resourcefulness. These four fundamental properties are also known as the "4 R's." Robustness and rapidity can be seen as the major measures of resilience-based design, while redundancy and resourcefulness are properties that may seem abstract but can decisively improve the system's resilience.

Robustness refers to the "strength", or "the ability of elements, systems, and other units of analysis to withstand a given level of stress or demand without suffering degradation or loss of function" [20]. In Fig. 9.3, the robustness is defined as the residual functionality of the system immediately after the event and is denoted as R. The actual robustness R of the system depends on numerous parameters and thus it can be considered probabilistically, i.e., with a mean value, standard deviation, and a certain distribution (e.g., uniform) [41].

Rapidity is "the capacity to meet priorities and achieve goals in a timely manner in order to contain losses and avoid future disruption" [20]. Using the plot of Fig. 9.1, the overall rapidity can be defined as the slope of a straight line that connects time instants t_0 and t_1. A local measure of rapidity is offered by the slope of the recovery function $dQ(t)/dt$ as shown in Fig. 9.3.

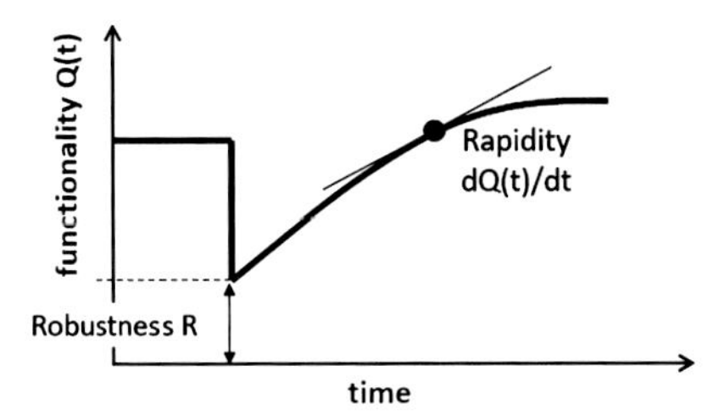

Figure 9.3 Definition of robustness and rapidity.

Redundancy refers to the capacity of the system to successfully serve the demand using alternative mechanisms or paths, redistributing the demand to its undamaged (or slightly damaged) elements. In the case of water networks, redundancy is the ability to have alternative paths that the water can use in order to reach the house connection of interest.

Resourcefulness is the capacity to mobilize and apply material and human resources in order to achieve goals when disruptions occur. For example, assume an earthquake that strikes a town near the sea. The fire extinguishing capacity could be considerably enhanced if pumps were used in order pump out water from the sea.

9.1.3 Resilience-Based Design

Resilience has several dimensions and thus developing a resilience-based design framework that consists of quantifiable and meaningful resilience objectives and metrics is always a challenge. Bruneau et al. [20] present 80 illustrative measures that relate the four dimensions of resilience (technical, organizational, social, and economic) with the four properties of resilience (4 R's) for different systems. For example, a social performance measure for a hospital system might be defined as "all human injuries must be treated in the first day" [23]. Therefore, resilience-based design requires to quantify resilience through identifying resilience objectives, e.g., network redundancy, integrity of the network components, access to critical services, minimum disruption of the financial activity and commerce. Furthermore, different levels for the resilience objectives (minimum, medium, etc.) should be established and a series of actions for reaching each level at a prespecified time following the event should be defined.

The first step towards developing a resilience-based design framework is to identity the primary performance objectives that the system should serve. Davis [48] identifies the various services offered by WDNs and dis-

Table 9.1 Resilience levels of a WDN [112].

Resilience parameter	Low	Medium	High
WDN redundancy	No redundancy. Single supply points for multiple demand points. A single pipeline break may shut down service to a large demand area	Redundancy to critical locations. Multiple supply points for the most critical demands (i.e., hospitals, government facilities, and schools)	Redundancy to all locations. Multiple supply points for all demands. System is able to operate during nearly all hazard events
Backup power and structural stability	No backup power to pumping or treatment facilities. Facilities are not able to withstand moderate disasters, resulting in interrupted service	Backup power to some pumping or treatment facilities. Facilities can withstand most moderate disasters, but extreme events result in interrupted service	Backup power to all pumping and treatment facilities. Facilities are able to withstand nearly all extreme disaster events

cusses the acceptable objectives, or performance levels, for each service category. The primary objectives of a water system are: (*i*) to provide water services, (*ii*) life safety (avoid injuries and casualties), and (*iii*) property protection. Safety objectives always come first, but the other two aspects are also important. Furthermore, the objective of water services has five dimensions, i.e., *water delivery*, *water quality*, *quantity*, *fire protection,* and *functionality*. Water delivery is the capacity of the network to provide water regardless of its quality or quantity (pressure). Fire protection is another critical aspect, since immediately after a disaster this service is critical. The different services may interact, while restoration actions may improve several dimensions of resilience.

A detailed quantification of the key aspects of water infrastructure resiliency is discussed by Matthews [112]. Table 9.1 lists the requirements for low, medium, and high resilience when the resilience parameters examined are the redundancy of the system and the backup power to pumping or treatment facilities. A WDN can be characterized as resilient or not, depending on which of the requirements of this Table can be met. The authorities should target at least to a medium resilience level and make proactive actions.

There are different ways to express the acceptance criteria of the system's functionality. In the simplest case, acceptance criteria can be set directly on the resilience index of Eq. (9.1). For example, R_{es} should exceed 80% for a system with high resilience. Alternatively, Chang and Shinozuka [23] proposed that the resilience should be defined as a minimum probability value that the resilience parameter is restored to an acceptable level and within an acceptable time period. Therefore, they proposed setting a limit value on the probability $P(Q \geq Q^* \cap t \geq t^*)$, where Q^* and t^* is the desired functionality level and recovery time, respectively.

9.1.4 Measuring Resilience

In order to measure resilience, we need to quantify the network's functionality Q (see Fig. 9.1). The functionality function $Q(t)$ depends on the loss function $L_{IM}(t)$ and the recovery function $f_{rec}(t)$ [41]. The recovery function provides the way that the system is restored, while the loss function gives information about the amount, or the consequences, of service loss. The subscript IM is used to denote that the loss function depends on the seismic intensity of the triggering event, while both parameters are functions of time t, taking into consideration the gradual restoration of the network. Before the catastrophic event and after the restoration of the system to a stable condition, the recovery function $f_{rec}(t)$ is constant. If the loss function is known, the functionality $Q(t)$ is defined as

$$Q(t) = 1 - L_{IM}(t). \tag{9.2}$$

If the recovery function is known, the system's resilience index is obtained as

$$R_{es} = \frac{1}{T_{RE}} \int_{t_0}^{t_0+T_{RE}} (1 - L_{IM}(t))\ dt. \tag{9.3}$$

$f_{rec}(t)$ does not appear in Eq. (9.3) but it has been explicitly considered in the integral since it controls the variation of $L_{IM}(t)$ in time.

Different recovery functions can be selected depending on the system and the preparedness of the society to respond [41]. The simplest recovery function is linear (see Fig. 9.4, grey dashed line). A linear function is commonly adopted and usually implies that there is no information regarding the resilience capacity of the network. A sinusoidal or an exponential function can be also used when the response is fast immediately after the event and its rapidity slows down near the end of the restoration phase

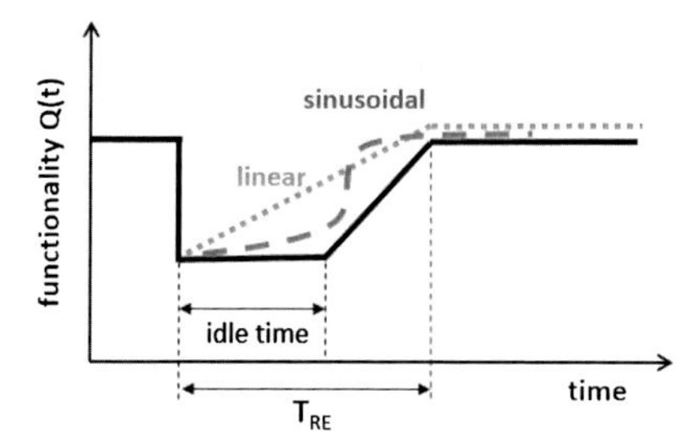

Figure 9.4 Use of different recovery functions f_{rec}.

(e.g., Fig. 9.4, blue (dark grey in print version) dashed curve). On several occasions, it is reasonable to assume some idle period before the restoration actions are initiated. Fig. 9.4 shows a case of an idle period followed by a linear recovery function (Fig. 9.4, solid black line).

In order to better understand the underling concepts, we focus on the resilience with respect to its capacity to provide water to its customers. In the case of water delivery, a simple loss function that considers the number of customers that will be left without water can be obtained with the aid of the serviceability ratio (*SR*) (see Section 5.4.1, p. 190). When the *SR* metric is adopted, the loss function becomes

$$L_{IM}(t) = \frac{\sum_j^N \omega_j X_j}{\omega_{tot}} \tag{9.4}$$

where ω_j is the number of users of each household that suffer from insufficient pressure or are left without water at time t, N is the number of households of the network and X_j is a binary parameter that denotes whether household j is accessible or not, i.e., water is able to move from the source/s to this household. ω_{tot} is the total number of users of the WDN.

Depending on the resilience objectives considered, different loss functions may be adopted. As extensively discussed by Davis [48], water quality is also a significant property that should be considered for the resilience-based design of a WDN. Typically, water quality indices measure the expected concentrations and are normalized with the aid of a concentration value that implies acceptable quality. Since water quality indices are beyond the scope of this book, we arbitrarily use Q_q as a water quality index which is compared with Q_q^*, a parameter that defines a level of acceptable quality.

The loss function for water quality can be defined as [43]

$$L_{IM}(t) = 1 - \frac{Q_q(t)}{Q_q^*}. \tag{9.5}$$

In practice several resilience indices must be met, and thus the system's resilience should be examined for every parameter separately. On some occasions it may be useful to derive a global index that combines the various indices. Such a global metric can be defined as follows:

$$R_{re}^{tot} = R_1 \cdots R_m = \prod_i^m R_i \tag{9.6}$$

where R_{re}^{tot} summarizes the performance of the WDN considering the m resilience indices that may represent various objectives (e.g., water delivery, water level in the tank, water quality). The multiplication of the metrics provides a global index that is sensitive to all different aspects of resilience considered. For example, a network may have high resilience with respect to its water delivery capacity, but the quality of the water may be low [43].

In order to calculate the integral of Eq. (9.3), we need to define also the recovery function f_{rec} and the recovery time T_{RE}. For example, for the case study network of [43], a reasonable assumption for the total recovery time (or "control time") is 48 hours. This time was decided as follows: the first emergency operations (e.g., isolate the zone where the pipes are damaged) are realized within 1 hour, while the repair operations are realized in maximum 12 hours (if the pipe diameter is less than 600 mm). 24 hours are additionally considered in order to inform the residents about the repair operations in advance. The remaining time is the idle time and some additional time which we add so that our estimate is on the safe side.

The recovery function can be considered as linear but this is often an oversimplifying assumption. Recovery times always depend on the properties of the network examined, and thus no general prescriptions exist. For an example, for the case of a WDN, two independent phases of a constant recovery function between them could be assumed. The first phase may combine some idle time and the time necessary to locate where the damage is, while a second phase would include all the repairing works during which the network may be shut down or remain operational but with reduced functionality.

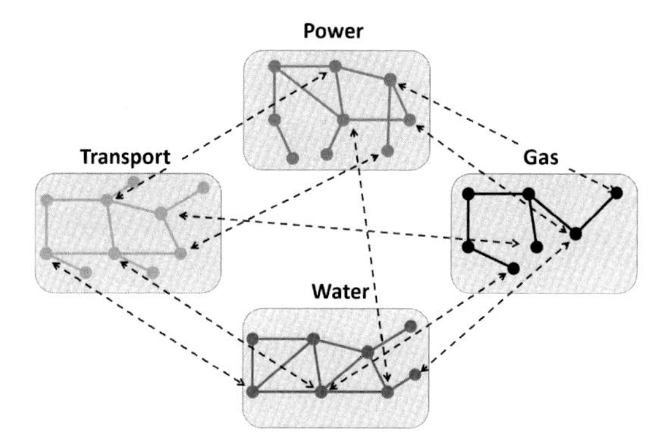

Figure 9.5 Interconnected infrastructure networks.

9.2 INTERCONNECTED NETWORKS

9.2.1 Properties of Interconnected Networks

Contemporary civilization has transitioned into societies and economies highly dependent on networks of varying nature and importance. Regardless of the nature of these networks, experience has shown us their interconnectedness and interdependence, and thus their effect on our lives. Their vulnerability is in effect our vulnerability as well. Interconnectivity is the state or quality of being connected together, and a supernetwork (or supernet) is a network that is formed from the combination of two or more networks (or subnets) interconnected networks. The decisions made by the users of the networks, in turn, affect not only the networks in question and their users but other networks as well. Hence, the understanding of the impacts of human decision-making on such networks is of paramount importance for resilience-based design.

Fig. 9.5 schematically shows the interconnectivity of various networks on which modern societies are increasingly reliant. Such networks may be either physical (i.e., transportation and logistical networks, communication, energy and power networks), or more abstract networks (i.e., financial networks, environmental networks, social, and knowledge networks). Interconnected networks (also termed "supernetworks" or "hypernetworks") are, in essence, the paradigm for the modeling, analysis, and solution of complex problems in the interlocked, internationalized and information-based network economy. Supernetworks are networks that are above and beyond existing networks, consisting of nodes, links, and flows, with nodes corresponding to locations in space, links to connections in the form of roads, cables, etc., and flows to vehicles, data, etc.

The supernetwork framework, captures, in a unified fashion, decision making facing a variety of decision-makers and its impacts on the component networks. The behavior of the individual decision-makers is modeled as well as their interactions on the complex network systems with the goal of minimizing the impacts on stakeholders. The ability to model and predict the various flows based on network topologies and interactions amongst the networks and amongst the decision-makers provide improved insight into the vulnerabilities of various linkages and network structures.

Regardless of the network type (albeit water, power, roadway, supernetwork), the concepts of fragility, vulnerability, and reliability discussed in previous chapters apply. Fragility is subject to numerous risk-of-failure metrics and may be calculated by, among others, survival analysis. Vulnerability refers to the inability of the network to withstand the effects of a hostile environment, while it is also subject to the network topology and connectivity. In all, interdependency cannot be overlooked should one desires the correct modeling and full understanding of the resilience of modern-time infrastructure networks.

9.2.2 Modeling Interconnectivity

In the case of supernetworks, a methodology similar to the one adopted for single water networks could be applied but with the extra degree of difficulty stemming from the interconnectivity of the constituent networks. One approach would be to first consider the supernetwork and determine the critical nodes/arcs of interconnectivity (Fig. 9.5), by use of either graph theory (e.g., the algebraic connectivity theorem), or Monte Carlo simulation. Another approach would be deducing the vulnerability of the supernet through simulating the interactions of the various subnets upon catastrophic events in one of them. The later approach is described in the lines below.

In order to update the network reliabilities due to interdependencies, we define the interoperability matrix $I_{W|P}$ between two networks, e.g., the water and the power network. $I_{W|P}$ captures the influence of the power network, denoted as P, on the water network (denoted as W), and thus the water network is the primary (or major) network and the power network is the dependent network. Note that $I_{P|W} \neq I_{W|P}$, while the size of $I_{W|P}$ depends on the number of vertices of the two networks. According to [51,139], the interoperability matrix is a weighted version of the adjacency matrix with weights that provide the strength of coupling between the vertices of the two interconnected networks. Therefore, $I_{W|P}(m, d)$ contains the failure probability of vertex m of the water network W, due to a failure at vertex d of the power network P. We may write

$$I_{W|P}(m, d) = P(W_m|P_d) \tag{9.7}$$

$I_{W|P}(m, d)$ equal to 0 or 1 implies complete independence and full dependence, respectively [139]. The derivation of $I_{W|P}$ is not always straightforward, but a simple procedure for calculating its terms can be borrowed for the concept of stiffness matrix used in structural analysis. We can simply apply some perturbation on vertex d of the power network (dependent network) and measure with our models the sensitivity on every vertex of the water network (primary network).

Once the interoperability matrix is known, the fragilities of each vertex of the primary network can be updated assuming that the network failure of vertex m is equal to either to failure of the water or the power network, thus $P(F_m) = P(W_m \cup P_m)$. $P(W_m)$ and $P(P_m)$ are the fragilities of vertex m for the two networks separately and are obviously independent, i.e., $P(W_m \cap P_m) = P(W_m)P(P_m)$. $P(W_m)$ is the fragility of vertex m or, in other words, it is the conditional failure probability $P(W_m|IM)$ of m and is obtained applying one of the approaches discussed in previous chapters. $P(P_d)$ is the fragility of the dependent network and its calculation is discussed in the lines below. The updated fragility of vertex m of the water network, considering interconnectivity will be

$$P(F_m) = P(W_m \cup P_m) = P(W_m) + P(P_m) - P(W_m)P(P_m). \tag{9.8}$$

$P(P_m)$ will be determined following the derivation proposed in [139]. A failure of the power (dependent) network at m will occur provided that always a failure at vertex d of the power network has also occurred and therefore $P(P_m \cap P_d) = P(P_m)$. Since P_m and P_d are statistically dependent we may write

$$P(P_m|P_d) = \frac{P(P_m \cap P_d)}{P(P_d)} = \frac{P(P_m)}{P(P_d)}. \tag{9.9}$$

$P(P_m|P_d)$ is the corresponding entry of the interoperability matrix $I_{W|P}(m, d)$. Rearranging the above expression will provide the updated fragility of vertex d as $P(P_m) = I_{W|P}(m, d) \cdot P(P_d)$. With the aid of Eq. (9.9), the fragility $P(F_m)$ will finally be

$$P(F_m) = P(W_k) + I_{W|P}(m, d)\ P(P_d) - P(W_m) I_{W|P}(m, d)\ P(P_d). \qquad (9.10)$$

The above expression combines only three quantities, the fragilities of the two vertices of each network (obtained separately) and the interoperability matrix.

9.3 ASSESSMENT OF SEISMIC RISK AND LOSS

Seismic risk and loss assessment combine fragility and vulnerability analysis with seismic hazard (Fragiadakis et al. [62]). Although many researchers use both fragility and vulnerability as two terms with the same meaning, in earthquake engineering, vulnerability has a broader meaning referring to the losses, while the fragility provides the probability that a damage-state (e.g., collapse) is exceeded. The vulnerability analysis of infrastructure requires first calculating the fragilities of the major components of the network (e.g., pipes) and then using them for calculating the reliability of the whole network. As already discussed, the network reliability depends on the topology of the network and also on its interdependencies with other networks [59].

Seismic risk assessment is a valuable decision-making tool, essential for prioritizing proactive actions and for mitigating losses and service disruption. Adopting the PEER (Pacific-Earthquake Engineering Research Center) formula, the seismic risk assessment can be seen as the mean annual frequency that a network performance metric exceeds a threshold value. Assuming that the network performance metric of interest is the serviceability ratio (SR, Section 5.4.1), the seismic risk may be obtained as

$$\lambda_{SR} = \int_{IM} P(SR \geq sr|IM)\ |d\lambda_{dIM}|\ dIM \qquad (9.11)$$

where λ_{SR} is the mean annual frequency that SR exceeds a threshold level sr, while $d\lambda_{IM}$ is the slope of the site's seismic hazard curve. $P(SR \geq sr|IM)$ is an abbreviation of $P(SR|IM)$ which is the global vulnerability function on the network, since SR is a global network performance metric. The integral of Eq. (9.11) is easily calculated numerically.

According to Eq. (9.11), the seismic risk is the convolution of the hazard curve and the network's vulnerability curve. The seismic hazard is obtained considering separately the probability of exceedance for various levels of seismic intensity. Seismic hazard models consider the seismicity of the site considered and depend on earthquake magnitude, the source-to-site distance, the local soil conditions, the fault mechanism, etc.

Apart from the relatively simple loss functions L_{IM} presented in Section 9.1.4 (e.g., Eq. (9.4)), seismic losses can be measured in terms of one of the 3 D's: Deaths, Dollars, and Downtime [41]. When this is the case, the loss function is determined by summing for every component the convolution of the component's cost function with its seismic vulnerability function. Using again the PEER framework, and adopting $P_{f,ij}$ as a local network performance metric (e.g., failure probability of water from vertex i reaching vertex j), the seismic losses may be calculated expanding Eq. (9.11) as follows:

$$\lambda_{DV} = \int_{IM} \int_{P_f} \int_{DM} P(DV|DM)\ dP(DM|P_{f,ij})\ dP(P_{f,ij}|IM)\ |d\lambda_{dIM}|\ dIM \tag{9.12}$$

where λ_{DV} is an abbreviation for $\lambda(DV \geq dv)$ which is the mean annual frequency that a decision variable (DV) exceeds a threshold value dv. The decision variable has units of one of the 3 D's and is the primary quantity of interest. Moreover, $P(DV|DM)$ is the loss curve, providing the probability of a decision variable being exceeded for the damage state (DM) examined (e.g., minor, major, total), while $P(DM|P_{f,ij})$ is the damage curve, i.e., the probability of a damage state being exceeded as function of $P_{f,ij}$. The numerical integration of Eq. (9.12) is performed over all components of the network and is preferably calculated with the aid of a Monte Carlo scheme [176]. Of course, the calculation of losses always depends strongly on the assumptions adopted. The loss curve $P(DV|DM)$ and the damage curve $P(DM|P_{f,ij})$ are often based on engineering judgement.

EPILOGUE

After all is said and done, more is said than done.
Aesop (620–564 BC).

As contemporary civilization has transitioned into societies and economies highly dependent on networks of varying nature and importance, so have societies become increasingly more reliant on networks, and on networks of networks.

Regardless of the network type (water, power, roadway, or hypernetwork), the concepts of fragility, vulnerability and reliability discussed in previous chapters apply. Further, regardless of the nature of these networks, albeit physical, knowledge or more abstract networks, experience has shown us their interconnectedness and interdependence, it has highlighted their effect on our lives and has transfused their vulnerability to a societal vulnerability at large.

Even though hypernetworks present topological and operational features quite different from those observed in isolated networks, the presence of links between the individual networks affects the way such interconnected networks behave. Granted that in the case of hypernetworks the same methodology as the one for WDNs could potentially be applied, the fact is that there is an extra degree of difficulty stemming from the interconnectivity of the constituent networks. This interconnectedness can not be overlooked though, should one desires the correct modeling and understanding of the behavior of contemporary infrastructure networks.

By extension, the decisions made by the managers of the constituent networks affect not only the networks in question and their users but other networks as well, in the greatest scheme of hypernetworks. Hence, the understanding of the impacts of human decision-making on such networks is of paramount importance to the study and the sustainable management of hypernetworks.

BIBLIOGRAPHY

[1] T. Adachi, B.R. Ellingwood, Serviceability assessment of a municipal water system under spatially correlated seismic intensities, Computer-Aided Civil and Infrastructure Engineering 24 (4) (2009) 237–248.

[2] N.F. Adnan, M.F. Ghazali, M.M. Amin, A.M.A. Hamat, Leak detection in gas pipeline by acoustic and signal processing – a review, in: IOP Conference Series: Materials Science and Engineering, vol. 100, IOP Publishing, 2015, p. 012013.

[3] A. Agathokleus, Sensor-Based Sustainable Management of Urban Water Distribution Networks Utilizing Survival Analysis Modeling, PhD Thesis, University of Cyprus, Cyprus, 2015.

[4] S.H. Ahmad, Simple enumeration of minimal cutsets of acyclic directed graph, IEEE Transactions on Reliability 37 (5) (1988) 484–487.

[5] American Lifelines Alliance (ALA), American Lifelines Alliance: Seismic Fragility Formulations for Water Systems—Guideline and Appendices, Technical Report, American Lifelines Alliance (ALA), Washington DC, USA, 2001.

[6] American Lifelines Alliance (ALA), American Lifelines Alliance: Seismic Guidelines for Water Pipelines, Technical Report, American Lifelines Alliance (ALA), Washington DC, USA, 2005.

[7] S.P. Andey, P.S. Kelkar, Influence of intermittent and continuous modes of water supply on domestic water consumption, Water Resources Management 23 (12) (January 2009) 2555–2566.

[8] S.A. Andreou, D.H. Marks, R.M. Clark, A new methodology for modelling break failure patterns in deteriorating water distribution systems: applications, Advances in Water Resources 10 (1) (1987) 11–20.

[9] S.A. Andreou, D.H. Marks, R.M. Clark, A new methodology for modelling break failure patterns in deteriorating water distribution systems: theory, Advances in Water Resources 10 (1) (1987) 2–10.

[10] J. Anthonisse, The Rush in a Directed Graph (Technical Report BN 9/71), Technical Report, Stichting Mathematisch Centrum, Amsterdam, Netherlands, 2001.

[11] T. Ariman, G.E. Muleski, A review of the response of buried pipelines under seismic excitations, Earthquake Engineering & Structural Dynamics 9 (2) (1981) 133–152.

[12] P. Aslani, Hazard Rate Modeling and Risk Analysis of Water Mains, MSc Thesis, Polytechnic University, USA, 2003.

[13] A. Athuraliya, P. Roberts, A. Brown, Yarra Valley Future Water – Residential Water Use Study, Technical Report, Stichting Mathematisch Centrum, Melbourne, Australia, 2012.

[14] N.Y. Aydin, L. Mays, T. Schmitt, Sustainability assessment of urban water distribution systems, Water Resources Management 28 (12) (2014) 4373–4384.

[15] D.B. Ballantyne, Earthquake Loss Estimation Modeling of the Seattle Water System, Kennedy/Jenks/Chilton, 1990.

[16] I. Bentes, L. Afonso, H. Varum, J. Pinto, J. Varajao, A. Duarte, J. Agarwal, A new tool to assess water pipe networks vulnerability and robustness, Engineering Failure Analysis 18 (7) (October 2011) 1637–1644.

[17] P. Bonacich, Power and centrality: a family of measures, American Journal of Sociology 92 (5) (1987) 1170–1182.

[18] S.P. Borgatti, Centrality and network flow, Social Networks 27 (1) (2005) 55–71.
[19] L. Breiman, J. Friedman, C.J. Stone, R.A. Olshen, Classification and Regression Trees, CRC Press, 1984.
[20] M. Bruneau, S.E. Chang, R.T. Eguchi, G.C. Lee, T.D. O'Rourke, A.M. Reinhorn, M. Shinozuka, K. Tierney, W.A. Wallace, D. Von Winterfeldt, A framework to quantitatively assess and enhance the seismic resilience of communities, Earthquake Spectra 19 (4) (2003) 733–752.
[21] A. Carrión, H. Solano, M.L. Gamiz, A. Debón, Evaluation of the reliability of a water supply network from right-censored and left-truncated break data, Water Resources Management 24 (12) (2010) 2917–2935.
[22] A.M. Cassa, J.E. Van Zyl, R.F. Laubscher, A numerical investigation into the effect of pressure on holes and cracks in water supply pipes, Urban Water Journal 7 (2) (2010) 109–120.
[23] S.E. Chang, M. Shinozuka, Measuring improvements in the disaster resilience of communities, Earthquake Spectra 20 (3) (2004) 739–755.
[24] D.-Y. Choi, S.-W. Kim, M.-A. Choi, Z.W. Geem, Adaptive Kalman filter based on adjustable sampling interval in burst detection for water distribution system, Water 8 (4) (2016) 142.
[25] K.W. Chou, G.Y. Liu, C.H. Yeh, C.W. Huang, Taiwan water supply network's seismic damage simulation applying negative pressure treatment, in: 5th International Conference on Advances in Experimental Structural Engineering 5AESE, Taipei, Taiwan, November 2013.
[26] S. Christodoulou, A. Agathokleous, B. Charalambous, A. Adamou, Proactive risk-based integrity assessment of water distribution networks, Water Resources Management 24 (13) (2010) 3715–3730.
[27] S.E. Christodoulou, Water network assessment and reliability analysis by use of survival analysis, Water Resources Management 25 (4) (2011) 1229–1238.
[28] S.E. Christodoulou, A. Agathokleous, A study on the effects of intermittent water supply on the vulnerability of urban water distribution networks, Water Science & Technology: Water Supply 12 (4) (2012) 523–530.
[29] S.E. Christodoulou, P. Aslani, A. Deligianni, Integrated GIS-based management of water distribution networks, in: 2006 ASCE International Conference on Computing and Decision Making in Civil and Building Engineering, Montreal, Canada, 2006.
[30] S.E. Christodoulou, P. Aslani, A. Vanrenterghem, A risk analysis framework for evaluating structural degradation of water mains in urban settings, using neurofuzzy systems and statistical modeling techniques, in: P. Bizier, P. DeBarry (Eds.), World Water and Environmental Resources Congress, American Society of Civil Engineers, Philadelphia, Pennsylvania, USA, June 2003, pp. 1–9.
[31] S.E. Christodoulou, B. Charalambous, A. Adamou, Managing the 'repair or replace' dilemma on water leakages, in: International Water Association's (IWA) Third Specialty Conference on Water Loss Reduction, Bucharest, Romania, September 2007.
[32] S.E. Christodoulou, C. Charalambous, A. Adamou, Rehabilitation and maintenance of water distribution network assets, Water Science & Technology: Water Supply 8 (2) (2008) 231–237.
[33] S.E. Christodoulou, A. Deligianni, A neurofuzzy decision framework for the management of water distribution networks, Water Resources Management 24 (1) (2010) 139–156.

[34] S.E. Christodoulou, G. Ellinas, Pipe routing through ant colony optimization, Journal of Infrastructure Systems 16 (2) (2010) 149–159.
[35] S.E. Christodoulou, M. Fragiadakis, Vulnerability assessment of water distribution networks considering performance data, Journal of Infrastructure Systems (2014).
[36] S.E. Christodoulou, E. Kourti, A. Agathokleous, Waterloss detection in water distribution networks using wavelet change-point detection, Water Resources Management 31 (3) (2017) 979–994.
[37] S.E. Christodoulou, A. Deligianni, P. Aslani, A. Agathokleous, Risk-based asset management of water piping networks using neurofuzzy systems, Computers, Environment and Urban Systems 33 (2) (2009) 138–149.
[38] S.E. Christodoulou, A. Agathokleous, B. Charalambous, A. Adamou, Proactive risk-based integrity assessment of water distribution networks, Water Resources Management 24 (13) (2010) 3715–3730.
[39] S.E. Christodoulou, A. Agathokleous, A. Kounoudes, M. Milis, Wireless sensor networks for water loss detection, European Water (30) (2010) 41–48.
[40] S.E. Christodoulou, A. Gagatsis, A. Agathokleous, S. Xanthos, S. Kranioti, Urban water distribution network asset management using spatio-temporal analysis of pipe-failure data, in: International Conference on Computing in Civil and Building Engineering, ICCBEE, Moscow, Russia, June 2012.
[41] G.P. Cimellaro, A. Reinhorn, M. Bruneau, Framework for the analytical quantification of disaster resilience, Engineering Structures 32 (2010) 3639–3649.
[42] G.P. Cimellaro, C. Renschler, L. Arendt, M. Bruneau, A.M. Reinhorn, Community resilience index for road network systems, in: G. De Roeck, G. Degrande, G. Lombaert, G. Müller (Eds.), EURODYN, 2011, pp. 370–376.
[43] G.P. Cimellaro, A. Tinebra, C. Renschler, M. Fragiadakis, New resilience index for urban water distribution networks, Journal of Structural Engineering 142 (8) (2016).
[44] R.M. Clark, C.L. Stafford, J.A. Goodrich, Water distribution systems: a spatial and cost evaluation, Journal of the Water Resources Planning and Management Division 108 (3) (1982) 140–147.
[45] D.I.C. Covas, A.C. Jacob, H.M. Ramos, Water losses' assessment in an urban water network, Water Practice & Technology 3 (3) (2008) 1–9.
[46] D.R. Cox, Regression models and life-tables, Journal of the Royal Statistical Society – Series B (Methodological) 34 (2) (1972) 187–220.
[47] T.K. Datta, Seismic response of buried pipelines: a state-of-the-art review, Nuclear Engineering and Design 192 (2) (1999) 271–284.
[48] C.A. Davis, Water system service categories, postearthquake interaction, and restoration strategies, Earthquake Spectra 30 (4) (2014) 1487–1509.
[49] O. Ditlevsen, Narrow reliability bounds for structural systems, Journal of Structural Mechanics 7 (4) (1979) 453–472.
[50] L. Dridi, A. Mailhot, M. Parizeau, J.P. Villeneuve, A strategy for optimal replacement of water pipes integrating structural and hydraulic indicators based on a statistical water pipe break model, in: Proceedings of the 8th International Conference on Computing and Control for the Water Industry, 2005, pp. 65–70.
[51] L. Duenas-Osorio, J.I. Craig, B.J. Goodno, Seismic response of critical interdependent networks, Earthquake Engineering and Structural Dynamics 32 (2) (2007) 285–306.
[52] D.G. Eliades, M.M. Polycarpou, Leakage fault detection in district metered areas of water distribution systems, Journal of Hydroinformatics 14 (4) (2012) 992–1005.

[53] S. Esposito, I. Iervolino, F. Silvestri, A. d'Onofrio, A. Santo, F. Cavalieri, P. Franchin, Seismic risk analysis of lifelines: preliminary results for the case-study of L'Aquila ENEL rete gas, in: 15th World Conference of Earthquake Engineering, Lisbon, Portugal, 2012, p. 2998.
[54] P. Fanner, Assessing real water losses: a practical approach, Water 21, International Water Association 6 (2) (April 2004) 49–50.
[55] M. Farley, Leakage Management and Control: A Best Practice Training Manual, Technical Report, World Health Organization, Geneva, Switzerland, 2001.
[56] M. Farley, S. Trow, Losses in Water Distribution Networks: A Practitioner's Guide to Assessment, Monitoring and Control, IWA, London, 2003.
[57] P. Fattahi, S. Fayyaz, A compromise programming model to integrated urban water management, Water Resources Management 24 (6) (2010) 1211–1227.
[58] C. Fitzpatrick, Tackling the global problem of water loss, Water 21, International Water Association 12 (2) (April 2010) 36–38.
[59] M. Fragiadakis, S.E. Christodoulou, Seismic reliability assessment of urban water networks, Earthquake Engineering and Structural Dynamics 43 (3) (2014) 357–374.
[60] M. Fragiadakis, S.E. Christodoulou, D. Vamvatsikos, Reliability assessment of urban water distribution networks under seismic loads, Water Resources Management 27 (10) (2013) 3739–3764.
[61] M. Fragiadakis, D. Vamvatsikos, S.E. Christodoulou, Reliability assessment of urban water networks, in: 15th World Conference on Earthquake Engineering, Lisbon, Portugal, 2012.
[62] M. Fragiadakis, D. Vamvatsikos, M.G. Karlaftis, N.D. Lagaros, M. Papadrakakis, Seismic assessment of structures and lifelines, Journal of Sound and Vibration 334 (2015) 29–56.
[63] L.C. Freeman, A set of measures of centrality based on betweenness, Sociometry 40 (1) (1977) 35–41.
[64] A. Gagatsis, S. Kranioti, S.E. Christodoulou, A. Agathokleous, S. Xanthos, An integrated software solution for identifying, monitoring and visualizing water leak incidents in water distribution networks, in: IWA's International Conference on New Developments in IT & Water, Amsterdam, Netherlands, November 2012.
[65] A. Gibbons, Algorithmic Graph Theory, Cambridge University Press, 1985.
[66] I.C. Goulter, A. Kazemi, Spatial and temporal groupings of water main pipe breakage in Winnipeg, Canadian Journal of Civil Engineering 15 (1) (1988) 91–97.
[67] N.S. Grigg, D.G. Fontane, J. Van Zyl, Water Distribution System Risk Tool for Investment Planning, 2013.
[68] Y.Y. Haimes, N.C. Matalas, J.H. Lambert, B.A. Jackson, J.F.R. Fellows, Reducing vulnerability of water supply systems to attack, Infrastructure Systems 4 (4) (1998) 164–177.
[69] E.J. Henley, R.A. Williams, Graph Theory in Modern Engineering: Computer Aided Design, Optimization, Reliability Analysis, Academic Press, Inc., 1973.
[70] J. Hintze, NCSS, PASS, and GESS, NCSS, Kaysville, Utah, 2006.
[71] J. Hintze, NCSS 2007, NCSS, LLC, 2007.
[72] H.P. Hong, E.N. Allouche, M. Trivedi, Optimal scheduling of replacement and rehabilitation of water distribution systems, Journal of Infrastructure Systems 12 (3) (2006) 184–191.
[73] D.W. Hosmer, S. Lemeshow, S. May, Applied Survival Analysis: Regression Modeling of Time-to-Event Data, Wiley Series in Probability and Statistics, 2nd edition, Wiley-Interscience, Hoboken, N.J., 2008.

[74] D.W. Hosmer, S. May, S. Lemeshow, Applied Survival Analysis, 2008.
[75] O. Hunaidi, P. Giamou, Ground penetrating radar for detection of leaks in buried plastic water distribution pipes, in: Seventh International Conference on Ground-Penetrating Radar, vol. 2, Lawrence, Kansas USA, May 27–30, University of Kansas, 1998, pp. 783–786.
[76] O. Hunaidi, W. Chu, A. Wang, W. Guan, Detecting leaks in plastic pipes, Journal of American Water Works Association 92 (2) (February 2000) 82–94.
[77] C. Hutton, Z. Kapelan, Real-time burst detection in water distribution systems using a Bayesian demand forecasting methodology, Procedia Engineering 119 (2015) 13–18.
[78] H.H.M. Hwang, H. Lin, M. Shinozuka, Seismic performance assessment of water delivery systems, Infrastructure Systems 4 (3) (1998) 118–125.
[79] R. Isoyama, E. Ishida, K. Yune, T. Shirozu, Seismic damage estimation procedure for water supply pipelines, Water Supply 18 (3) (2000) 63–68.
[80] A. Jacobson, M. Grigoriu, Fragility Analysis of Water Supply Systems, Technical Report, Multidisciplinary Center for Earthquake Engineering Research, 2008.
[81] M.B. Javanbarg, C. Scawthorn, J. Kiyono, Y. Ono, Minimal path sets seismic reliability evaluation of lifeline networks with link and node failures, in: Proc. Lifeline Earthquake Engineering in a Multihazard Environment, 2009.
[82] D. Johnston, M. Bracken, J. Thornton, Assessing pipe condition, useful remaining life and maximum operating and surge pressures: when to change the pipe or control the pressure, Water 21, International Water Association 14 (5) (October 2012) 43–44.
[83] H. Jun, G.V. Loganathan, J.H. Kim, S. Park, Identifying pipes and valves of high importance for efficient operation and maintenance of water distribution systems, Water Resources Management 22 (6) (2008) 719–736.
[84] V. Kanakoudis, S. Tsitsifli, Water pipe network reliability assessment using the DAC method, Desalination and Water Treatment 33 (1–3) (2011) 97–106.
[85] D. Jung, K. Lansey, Water distribution system burst detection using a nonlinear Kalman filter, Water Resources Planning and Management 141 (5) (2014) 04014070.
[86] V.K. Kanakoudis, Vulnerability based management of water resources systems, Journal of Hydroinformatics 6 (2) (2004) 133–155.
[87] V.K. Kanakoudis, D.K. Tolikas, Assessing the performance level of a water system, Water, Air, & Soil Pollution: Focus 4 (4/5) (2004) 307–318.
[88] V.K. Kanakoudis, S. Tsitsifli, Reliability assessment and data classification using discriminant functions and factor analysis, in: International Water Association's (IWA) Third Specialty Conference on Water Loss Reduction, vol. 3, Bucharest, Romania, September 2007, pp. 836–845.
[89] E.L. Kaplan, P. Meier, Nonparametric estimation from incomplete observations, Journal of the American Statistical Association 53 (282) (1958) 457.
[90] A.J. Kettler, I.C. Goulter, An analysis of pipe breakage in urban water distribution networks, Canadian Journal of Civil Engineering 12 (2) (1985) 286–293.
[91] A. Khan, P.D. Widdop, A.J. Day, A.S. Wood, S.R. Mounce, J. Machell, Artificial neural network model for a low cost failure sensor: performance assessment in pipeline distribution, Transactions on Engineering, Computing and Technology 15 (2006) 195–201.
[92] Y. Kim, S.J. Lee, T. Park, G. Lee, J.C. Suh, J.M. Lee, Robust leak detection and its localization using interval estimation for water distribution network, Computers & Chemical Engineering 92 (2016) 1–17.

[93] J.P. Klein, M.L. Moeschberger, Survival Analysis: Techniques for Censored and Truncated Data, Springer, New York, 1997.
[94] D.G. Kleinbaum, M. Klein, Survival Analysis: A Self-Learning Text, Statistics for Biology and Health, 3rd edition, Springer, New York, 2012.
[95] Y. Kleiner, B.J. Adams, J.S. Rogers, Long-term planning methodology for water distribution system rehabilitation, Water Resources Research 34 (8) (1998) 2039–2051.
[96] Y. Kleiner, B. Rajani, Using limited data to assess future needs, American Water Works Association 91 (7) (1999) 47–61.
[97] Y. Kleiner, B. Rajani, Comprehensive review of structural deterioration of water mains: statistical models, Urban Water 3 (3) (2001) 131–150.
[98] N.T. Kottegoda, R. Rosso, Probability, Statistics, and Reliability for Civil and Environmental Engineers, 1997.
[99] E.M. Laatsch, D.G. Honegger, T.D. Sheckler, The American Lifelines Alliance Approach: Four Years of Progress and Future Directions, American Society of Civil Engineers, July 2003, pp. 359–367.
[100] Z.M. Lahlou, Leak Detection and Water Loss Control, May 2001, http://www.nesc.wvu.edu/ndwc/pdf/ot/tb/tb_leakdetection.pdf.
[101] D. Laucelli, M. Romano, D. Savić, O. Giustolisi, Detecting anomalies in water distribution networks using EPR modelling paradigm, Journal of Hydroinformatics 18 (3) (2016) 409–427.
[102] E.T. Lee, J.W. Wang, Statistical Methods for Survival Data Analysis, Wiley Series in Probability and Statistics, 3rd edition, J. Wiley, New York, 2003.
[103] J.R. Lee, H. Tsuda, Fiber optic liquid leak detection technique with an ultrasonic actuator and a fiber Bragg grating, Optics Letters 30 (24) (December 2005) 3293–3295.
[104] R. Liemberger, Manila's remarkable performance offers leakage lessons for Asia, Water 21, International Water Association 13 (3) (June 2011) 43–44.
[105] A. Lindhe, Risk Assessment and Decision Support for Managing Drinking Water Systems, PhD Thesis, Chalmers University of Technology, Gothenburg, Sweden, 2010.
[106] S. Liu, M. Yamada, N. Collier, M. Sugiyama, Change-point detection in time-series data by relative density-ratio estimation, Neural Networks 43 (2013) 72–83.
[107] D. Loureiro, H. Alegre, S.T. Coelho, Controlling real losses: how much of minimum night flow is due to household consumption?, Water 21, International Water Association 14 (6) (December 2012) 28–30.
[108] D. Loureiro, C. Amado, A. Martins, D. Vitorino, A. Mamade, S.T. Coelho, Water distribution systems flow monitoring and anomalous event detection: a practical approach, Urban Water Journal 13 (3) (2016) 242–252.
[109] A. Mailhot, A. Poulin, J-P. Villeneuve, Optimal replacement of water pipes, Water Resources Research 39 (5) (2003).
[110] J.B. Martínez-Rodríguez, I. Montalvo, J. Izquierdo, R. Pérez-García, Reliability and tolerance comparison in water supply networks, Water Resources Management 25 (5) (2011) 1437–1448.
[111] J.C. Matthews, Disaster resilience of critical water infrastructure systems, Journal of Structural Engineering 142 (8) (2016) 14–24.
[112] J.C. Matthews, Disaster resilience of critical water infrastructure systems, Journal of Structural Engineering 142 (2016).
[113] R.E. Melchers, Structural Reliability, Horwood, 1987.
[114] D. Michaud, G.E. Apostolakis, Methodology for ranking the elements of water-supply networks, Infrastructure Systems 12 (4) (2006) 230–242.

[115] M. Moghtaderizadeh, R.K. Wood, A. Der Kiureghian, R.E. Barlow, Seismic reliability of lifeline networks, Journal of the Technical Councils of ASCE 108 (1) (1982) 60–78.
[116] J. Morrison, S. Tooms, D. Rogers, Guidance Notes on District Metered Areas, 2007.
[117] J. Morrison, S. Tooms, D. Rogers, Guidance Notes on District Metered Areas, 2007.
[118] R. Mossad, H. Aral, Numerical modeling of flow in a horizontal sand filter, American Journal of Engineering and Applied Sciences 3 (2) (2010) 286–292.
[119] MRI, HAZUS-MH, Multi Hazard Loss Estimation Methodology: Earthquake Model, Technical Report, Department of Homeland Security, FEMA, Washington, DC, 2003.
[120] NRC-CNRC, Deterioration and Inspection of Water Distribution Systems: A Best Practice, Technical Report, Federation of Canadian Municipalities and National Research Council, Canada, 2003.
[121] C. Nuti, A. Rasulo, I. Vanzi, Seismic safety of network structures and infrastructures, Structure and Infrastructure Engineering 6 (1–2) (2010) 95–110.
[122] D.K. O'Day, Organizing and analysing leak and break data for making main replacement decisions, American Water Works Association 74 (11) (1982) 589–594.
[123] T. Opsahl, F. Agneessens, J. Skvoretz, Node centrality in weighted networks: generalizing degree and shortest paths, Social Networks 32 (2010) 245–251.
[124] T.D. O'Rourke, M.D. Grigoriu, M.M. Khater, A state of the art review: seismic response of buried pipelines, in: Pressure Vessel and Piping Technology–1985: A Decade of Progress, ASCE, New York, NY, 1985, pp. 281–323.
[125] A. Ostfeld, J.G. Uber, E. Salomons, J.W. Berry, W.E. Hart, C.A. Phillips, J.-P. Watson, G. Dorini, P. Jonkergouw, Z. Kapelan, et al., The battle of the water sensor networks (BWSN): a design challenge for engineers and algorithms, Water Resources Planning and Management 134 (6) (2008) 556–568.
[126] S. Panguluri, W.M. Grayman, R.M. Clark, L.M. Garner, R. Haught, Water Distribution System Analysis: Field Studies, Modeling and Management, Report, U.S. Environmental Protection Agency, Cincinnati, Ohio, USA, 2005.
[127] S. Park, An Optimal Replacement Scheduling Model for Water Distribution Systems, MSc Thesis, Virginia Polytechnic Institute and State University, USA, 2000.
[128] S. Park, Identifying the hazard characteristics of pipes in water distribution systems by using the proportional hazards model: 1. Theory, KSCE Journal of Civil Engineering 8 (6) (2004) 663–668.
[129] S. Park, Identifying the hazard characteristics of pipes in water distribution systems by using the proportional hazards model: 2. Applications, KSCE Journal of Civil Engineering 8 (6) (2004) 669–677.
[130] S. Park, Estimating the timing of the economical replacement of water mains based on the predicted pipe break times using the proportional hazards models, Water Resources Management 25 (10) (2011) 2509–2524.
[131] S. Park, H. Jun, B.J. Kim, G.C. Im, Modeling of water main failure rates using the log-linear ROCOF and the power law process, Water Resources Management 22 (9) (2008) 1311–1324.
[132] S. Park, C.L. Choi, J.H. Kim, C.H. Bae, Evaluating the economic residual life of water pipes using the proportional hazards model, Water Resources Management 24 (12) (2010) 3195–3217.
[133] S. Park, R. Vega, Z. Choto, M. Grewe, Risk-based asset prioritization of water transmission/distribution pipes for the City of Tampa, Florida Water Resources (2010) 22–28.

[134] S. Park, H. Jun, N. Agbenowosi, B.J. Kim, K. Lim, The proportional hazards modeling of water main failure data incorporating the time-dependent effects of covariates, Water Resources Management 25 (1) (2011) 1–19.
[135] R. Pilcher, S. Hamilton, H. Chapman, D. Field, B. Ristovski, Guidance Notes on Leak Detection and Repair, 2007.
[136] O.A. Pineda-Porras, M. Ordaz, Seismic damage estimation in buried pipelines due to future earthquakes – the case of the Mexico City water system, in: A. Moustafa (Ed.), Earthquake-Resistant Structures – Design, Assessment and Rehabilitation, InTech, February 2012.
[137] J. Pinto, H. Varum, I. Bentes, J. Agarwal, A theory of vulnerability of water pipe network (TVWPN), Water Resources Management 24 (15) (2010) 4237–4254.
[138] K. Pitilakis, M. Alexoudi, S. Argyroudis, O. Monge, C. Martin, Earthquake risk assessment of lifelines, Bulletin of Earthquake Engineering 4 (4) (2006) 365–390.
[139] K. Poljanšek, F. Bono, E. Gutiérrez, Seismic risk assessment of interdependent critical infrastructure systems: the case of European gas and electricity networks, Earthquake Engineering and Structural Dynamics 41 (1) (2012) 61–79.
[140] T.D. Prasad, S.-H. Hong, N. Park, Reliability based design of water distribution networks using multi-objective genetic algorithms, KSCE Journal of Civil Engineering 7 (3) (2003) 351–361.
[141] R. Puust, Z. Kapelan, D.A. Savic, T. Koppel, A review of methods for leakage management in pipe networks, Urban Water Journal 7 (1) (February 2010) 25–45.
[142] S. Rajah, S. Nasr, K. Ferguson, J. Ey, Seismic fragility evaluation of water pipelines: Part 1–Current practice, in: Pipelines 2012, American Society of Civil Engineers, 2012, pp. 257–267.
[143] K. Ramachandran, System reliability bounds: a new look with improvements, Civil Engineering and Environmental Systems 21 (4) (2004) 265–278.
[144] J.E. Ramirez-Marquez, D.W. Coit, A Monte-Carlo simulation approach for approximating multi-state two-terminal reliability, Reliability Engineering & System Safety 87 (2) (2005) 253–264.
[145] M. Romano, Z. Kapelan, D.A. Savić, Automated detection of pipe bursts and other events in water distribution systems, Water Resources Planning and Management 140 (4) (2012) 457–467.
[146] N. Romero, T.D. O'Rourke, L.K. Nozick, C.A. Davis, Seismic hazards and water supply performance, Journal of Earthquake Engineering 14 (7) (2010) 1022–1043.
[147] L.A. Rossman, The EPANET Water Quality Model, Technical Report, Environmental Protection Agency, Cincinnati, OH (United States), 1995.
[148] R. Rupakhety, S.U. Sigurdsson, A.S. Papageorgiou, R. Sigbjörnsson, Quantification of ground-motion parameters and response spectra in the near-fault region, Bulletin of Earthquake Engineering 9 (4) (2011) 893–930.
[149] G.Z. Salehpour, Risk-Based Asset Management of Potable Water Distribution Systems: Case Study Göteborg, MSc Thesis, Chalmers University of Technology, Gothenburg, Sweden, 2009.
[150] H. Sallm, N. Manurung, Bottom-up water balance estimation – applications and limitations, in: Water Loss 2012, International Water Association, Manila, Philippines, February 2012, pp. 1–8.
[151] D.A. Savic, J.B. Boxall, B. Ulanicki, Z. Kapelan, C. Makropoulos, R. Fenner, K. Soga, I.W. Marshall, C. Maksimovic, I. Postlethwaite, R. Ashley, N. Graham, Project

Neptune: improved operation of water distribution networks, in: K. Van Zyl (Ed.), Water Distribution Systems Analysis, American Society of Civil Engineers, Kruger National Park, South Africa, 2009, pp. 1–16.
[152] U. Shamir, C.D.D. Howard, An analytic approach to scheduling pipe replacement, Journal of the American Water Works Association 71 (5) (1978).
[153] Q. Shuang, M. Zhang, Y. Yuan, Node vulnerability of water distribution networks under cascading failures, Reliability Engineering & System Safety 124 (2014) 132–141.
[154] J. Song, A. Der Kiureghian, Bounds on system reliability by linear programming, Journal of Engineering Mechanics 129 (6) (2003) 627–636.
[155] S. Stone, E.J. Dzuray, D. Meisegeier, A. Dahlborg, M. Erickson, Decision-support tools for predicting the performance of water distribution and wastewater collection systems, Report 95538, National Risk Management Research Laboratory, U.S. Environmental Protection Agency, Cincinnati, Ohio, USA, 2002.
[156] M. Tabesh, J. Soltani, R. Farmani, D. Savic, Assessing pipe failure rate and mechanical reliability of water distribution networks using data-driven modeling, Journal of Hydroinformatics 11 (1) (2009) 1–17.
[157] T.T. Tanyimboh, M. Tabesh, R. Burrows, Appraisal of source head methods for calculating reliability of water distribution networks, Water Resources Planning and Management 127 (4) (2001) 206–213.
[158] J. Thornton, A. Lambert, Managing Pressures to Reduce New Break Frequencies and Improve Infrastructure Management, February 2007.
[159] S. Tsitsifli, V. Kanakoudis, Predicting the behavior of a pipe network using the "critical Z-score" as its performance indicator, Desalination 250 (1) (2010) 258–265.
[160] S. Tsitsifli, V. Kanakoudis, I. Bakouros, Pipe networks risk assessment based on survival analysis, Water Resources Management 25 (14) (2011) 3729–3746.
[161] L. Tuhovcak, J. Rucka, T. Juhanak, Risk analysis of water distribution systems, in: Security of Water Supply Systems: From Source to Tap, Springer, 2006, pp. 169–182.
[162] L. Tuhovcak, J. Rucka, T. Juhanak, Risk analysis of water distribution systems, in: Security of Water Supply Systems: From Source to Tap, in: NATO Security Through Science Series, vol. 8, Kluwer Academic Publishers, Dordrecht, 2006, pp. 169–182.
[163] H. Unen, A. Elnashai, M. Sahin, Seismic performance assessment of interdependent utility network systems, TCLEE (2009).
[164] K. Vafai, Handbook of Porous Media, 2 edition, 2005.
[165] K. Vairavamoorthy, J. Yan, H.M. Galgale, S.D. Gorantiwar, IRA-WDS: a GIS-based risk analysis tool for water distribution systems, Environmental Modelling & Software 22 (7) (2007) 951–965.
[166] J.E. Van Zyl, M.O.A. Alsaydalani, C.R.I. Clayton, T. Bird, A. Dennis, Soil fluidisation outside leaks in water distribution pipes-preliminary observations, Proceedings of the Institution of Civil Engineers 166 (10) (2013) 546.
[167] A. Vanrenterghem, P. Eisenbeis, I. Juran, S.E. Christodoulou, Statistical Modeling of the Structural Degradation of an Urban Water Distribution System: Case Study of New York City, American Society of Civil Engineers, 2003, pp. 1–10.
[168] T.M. Walski, A. Pelliccia, Economic analysis of water main breaks, American Water Works Association 74 (3) (January 1982) 140–147.
[169] Y. Wang, S.-K. Au, Q. Fu, Seismic risk assessment and mitigation of water supply systems, Earthquake Spectra 26 (1) (2010) 257–274.

[170] Y.-K. Wen, Probabilistic aspects of earthquake engineering, in: Earthquake Engineering: From Engineering Seismology to Performance-Based Engineering, CRC Press, 2004.
[171] C. Xu, I.C. Goulter, Reliability based optimal design of water distribution networks, Water Resources Planning and Management 125 (6) (1999) 352–362.
[172] Q. Xu, Q. Chen, J. Ma, K. Blanckaert, Optimal pipe replacement strategy based on break rate prediction through genetic programming for water distribution network, Journal of Hydro-Environment Research 7 (2) (2013) 134–140.
[173] M. Yamada, T. Suzuki, T. Kanamori, H. Hachiya, M. Sugiyama, Relative density-ratio estimation for robust distribution comparison, Neural Computation 25 (5) (2013) 1324–1370.
[174] H. Yang, S. Shaoping, C. Yuping, Some parameters for seismic design and analysis of buried pipelines, in: Pipeline Engineering and Construction International Conference, 2003.
[175] S.-L. Yang, N.-S. Hsu, P.W.F. Louie, W.W.G. Yeh, Water distribution network reliability: connectivity analysis, Infrastructure Systems 2 (2) (1996) 54–64.
[176] T.Y. Yang, J. Moehle, B. Stojadinovic, A. Der Kiureghian, Seismic performance evaluation of facilities: methodology and implementation, Journal of Structural Engineering 135 (2009) 1146–1154.
[177] S. Yannopoulos, M. Spiliotis, Water distribution system reliability based on minimum cut–set approach and the hydraulic availability, Water Resources Management 27 (6) (2013) 1821–1836.
[178] A. Yazdani, P. Jeffrey, Robustness and vulnerability analysis of water distribution networks using graph theoretic and complex network principles, in: Water Distribution Systems Analysis, American Society of Civil Engineers, 2010, pp. 933–945.
[179] S. Yazdekhasti, K.R. Piratla, S. Atamturktur, A.A. Khan, Novel vibration-based technique for detecting water pipeline leakage, Structure and Infrastructure Engineering (2016) 1–12.
[180] Y.C. Zhang, High-order reliability bounds for series systems and application to structural systems, Computers & Structures 46 (2) (1993) 381–386.
[181] E. Zio, Challenges in the vulnerability and risk analysis of critical infrastructures, Reliability Engineering and System Safety 152 (2016) 137–150.

INDEX

Printed in the United States
By Bookmasters